EMBRACING AND MANAGING CHANGE IN TOURISM

Growth in international tourism in the second half of the twentieth century is a reflection of a range of changes taking place in the global economic, social and political environments. At the local scale, the development of tourism infrastructure and tourist activities also causes change to occur. As the case studies assembled in this volume indicate, there are myriad events and developments in relation to which tourism has represented either a response to change, or it has been an agent of change.

Embracing and Managing Change in Tourism examines management responses to the major changes taking place in international tourism and considers tourism itself as an agent of change. Including 22 detailed case studies from the developed and developing worlds, this book explores two key principles: first, that change is inevitable and, if effectively managed, has the potential to benefit all those living in, working in and visiting the destination; second, that there are no universal prescriptions for the effective management of change in tourism, since each destination has distinguishing characteristics and the nature of the problems facing it change over time.

Embracing and Managing Change in Tourism presents a unique collection of studies of tourism management at a time of increasing change in the industry. Analysing a wide range of methods of change management adopted by the diverse places featured in this book, the authors demonstrate that adaptive change management strategies must be implemented if the complexities and turbulence of the industry and its effects are to be dealt with successfully as the international tourism industry continues to change in the third millennium.

Eric Laws is Senior Lecturer in Marketing and International Business, Queensland University of Technology, **Bill Faulkner** is Director of the Centre for Tourism and Hotel Management Research, Griffith University and **Gianna Moscardo** is Research Fellow in Tourism, James Cook University, all in Australia.

EMBRACING AND MANAGING CHANGE IN TOURISM

International case studies

Eric Laws, Bill Faulkner and Gianna Moscardo

LONDON AND NEW YORK

First published 1998
by Routledge
2 Park Square, Milton Park, Abingdon, Oxfordshire OX14 4RN

Simultaneously published in the USA and Canada
by Routledge
711 Third Avenue, New York, NY 10017

First issued in paperback 2014

Routledge is an imprint of the Taylor & Francis Group, an informa business

British Library Cataloguing in Publication Data
A catalogue record for this book is available from the British Library

Library of Congress Cataloguing in Publication Data
Embracing and managing change in tourism : international case studies/ [edited by]
Eric Laws, Bill Faulkner, and Gianna Moscardo.
p.   cm.
Includes bibliographical references and index.
1. Tourist trade. I. Laws, Eric II. Faulkner, H. W. (Herbert William)
III. Moscardo, Gianna.
G155.A1E4285   1998
338.4'791–dc21   97–35290

ISBN 13 : 978-1-138-88132-7 (pbk)
ISBN 13 : 978-0-415-15998-2 (hbk)

Publisher's Note
The publisher has gone to great lengths to ensure the quality of this reprint
but points out that some imperfections in the original may be apparent

CONTENTS

CONTENTS

CONTENTS

FIGURES

FIGURES

TABLES

TABLES

CHAPTER SUMMARIES

Embracing and managing change in tourism

Eric Laws, Bill Faulkner and Gianni Moscardo

Change constantly challenges tourism planners and managers, yet in all the literature about tourism research, little attention has been paid to the nature and implications of change for the industry. In introducing this book of case studies, Chapter 1 reviews the change contexts to tourism management and the complexity of tourism management responses to change.

NAFTA and tourism development policy in North America

Ginger Smith and Abraham Pizam

The creation of international trading associations in various regions of the world have the potential to affect tourism through such impacts as increased ease of travel, investment and closer commercial ties. The North American Free Trade Agreement (NAFTA), involving USA, Mexico and Canada, is an example of this phenomenon. This chapter endeavours to assess the impact of NAFTA on the US economy in general, and its tourism industry in particular. Our efforts in this regard are complicated by the fact that many of the impacts of NAFTA are long term and are therefore not yet discernible. Furthermore, isolating NAFTA impacts from those of other factors is difficult, particularly when NAFTA-related changes have coincided with events such as the pesos devaluation and the resulting Mexican economic crisis. However, an analysis of the arrivals/departures data shows that, while Mexico has seen an increase in average annual arrivals from the US and Canada in the post-NAFTA years, the US and Canada have seen a decline in average annual arrivals or receipts from Mexico.

Regional tourism in Africa: scope and critical issues

Peter U.C. Dieke

The purpose of this chapter is to assess the prospects for regional cooperation in tourism development in Africa. The first section of the chapter summarises the inherent economic benefits and challenges related to the development of tourism, and the next outlines Africa's place in the global tourism market. The third section considers the case for intra-regional tourism in Africa, by examining some of the issues which are of concern in the region. The final section suggests ways in which intra-African tourism might provide a conduit for future development in the continent.

Tourism planning and implementation in the Polish city of Kalisz

Gary Akehurst

This chapter examines an urban development project in the Polish City of Kalisz, which was funded by the Foreign and Commonwealth Office of the United Kingdom Government (as part of its Local Authority Technical Links Scheme), the Southampton City Council and the City of Kalisz Council. It highlights the realities of tourism planning in a developing Eastern European economy, and illustrates the comprehensive nature of tourism planning, which of necessity, covers a wide range of local government and commercial activities. Co-ordinating these diverse activities in order to generate tourism outcomes which are beneficial to both local citizens and tourists, and securing local support has been a major challenge for the planning team. This challenge has been compounded by the need to apply the planning principles, concepts and approaches which have been developed in free market systems to the unique circumstances of a city within a state emerging from over 40 years of communist central planning and control.

A stakeholder-benefits approach to tourism management in a historic city centre: the Canterbury City Centre Initiative

Barbara Le Pelley and Eric Laws

This chapter reviews the general problems posed for Europe's small historic cities by increased tourist activity, then examines the background, purposes, progress and future direction of one historic city's approach to tackling this issue, the Canterbury City Centre Initiative (CCCI)).

The initial phase of CCCI introduced two concepts: the PATHS Initiative (Proactive Approaches to Tourism Hosting Strategies) – whose purpose is to involve the travel trade in adopting a sustainable approach towards methods of visiting Europe's historic cities, beginning before the tourist has left home, and second, a partnership within Canterbury conceived as a self-financing public/private sector collaboration involving the local authorities (district and county councils), key members of the community and the tourism industry. Both approaches bring into play local Agenda 21 and sustainability issues, and encourage innovation through targeting the private sector's business expertise. The main objectives are to manage the flow of visitors within Canterbury, in order to reduce congestion and enhance the appreciation of the aesthetic values of the walled town, while improving the environment for residents, businesses and visitors and helping support the local economy. Both concepts involve partnership approaches to destination policy, and this chapter considers some of the benefits and problems experienced.

An open, soft systems model is used to clarify the differing significance of the initiative for selected stakeholder groups with interests in the city, its society, economy, environment and ecology. The monitoring process, which is an integral part of the project, is also discussed, and proposals to develop a second phase of CCCI are considered.

Reliving the destination life cycle in Coolangatta: an historical perspective on the rise, decline and rejuvenation of an Australian seaside resort

Roslyn Russell and Bill Faulkner

Coolangatta is one of Australia's oldest seaside resorts. It is located at the southern end of the Gold Coast, which is the country's leading tourist destination outside metropolitan centres such as Sydney and Melbourne. As a mature destination, Coolangatta has recently experienced the decline which Butler's Destination Life Cycle model suggests is inevitable in the absence of management and planning measures that might otherwise rejuvenate such destinations. While the evolution of tourism at Coolangatta exhibits many of the features of the Butler scenario, it also deviates from the classical pattern in some respects. Such factors as the sudden growth of nearby more modern resorts, the failure to adapt to change and geographical constraints have contributed to this and it is these factors, in particular, that warrant attention in the development of a rejuvenation strategy. The analysis contained in this chapter highlights the irony of the area's current situation and the dilemma it faces in the future. On the one hand, Coolangatta's history has been instrumental in creating the problems its tourism industry now faces. In particular, the inertia of tradition and the failure to modernise have reduced the area's competitiveness *vis-à-vis* more modern destinations elsewhere on the Gold Coast. On the other hand, the area's history also distinguishes it from neighbouring resorts and, in this sense, it is a potential asset which could provide the basis of a tourism recovery strategy.

Tourism marketing and the small island environment: cases from the periphery

Tom Baum

This chapter considers tourism development in the context of peripheral island destinations and the marketing issues that are the result of peripherality and 'islandness'. The key tourism concerns of peripheral cold-water island locations are discussed and the chapter uses Butler's Tourism Area Life Cycle as a market analysis model and the means by which to undertake a comparative analysis of seven case-study northern islands, focusing on their tourism history and potential for future development.

Planning for stability and managing chaos: the case of Alpine ski resorts

Richard D. Lewis and Suzan Green

This chapter contrasts a traditional management perspective grounded in steady state theory with an emergent management perspective grounded in chaos/ complexity theory. These two perspectives are operationalised through different yet complementary approaches applied to the management process, and are further examined in relation to the actions of resort planners and entrepreneurs in winter alpine resorts. Using case study material drawn from the example of 'white tourism' in the European Alps, a comparative analysis of resort development, service system delivery and market development provides evidence of these two perspectives in action.

A close examination of further case material reveals that apparently rational (bounded) decision-making in alpine resorts is frequently based on (unbounded) chaotic/complex systems. Resorts thus progress in an ad hoc manner with periods of relative stability preceded and superceded by chaotic and turbulent episodes. In an ever-changing world, resorts can only ever manage and control small, tightly circumscribed, parts of the vast, complex whole.

In conclusion, a paradox is revealed, when the compromise is sought between steady state and chaos. In dynamic systems where innovations, phase-shifts and step changes are common, resort planners and managers alike are faced with making decisions for a future which they are unable to envision or control.

Ecotourism and its potential contribution to the economic revitalisation of Saskatchewan and Manitoba

David B. Weaver

This chapter posits that ecotourism, liberally defined, can play a useful secondary role in the economic rehabilitation of marginal hinterland jurisdictions such as Saskatchewan and Manitoba. Current ecotourism planning in the two provinces has been focused upon inventories of isolated attractions, and the shortcoming of this approach is the failure to take into account the influence and opportunities of the spatial context which surrounds each of these nodes. The inventories, therefore, can be usefully augmented by the introduction of *ecotourism context zones* (ECZs), or categorisations of the environments within which ecotourism can occur. These range from *urban* areas to the *agricultural heartland*, the *agricultural frontier*, the *resource frontier* and the *wilderness*. Urban areas and the agricultural frontier, it is argued, are the crucial future areas of ecotourism within the two provinces. This is because the first ECZ provides accessible opportunities to most of the population and functions as a recruitment ground for more involved forms of ecotourism, while the second ECZ constitutes the core of an extensive ecotourism industry unchallenged by agriculture or other primary activities.

Village-based tourism in the Soloman Islands: impediments and impacts

Adrian J.H. Lipscomb

Conventional forms of tourism have been introduced into many pacific nations over the last three decades. Although frequently profitable, they have impacted significantly upon the societies in which they have been established. Village-based tourism has been proposed by some commentators as a more appropriate form of tourism in the South Pacific, and a way of bridging the gap between subsistence and formal economies – a factor of some importance to the Solomon Islands where 85 per cent of the inhabitants live in villages. But the benefits of village-based tourism can only be realised if, first, the impediments to its introduction are adequately addressed, and second, if the resultant cultural and environmental impacts are minimised. Its successful introduction also requires a balanced and co-ordinated approach to the four essential requirements of any developing tourism industry: product development, marketing, training and infrastructure development. Significant changes are consequently required at both the micro- and the macro-level, and it is questionable whether the Solomon Islands is capable of such change. If it is not, then the country runs the risk of reinforcing the duality of its economy. Such a scenario is more likely to engender economic imperialism and social and economic elites than development in the true sense of the word.

Tourism development

Bill Faulkner

After identifying tourism as a potential catalyst for economic growth and development, the Indonesian Government has featured this sector in its Sixth Five Year (national development) Plan and established a target for international visitor arrivals of between 6 and 6.5 million in 1998. While the growth of other nearby Asian economies and the increasing affluence of Indonesia's own population suggests that these aspirations for the tourism industry are not necessarily unrealistic, they presuppose an emphasis on mass tourism that is not consistent with the long term sustainability of the industry and, as a consequence, the wellbeing of individual communities. This chapter provides an overview of the principles of sustainable tourism development as a starting point for identifying options for tourism development in Indonesia. It is concluded that, just as (old) mass tourism is not necessarily unsustainable in all circumstances, new or alternative forms of tourism are not inevitably a panacea in all situations. Indeed, neither form of tourism is sustainable unless an appropriate planning and management regime is in place. The case of the Sodong Agro-Tourism Project in Central Java is examined to elaborate on this point and illustrate how a relatively small scale tourism project can be instrumental in achieving the dual objectives of contributing to regional economic development and preserving the national cultural heritage. The latter outcome is of particular significance when it is considered that traditional lifestyles in Indonesia are being increasingly marginalised by economic growth and urbanisation. As the Sodong case illustrates, however, the good intentions of planners are often distorted at the implementation stage by the pressures of the commercial environment.

Local participation in tourism in the West Indian islands

Lesley France

Participation and empowerment of local populations in the tourism management process has been recognised as a cornerstone of sustainable tourism development. This chapter examines local participation in the tourism industry of the Caribbean islands by adopting Pretty's model of participation to tourism and by extending it backwards, to include an earlier 'plantation' phase which was common in the Caribbean islands. The processes that move the model on from one stage to the next are complex but include: ownership and management patterns; the extent of autonomy; food; different types of tourist and of tourism. These processes are used as criteria by which the stages of the model are identified, thereby establishing a typology. Examples from the West Indies demonstrate the extent to which participation and empowerment have emerged in tourism and its management within the islands.

The economic evaluation of the role of conservation and tourism in the regeneration of historic urban destinations

Mike Stabler

From a largely economic perspective, the chapter investigates the extent to which conservation of the historic quarters of cities and town acts as a catalyst for their regeneration, defined to embrace physical and economic improvement and revitalisation of socio-cultural life, and assesses the complementary role tourism plays. The contribution which economics can make to appraising both the rationale for conservation and the process is explored holistically, applying the analytical approach of Cost-Benefit Analysis within which framework the benefits and costs of regeneration initiatives are identified and evaluated. Particular attention is paid to methods for measuring the benefits of the historic built environment and the socio-cultural factors, which although not traded in the market, i.e. are not priced, impart added value to goods and services supplied in tourism destinations and enhance both residents' and tourists' welfare. A key feature of economic analysis is that it can show that public sector grants and subsidies are justified because they initiate and sustain regeneration and revitalisation and give rise to benefits greater than their cost.

The case study of Temple Bar, Dublin is of interest because it not only exemplifies what might be termed second rank tourist attractions, but it also demonstrates that successful regeneration schemes must primarily meet the needs of local communities. By doing so spillover effects occur which benefit the wider economy of the city, as well as acting as a draw to tourism.

In a concluding section, more practical issues are examined, such as the need for a strong commitment by and cooperation between the public, business and voluntary sectors. Some general desiderata for conservation schemes which trigger successful regeneration are identified and the need for further research to establish more firmly how the process works and the role of conservation and tourism in it is emphasised.

Conservation and regeneration: two case studies in the Arab world

Martin Crookston

This chapter examines two tourism and conservation projects in the Arab world – in the city of Salt in Jordan, and in the historic Bastakia district of Dubai in the United Arab Emirates. In Salt, a USAID-funded study developed an integrated strategy using tourism as a vital component in the economic regeneration of the central area of Jordan's last intact traditional Ottoman city, with practical priorities for restoration and conservation of the built heritage and recommendations for change to planning and development policies. In Bastakia, a study for

Dubai Municipality defined the conservation strategy for the historic Wind Tower district alongside the Creek, recommending visitor attractions as part of the mix of activities that would rejuvenate the area in an authentic, self-sustaining way.

These two case studies provide insight into the potential urban tourism role and limitations in the regeneration of historic areas and economic development in the developing world. The need for a careful 'fit' of the potential demand to the actual heritage resource and the social setting is identified, while some of the policy and implementation approaches associated with tourism-led regeneration in the developing world is discussed.

Dolphins, canoes and *marae* – ecotourism products in New Zealand

Chris Ryan

This chapter considers ecotourism developments in New Zealand. It initially provides data on numbers of overseas visitors and operators which indicates that interest in outdoor products is a fast-growing sector within New Zealand tourism. Second, the chapter discusses the economic problems faced by small operators offering what may be termed 'ecotourism' products. It does this by reference to three short case studies involving dolphin watching, canoe trips and tramping and *marae* visits, and discusses resultant environmental and economic issues. It concludes that the economic viability of many New Zealand ecotourism businesses is vulnerable to extraneous factors including adverse weather as well as problems of limited capital and ease of entry by potential competitors. This last issue may be 'solved' by the creation of local monopolies through licensing by the Department of Conservation as one way of sustaining business and controlling over-use of resources. However government 'market-led' policies mean that the Department has to seek more revenues from tourism, thereby making the application of this policy more problematic as the Department seeks additional revenue from tourism. Additionally a tension exists within New Zealand eco-tourism operators. First there is a wish to remain small while offering high quality, sustainable tourism experiences. Second there is a contrary need, which is to consider growth to remain economically viable and to deter competition. While in many areas of consumer products competition is perceived to be beneficial, in this instance it may not only reduce the economic viability of existing operators, but lead to an over-use of natural resources, and a reduction in the quality of experience sought by tourists. A political problem exists in New Zealand for conservation by policies based on restraints on trade are contrary to the free market ideology exposed by the present government.

Managing tourism in the Wet Tropics World Heritage Area: interpretation and the experience of visitors on Skyrail

Gianna Moscardo and Barbara Woods

It can be argued that much of the discussion of sustainable tourism has concentrated on the biophysical aspects of tourism development. This case study seeks to redress this imbalance by examining the potential value of interpretation as a management tool for sustainable tourism. Interpretation involves communicating with visitors, informing them about the significance of a place, providing them with information to encourage more sustainable behaviours and creating a positive experience. The specific case examined is Skyrail, a rainforest cableway operating in the World Heritage tropical rainforests of North Queensland, Australia. The case uses data collected from surveys and observations of Skyrail visitors to investigate the effectiveness of the Skyrail interpretation program in educating visitors about the rainforest and the value of the interpretation for the visitors' experiences. The results demonstrated that the interpretation was effective in improving tourists' rainforest knowledge and that it was a valuable component in visitors' satisfaction with their Skyrail experience.

Competition or co-operation? Small and medium sized tourism enterprises at the destination

Dimitrios Buhalis and Chris Cooper

Small businesses are a key ingredient of almost every tourism destination in the world. Yet, these enterprises provide both advantages and disadvantages to the destination. Whilst they represent a flexible form of supply and provide a rapid injection of cash into local economies, they also may lack the marketing and managerial skills to deliver the tourism product in a professional manner. At the same time, changes in the trends of both tourism demand and supply are creating real threats to the continued survival of small businesses in tourism. This chapter begins with an analysis of these issues and the contribution of small businesses to the tourism destination. A conceptual framework is developed which provides a way of thinking about competition at the destination and wider levels. This then forms the basis for a case study of Greek Aegean tourism through a SWOT analysis of small businesses on the Aegean islands. The findings clearly point to a sector in need of assistance and the chapter closes with the identification of a solution in the form of the establishment of a supporting agency for small businesses. The chapter concludes that the overall approach adopted in the case study is also applicable to many other tourism destinations around the world.

Facilitating tourist–host social interaction: an overview and assessment of the culture assimilator

Philip L. Pearce, Edward Kim and Syamsul Lussa

An enduring concern of tourism is the interpersonal impacts of tourists on hosts and hosts on tourists. This concern is embedded in the larger framework of social impacts generally and the sustainability of tourism in socio-cultural terms. This chapter suggests that culture shock, the phenomenon of disorientation, confusion, anxiety and malaise which accompanies novel and difficult social encounters is a significant component of the social problems of tourism.

Culture shock is viewed here as reciprocal, applying both to hosts and guests. In order to understand the dimensions and nature of culture shock two conceptual approaches are explored. The Co-ordinated Management of Meaning perspective is used to provide a framework to identify the diversity of cross-cultural factors which may generate communication and interaction difficulty. Subsequently a social situations analysis approach is employed as a technique which identifies some of the precise areas for change when difficulties occur. The value of Culture Assimilator training is assessed, at least in a preliminary way, by reporting on an evaluation of a Culture Assimilator trial for Australians travelling to Indonesia. The text-based instructional technique was shown to have an initial value in reducing self reported interaction difficulties. It was seen as useful by the small sample of users but modifications for the construction of future assimilators were also noted. Importantly, control group travellers who did not use the assimilator reported a broader range of problems and proportionately more problems than travellers who read the assimilator booklet thoroughly. Culture training, whether through Culture Assimilators or a combination of other approaches, is a potentially valuable activity to promote host-guest interaction.

A globalised theme park market? The case of Disney in Europe

Greg Richards and Bill Richards

Theme parks have become a global phenomenon, spreading rapidly from their origin in the American amusement park to Europe, the Pacific Rim and beyond. Major transnational leisure corporations such as Walt Disney and Warner Brothers have played a leading role in these developments. Despite considerable success in their home markets, recent problems with transplanting theme parks in Europe have underlined the importance of cultural factors in theme park management. The problems of developing theme parks transnationally are illustrated through an analysis of Disneyland Paris. The problems which Disney encountered in opening their new European park are examined in terms of locational, cultural, managerial and financial terms. It is argued that a basic lack of understanding of European consumer markets was a key cause of these problems.

Strategies for optimising revenues from short breaks: lessons from the Scottish hotel markets

David A. Edgar

Short break markets are widely recognised as a continual growth market. What is often not documented is that such growth is in terms of market share and as such results in highly monopolistic competition characteristics. Such a market therefore requires operators to adopt specialist and sophisticated marketing strategies to compete. Hotel groups and associated distributors have been at the forefront of short break development and as such can make a considerable contribution to the understanding of short breaks and the strategies for success. Organisations charged with the management of destinations and tourism markets can benefit considerably from a review of developments in the management of short breaks in Scottish hotel markets.

The chapter explores the strategies adopted by hotel groups operating in Scotland and examines the range of structural and strategic dimensions to the commercial elements of the short break market. The characteristics established are related to the performance of hotel operators and the overall paradigm applied to the management of short breaks in destinations, suggesting how tourism management organisations may utilise principles employed by hotel operators to maximise and manage revenues from short breaks.

An investigation of factors influencing the decision to travel: a case study of Japanese pleasure travellers

Sheauhsing Hsieh and Joseph T. O'Leary

A better understanding of decisions made by the travelling public can lead to more potential opportunities generated by efficient market planning and promotional strategies. The purpose of this study is to develop a travel behaviour choice model using sociodemographics, travel characteristics and psychographic factors for the Japanese travel market. A secondary analysis of data from the Pleasure Travel Markets Survey for Japan in 1989 was used. The major findings indicated that a model of travel behaviour choice which includes socio-demographics, travel characteristics and psychographics variables fitted the observed data very well. Sociodemographics and travel characteristics were significantly related to psychographic variables and further predicted travel decisions. Implications of the findings for marketing are discussed and future research directions are suggested for this important area.

An analysis of the promotion of marine tourism in Far North Queensland, Australia

Diane Burns and Laurie Murphy

The aim of this case study was to investigate options for improving the marketing of tours to the Great Barrier Reef in Far North Queensland. The study was based on the assumption that providing a good match to visitor expectations is a core element in successful tourism. As one of the sources of visitor expectations is the advertising provided by operators, this case study presents an analysis of reef operator and regional brochures. This content analysis was compared to the actual visitor experience of the product as assessed by a survey. Examination of the similarities and inconsistencies between the image and the reality focused the study on achieving a better understanding of the present situation and providing directions for future marketing. High levels of satisfaction with the reef trip and willingness to recommend the trip indicated that visitors' expectations were largely being met. Some differences, however, were found between the market survey results and the analysis of promotional material and the marketing implications of these are discussed. The chapter also demonstrates the usefulness of content analysis as a systematic means of analysing brochures promoting a tourism product, especially when the results are used in a comparison with visitor characteristics.

NOTES ON CONTRIBUTORS

Gary Akehurst is Professor of Marketing at the University of Portsmouth Business School. He has a first degree in economics, a Masters in management and a PhD in marketing. Previously he was with the Greene Belfield-Smith Division of Deloitte & Touche Consulting Group, during which time he led a team which prepared a tourism plan for the Government of Albania (funded by the European Bank for Reconstruction and Development) and a review of EU member states' tourism policies for the British Government Department of Employment. He is the founding Editor of the Service Industries Journal, the first academic journal in the world devoted to the study of the service sector.

Tom Baum PhD is Professor of International Hospitality Management in the Scottish Hotel School, University of Strathclyde. Prior to moving to Glasgow, he was responsible for the development of tourism and hospitality programmes at the University of Buckingham. He has extensive international tourism experience as researcher and consultant to governments and the private sector. His research interests are in the strategic human resource management within international tourism and in the tourism dimensions of small island economic development, in relation to which he is undertaking a major comparative project, outlined in his chapter. He is co-editor of a key book on island tourism and has written extensively on this and a number of other tourism and hospitality themes.

Dimitrios Buhalis is Senior Lecturer in Tourism at the Dept of Tourism, University of Westminster in London. Dr Buhalis is also Adjunct Professor at the Institut de Management Hotelier International (Cornell University-Ecole Superieure des Sciences Economiques et Commerciales ESSEC) in Paris. Other professional activities include: Assistant Editor of the *Progress in Tourism and Hospitality Research*; Books Editor for the *IT and Tourism: Applications-Methodologies and Techniques Journal*; Associate Editor for the *Encyclopaedia of Tourism*; Mediterranean Editor for *Anatolia International Journal of Tourism and Hospitality Research*; Book Editor for *Tourism*, the Tourism Society journal. Dimitrios is Chairman of the Association of Tourism Teachers and Trainers

(ATTT); executive committee member of the International Federation of Information Technology and Tourism (IFITT); Council Member of the Tourism Society; and Co-Chair for the ENTER '98 and '99 conferences on Tourism and Technology.

Diane Burns is a PhD student in the Department of Tourism, James Cook University and was employed by the Cooperative Research Centre for Ecologically Sustainable Development of the Reef (CRC Reef Research) as a research assistant on the Reef Tourism 2005 Project. Her PhD research is concerned with understanding community and tourist perceptions of user-pays strategies for environmental management.

Martin Crookston BSc(Econ) DipTRP MTS is an economist (LSE) and planner (Glasgow University) who has specialised in tourism policy and development, particularly as a component of economic regeneration strategies. He is a partner at Llewelyn-Davies, an international consultancy in the fields of planning, architecture, health and economics, and has worked in Britain, France, North Africa, and the Near and Middle East. He was a co-author, with Professor Peter Hall, Charles Landry and Nick Banks, of the *Four World Cities* comparative study for the Government Office for London, to which he contributed chapters on tourism, transport and information technology.

Chris Cooper received his geography degree and doctorate from University College London. Since then he was worked in market planning and research for both a major tour operator and retail company. He was in the University of Surrey's tourism team for ten years, before he joined Bournemouth University in 1996 as Chair in Tourism Management and with colleagues established the International Centre for Tourism and Hospitality Research.

He conceived, launched and edited *Progress in Tourism Recreation and Hospitality Management. Progress* is an annual, authoritative international review of research in the subject area and has run to six volumes. In 1995 *Progress* became a quarterly, fully international journal – *Progress in Tourism and Hospitality Research.* In addition Chris Cooper has jointly written *Tourism Principles and Practice* – a text which is now basic reading on many tourism courses around the world. He has published a second edition of *The Geography of Travel and Tourism* as well as many academic and conference papers.

Dr Peter U.C. Dieke is Senior Lecturer in Tourism at the Hotel School, University of Strathclyde. He took his BA in History and Archaeology, at the University of Nigeria in 1975 and an MSc in Hotel, Restaurant and Travel Administration at the University of Massachusetts, Amherst, USA, in 1981. His research and teaching interests are in the developmental aspects of tourism in less developed countries generally and, in particular, sub-Saharan Africa. He is currently completing two books on the latter subject, namely *The Political Economy of Tourism in Africa* (Cognizant Communication Corporation, New York) and *Tourism and Economic Development in Sub-Saharan Africa* (CABI,

Oxford). He has published several articles in *Annals of Tourism Research, World Development,* Journal of Tourism Studies and the *Review of African Political Economy.*

David Edgar is currently a lecturer in the Department of Marketing and Strategic Management, at Glasgow Caledonian University, Glasgow, following a move in July 1997 from a similar post in the Department of Hospitality and Tourism Management at Napier University, Edinburgh. He is an active researcher in strategic management with particular interests in the multiple linkages between market structure, strategy and performance. Most of his current research has revolved around chaos management, social auditing and the revision of research approaches in hospitality – particularly strategic and marketing management.

Bill (H.W.) Faulkner is the Director of the Centre for Tourism and Hotel Management Research at Griffith University (Gold Coast). Prior to taking up this position, he was the founding Director of the Australian Bureau of Tourism Research. His earlier career has included senior policy research positions in the Australian Government's tourism and transport administrations, and an academic position at Wollongong University. His PhD research was carried out in the Research School of Pacific Studies at the Australian National University, while undergraduate studies were completed at the University of New England.

Lesley France was a senior lecturer in geography, specialising in tourism, at the University of Northumbria at Newcastle for 20 years until she retired on the grounds of ill-health in January 1996. The main focus of her interests has been upon the evolution, impacts and planning of tourism. In recent years she has begun to explore aspects of sustainability as they relate to tourism. Most of her research has been undertaken in northern England; the Costa del Sol, Spain; and in the West Indies.

Suzan Green see Richard D. Lewis.

Sheauhsing Hsieh is an adjunct professor at Cleveland State University, Ohio, USA. Her research involves the study of travel behaviour and market segmentation using more than 20 different data sets in the travel and tourism area. Her research has included market segmentation studies that focused on activities, information channels, vacation arrangement and travel decision patterns; she also studies travel choice models in terms of travel expenditure and package vacations.

Edward Kim is a Research Officer at the Pai Chai Korean–Australian Study Centre, James Cook University, Cairns, Australia. He holds a first degree from Korea and a Graduate Diploma in Tourism from James Cook University. He has recently completed Doctoral work in Tourism. His principal interests are in tourism marketing, tourist motivation and cross-cultural interaction.

Eric Laws lectures in the School of Marketing and International Business, Queensland University of Technology, Brisbane. Before joining QUT, Eric lectured on tourism management in a number of British universities, and was Chairman of the Association of Teachers and Trainers in Tourism. He is the author of *Tourism Marketing, Service and Quality Management Perspectives* (Stanley Thornes, Cheltenham, 1991), *Tourist Destination Management: Issues, Analysis and Policies* (Routledge, London, 1995) and *Managing Packaged Tourism: Relationships, Responsibilities, and Service Quality in the Inclusive Holiday Industry* (Thomson International Business Press, London, 1997).

Barbara Le Pelley is a Forward Planning Officer with the States of Guernsey Island Development Committee. Formerly, she was a Senior Policy and Economic Development Officer with Canterbury City Council and for twenty-three years was closely involved with the development of tourism policy in Canterbury, culminating in devising, setting up and supporting Canterbury City Centre Initiative. Barbara is the co-author of more than twenty articles on the role of the local authority in tourism and visitor management and is a member of the English Historic Towns Forum Tourism Policy Group which published *Getting it Right . . . A Guide to Visitor Management in Historic Towns*.

Richard D. Lewis and **Suzan Green** are currently both Senior Lecturers in the School of Leisure Management at Sheffield Hallam University, England. Both have lived, worked and travelled extensively in alpine environments worldwide; and have now been actively engaged in an ongoing programme of research into aspects of white tourism for almost a decade.

In career terms, both have a combination of practical management and academic experience in the UK and abroad. They have also acted as freelance consultants in business and leisure and dabbled in entrepreneurship, having launched a couple of small businesses. Between them they possess a number of degrees in subjects as diverse as Zoology, Management Science, Education and Outdoor Pursuits.

Adrian Lipscomb is a tourism industry consultant and travel writer based in Bellingen on the east coast of Australia; his main interest in tourism lies in its potential to assist the development of less developed countries (LDCs). He has an academic background in Human Geography and has travelled widely in less developed parts of the world. His career has spanned international policy and intelligence with the Australian Department of Defence in Canberra, freelance journalism, small business operation, university lecturing in Tourism Studies, and tourism planning. In recent years he has been Tourism Adviser to the Western Province of the Solomon Islands, under the auspices of Australia's Overseas Service Bureau (OSB) (1994–5), Associate Lecturer in Tourism at Southern Cross University in Coffs Harbour, Australia (1996), and, most recently, he has updated the *Lonely Planet* guidebook to Papua New Guinea (1997).

Syamsul Lussa holds a first degree from Indonesia and a Graduate Diploma of Tourism and a Master of Administration (Tourism) by thesis degree from James Cook University, Australia. He is currently working in the Indonesian government in Jakarta in the area of tourism, public relations, tourism policy and cultural tourism. He has research interests in tourism, cross-cultural contact and international tourism development.

Gianna Moscardo is a Research Fellow with the Cooperative Research Centres for Reef Research and for Tropical Rainforest Ecology and Management (CRC TREM). Its aims include enhancing ecological and management studies of Australia's tropical rainforests and contributing to the well-being of the world's tropical rainforests through research, education and technology transfer. She is part of a team studying visitors to the North Queensland region to assist both public managers and tourism operators to develop more sustainable tourism in this area. The research program includes investigations of patterns of visitation, survey research to describe and segment visitors and evaluate visitor experiences, motives and expectations, and studies evaluating interpretive practices in the area.

Laurie Murphy is a lecturer in Tourism Marketing at James Cook University and a member of the (CRC Reef Research) team. Her research interests include tourism marketing and decision-making models and she has concentrated much of her research on understanding young independent long-stay travellers.

Joseph T. O'Leary is a professor in the Department of Forestry and Natural Resources at Purdue University. He is interested in the social behaviour and travel patterns of domestic and international recreation consumers, and secondary analysis of major regular national data sets and longitudinal travel and recreation-related data.

Philip Pearce holds a DPhil in Psychology from the University of Oxford, UK and a BA and Diploma in Education from the University of Adelaide. He has published widely in the area of tourist behaviour and is currently Foundation Professor and Head of Department of Tourism at James Cook University, Townsville and Cairns, Australia. He is involved in the research programs of two Cooperative Research Centres – CRC-Tropical Rainforest Ecology and Management and CRC-Reef Research. His principal interests are tourist motivation, tourism education, visitor behaviour and tourism community relationships.

Abraham Pizam is Professor of Tourism Management in the Department of Hospitality Management, University of Central Florida, Orlando, USA.

Professor Pizam is widely known in the field of Hospitality and Tourism Management and has conducted research projects, lectured and served as a consultant in more than thirty countries. He has held various academic positions, in the USA, UK, France, Austria, Australia, New Zealand, Singapore, Israel and Switzerland, is the author of more than 100 scientific publications and four books, and is on the editorial staff of ten academic journals in the field of tourism/hospitality management. Professor Pizam has conducted consulting and research projects for a variety of tourism organisations. Professor Pizam holds a Master's degree from New York University and a PhD from Cornell University, and is the recipient of several tourism academic awards.

Bill Richards is Senior Partner in Tourism Research and Marketing, a specialist tourism consultancy with offices in London, Amsterdam and Vienna. Tourism Research and Marketing is a member of the International Association of Amusement Parks and Attractions (IAAPA) and has produced a series of reports on the UK and international theme park markets.

Greg Richards obtained a PhD in Geography from University College London in 1982, and entered market research with RPA Marketing Communications. In 1984 he became a partner in Tourism Research and Marketing (TRAM), a consultancy specialising in tourism and event marketing.

Greg was Director of the Centre for Leisure and Tourism Studies (CELTS) at the University of North London from 1990 to 1993. He conducted research and consultancy projects in tourism, leisure and culture for the European Commission, the Department of National Heritage, the Exhibition Industry Federation, Spectator Management Group International, and Thomson Holidays. Greg is Co-ordinator of the European Association for Tourism and Leisure Education (ATLAS) and has directed a number of ATLAS projects for the European Commission on topics including cultural tourism, sustainable tourism and tourism education. He is currently directing a second European Cultural Tourism Survey, which covered some 30 cultural attractions and events in 12 European countries in 1997. ATLAS is also undertaking a project for DGXVI of the European Commission on crafts tourism, aiming to support local crafts and regional cultures through the promotion of crafts products to tourists.

Roslyn Russell is a PhD student with the Centre for Tourism and Hotel Management Research at Griffith University (Gold Coast). Her current research involves the application of Chaos Theory to the tourism field using the Gold Coast as a case study. Previous research (on which this chapter is based) includes the historical analysis of the evolution of tourism at Coolangatta. She completed her undergraduate studies in Business and the Honours programme in tourism within the Faculty of Business and Hotel Management at Griffith University (Gold Coast).

Chris Ryan is Professor of Tourism at Northern Territory University, Australia, and a visiting member of staff of the Nottingham Business School, UK, where he was formerly Head of the Tourism and Recreation Research Unit. He is the author of *Recreational Tourism – A Social Science Perspective, Researching Tourist Satisfaction* and *The Tourist Experience – A New Approach* and has written about 70 articles and papers about tourism. He has also taught at the University of Saskatchewan, Canada, and Massey University, New Zealand. He has undertaken research and written about tourism in each of these countries. His current research relates to tourist behaviours and impacts. He is the editor of the journal, *Tourism Management.*

Ginger Smith PhD, is Dean of the International School of Tourism and Hotel Management which opened in January 1997 at the Colegio Universitario del Este in Carolina, Puerto Rico, part of the Ana G. Mendez University System. She has also served as Senior International Policy Analyst, US Travel and Tourism Administration, US Department of Commerce, Washington, DC Dr Smith holds a BA in English from Cornell University with studies in honors humanities and comparative literature at Stanford University, an MA in public relations from the University of Delaware, and PhD in international relations and international communication from the School of International Service, The American University. Her PhD also included a concentration in international business and finance at the Landegger Program in International Business Diplomacy, School of Foreign Service, Georgetown University. Dr Smith's research interests include such areas as tourism and the environment, traveller safety and security, hospitality education and training, tourism and information technologies, international tourism policy and planning, and international organizations and free trade agreements (APEC, NAFTA, GATS, etc.). Upcoming publications include: 'Addressing Traveler Safety and Security Issues in the United States: A Follow-up on the 1995 White House Conference on Travel and Tourism,' 'NAFTA and the US Tourism Industry: Three Years After Implementation' (with A. Pizam), and 'Is Ecotourism an Appropriate Answer to Tourism's Environmental Concerns?' (with F. Dimanche).

Michael Stabler worked in local government and commercial management before becoming an academic. He is a senior research fellow in Economics in the Centre for Spatial and Real Estate Economics at The University of Reading where he is also Joint Director of the Tourism Research and Policy Unit and a member of the School of Planning Studies. In addition to his long-term interest in environment, leisure and land-use policy issues, he has investigated the economics of tourism, including its economic, environmental and socio-cultural impact. His research on these topics has been published in books, international journals, conference proceedings and research reports.

Dr David Weaver was based at the University of Regina, in Saskatchewan, for ten years after completing his PhD in geography at the University of Western Ontario. Since 1996, he has been a Senior Lecturer in the School of Tourism and Hotel Management at Griffith University, Gold Coast campus, in

Queensland, Australia. His primary interests are in the areas of sustainable tourism, ecotourism, resort cycles and tourism in peripheral regions.

Barbara Woods is a research officer with the Cooperative Research Centres for Reef Research and for Tropical Rainforest Ecology and Management. She has recently completed a Masters degree in Tourism at James Cook University focusing on the design of effective interpretive signs at tourist sites. Her research interests include evaluating interpretation, ecotourism and visitor management.

1

EMBRACING AND MANAGING CHANGE IN TOURISM

International case studies

Eric Laws, Bill Faulkner and Gianna Moscardo

Change contexts to tourism management

Change, and the effective management of change, are fundamental to contemporary management thinking. The one certainty in modern societies is that, whatever we do and experience today, the near future will be different. We live in a changing world, where individuals and organisations are constantly endeavouring to anticipate, respond to and influence events that impact on their existence in one way or another. Regardless of how much control we have over these events there will always be surprises that throw established systems into disarray until a new configuration of adaptive responses is established.

Not only do we live in a changing world, but also it has been suggested that the pace of change is accelerating. Thus, social commentators such as Toffler (1970) emphasised how the serendipitous effect of technology-driven change, and the consequent destabilisation of social, economic and political relationships has resulted in mounting pressure on the adaptive capabilities at both the individual and institutional levels. If anything, this process has intensified since the 1970s. Disequilibrium, instability and change are therefore as much a part of the modern economic, social and political landscape as the stable systems that we identify as organisational frameworks which structure relationships in these domains.

In tourism, change constantly challenges public sector planners and policy makers, but also provides opportunities for the development and expansion of operators' businesses. Without change there would be few opportunities for the creativeness of successful entrepreneurs, while for tourists, the opportunity to enjoy and experience a temporary change of place, culture or the pace of daily life is a prime motive for travel. Yet change can also be stressful and difficult to deal with, although it is clear from research in psychology that the more people avoid change the less they are able to cope with it (Langer, 1989). Langer distinguished mindful and mindless ways of dealing with change. Mindfulness is about actively seeking change and creating new conceptualisations of situations

1

and new routines or management approaches. She defined mindlessness in business as 'the application of yesterday's business solutions to today's problems.' (1989: 152). De Bono (1987) has provided us with the concept of lateral thinking which can be defined as a way of thinking about problems which does not rely upon one logical train of thought but rather seeks to associate and assimilate ideas from very different areas. The approaches of De Bono and Langer share two major principles. The first is that people and organisations need to actively seek and manage change rather than avoiding it. The second is that one of the keys to successfully managing change is to manage information. Successful managers are those who continuously seek information about the core components of their business and who monitor and evaluate their performance on a regular basis.

The emergence of tourism as a sphere of human activity and a significant element of global trade is itself a manifestation of change. World Tourism Organisation (1992) estimates reveal that, in the 1980s, tourism was the fastest growing sector of world trade, with an annual growth rate of 9.6 per cent. This compares with growth rates of 5.5 and 7.5 per cent for merchandise exports and commercial services respectively. From 1950 to 1970 international visitor arrivals increased more than six-fold from 25 million to 160 million, while this figure had increased a further three-fold to 450 million in 1991. It is anticipated that visitor arrivals will increase by a further 200 million, or 36 per cent, by the year 2000.

Growth in international tourism in the second half of the twentieth century is a reflection of a range of changes taking place in the global economic, social and political environments. Foremost among these changes is the widespread and rapid economic growth that has made travel more feasible for an increasing number of the world's population and the improvements in transport and communication technology which have made high-volume leisure or business travel possible to distant destinations. This may well come to be seen as a major change in human society, as for the first time in human history tourism has enabled large numbers of people to experience at first hand and on a temporary basis different cultures and remote, exotic environments (Laws, 1997).

As the case studies assembled in this volume indicate, there are myriad other (often interrelated) events and developments in relation to which tourism has represented either a response to change, or it has been an agent of change. A relatively comprehensive analysis of changes affecting tourism at the global level was carried out in 1990 at the First International Tourism Policy Forum at George Washington University (Brent Ritchie, 1992; Hawkins, 1993). Among the changes discussed by these authors, the following stand out as being most relevant to the cases referred to in the following chapters.

- The dismantling of the Soviet block and the spread of the private enterprise, free market system as a basis for economic organisation in communist countries has resulted in these countries being opened up for tourism purposes. While the evaporation of Cold War tensions has initiated a period of peace and co-operation at the global level, political conflicts nevertheless continue to erupt into violent confrontations at a more localised level and tourism-related activities are periodically targeted by acts of terrorism. Meanwhile, tourism is envisaged by some to be an instrument of world peace by virtue of its perceived impacts on inter-cultural tolerance and under-standing.

The opening up of economies to free trade and the reduction of restrictions on international investment have eroded the significance of national boun-daries in economic organisation. In association with these trends, multi-national corporations are driving tourism development in many parts of the world and local interests are often disenfranchised in this process.

The emergence of regional trading blocks, whereby neighbouring countries enter into arrangements aimed at establishing closer economic, political and cultural relationships, have the potential to influence tourism flows between the countries concerned to the extent that more intensive relationships in these areas stimulate travel. Such alliances also have the potential to facilitate greater co-ordination in tourism development strategies at the regional level.

Underpinning many of the above changes has been the pervasive shift towards a non-interventionist approach to government policy. Thus, many governments are moving towards a deregulated regime, government enterprises are being privatised and subsidised industry development is becoming less common. Within this environment, the parameters of tourism infrastructure development and service delivery are changing towards a more fundamental reliance on market forces.

With the changes referred to above, previously under-developed countries are being placed under increasing pressure to 'modernise' and become a part of the global economy. So far, the so-called 'Asian Tiger Economies' have responded to this pressure most effectively and, in the process, they have precipitated a major shift in the global economic order. This has been accompanied by a global redistribution in the focus of tourism activity, with new growth focuses in both tourism demand and the supply emerging in this region.

As mentioned previously, the emergence of the global economy has diluted local control of the means of production. Another implication of this trend is that residents of various countries have been exposed to the homogenising effects of corporate cultures and market expectations of consistent global standards. This, along with innovations in electronic media communication systems which are rapidly making McLuhan's global village a reality (McLuhan and Powers, 1989), is resulting in cultural distinctions becoming increasingly blurred. On the one hand, tourism has the potential to act as the vanguard of 'modernisation', with the homogenisation process being reinforced through the impacts of economic colonialism and demonstration effects. On the other hand, pockets of resistance to homogenisation are being encountered as the value of preserving the cultural heritage and traditions

3

of specific areas are being recognised, and as tourism is used as a mechanism for making the preservation of this heritage economically viable.

- Travel has become increasingly 'democratised' (Hawkins, 1993) in the sense that, in wealthy countries at least, opportunities for travel are no longer exclusively reserved for the wealthy elite. Travel is increasingly being seen as an entitlement, rather than a privilege, and as an essential ingredient of lifelong education, rather than as simply a leisure activity. This, along with the diversification of lifestyles, means that destinations and tourism organisations now have more varied and more demanding markets.

- As a response to the effects of many of these changes, an increasing environmental and conservation consciousness has emerged and is being expressed both in the market place and in the political process. The increasing demand for ecotourism product and the move towards the adoption of sustainability as a guiding principle for tourism development and operations are reflections of this trend.

- Cities are changing in response to the effects of growth and the impacts of innovations in transport and communications technology on the spatial distribution of people and activities. The growth of cities, and the pressure to adapt their infrastructure to changing land use demands, often means that the older inner city areas, where cultural and historical heritage assets are concentrated, either become degraded or increasingly inaccessible. The enhancement of these areas for tourism purposes is increasingly being recognised as a strategy for financially underwriting their preservation.

- Advances in information technology are revolutionising the tourism industry, both in terms of consumer access to product and the conduct of tourism businesses. The discerning taste of the emerging 'new tourism' market (Poon, 1993) is being reinforced by direct consumer access to product databases and destination information, particularly via the Internet. Meanwhile, innovations such as computerised reservation systems and automated systems for various aspects of office functions are supporting rising benchmarks in service levels and efficiency.

The complexity of tourism management responses to change

The purpose of this book is to examine tourism management responses to this changing global environment and to consider tourism itself as an agent of change. As Butler (in Hall, 1995: 102) has observed, 'We have tended to ignore the dynamic element of tourism', and Hall suggests 'Tourism is clearly a far more complex structure than the majority of present tools and frameworks of analysis of its impacts would have us believe' (Hall, 1995: 101).

Argyris and Schon (1978) investigating the ways in which organisational systems respond to external events, distinguished between single- and double-loop learning. The former is exemplified by the thermostat which turns heating on or off as a corrective response to information which it is programmed to sense and process. Organisations function in single-loop mode in periods and circumstances which can be characterised as routine and normal: decisions are correctly taken within the established paradigms of action. More radical action is required when the underlying conditions change. The result may be the formulation of a new mission, often carried forward by a new management team with new strategies and control systems. Argyris and Schon termed this double-loop learning, representing a paradigm shift, underpinned by new value systems.

The processes underlying double-loop learning have parallels with the concepts of positive and negative feedback which are key concepts in general systems theory (Bertalanffy, 1968). This argues that selected inputs are combined in a series of processes with the intention of producing specified outputs, each process stage adding value to the service. Efficiency in the system's operation can therefore be evaluated by measuring outputs against the inputs required to produce them, by examining the quality of those outputs, and by considering the way each process contributes to the overall service. Kirk (1995) has noted that although systems thinking has mainly been applied to 'hard' engineering situations, where outcomes are unambiguous and highly predictable, the concept can be applied in situations where human behaviour is a significant factor in business activities which combine social and technical processes (Checkland and Scholes, 1990).

Parallels to Argyris and Schon's single-/double-loop learning dichotomy can also be found in chaos theory, where a general distinction is drawn between stable (steady state) systems at one extreme and dynamic chaos and turbulent systems at the other (Nilson, 1995). The single-loop, negative-feedback based model for decision-making is more applicable to the former situation, while the double-loop, positive-feedback process dominates in the latter. Even though some systems can be in a virtually permanent steady state (e.g. the solar system), while others are perpetually dynamic (e.g. weather systems), many systems pass through phases involving transitions from one state to another. It would seem that tourism systems, whether they are viewed at the global, regional or individual destination levels, are more likely to be in the latter category. This, to some extent, explains our limited understanding of change phenomena. The conventional methods of tourism research which have dominated the field to date are more attuned to the analysis of stable systems, leaving a huge void in our understanding of dynamic chaos, or the turbulent phases in tourism development.

These lines of thought have been extended by Faulkner and Russell (1997), who explore the applicability of the chaos/complexity theory to tourism phenomena. In its present form, the methodological foundations of the tourism management field is largely derived from traditional social science disciplines (psychology, sociology, anthropology, economics, geography), which are themselves often based on the classical (Newtonian/Cartesian) paradigm of science. This paradigm adopts a reductionist view of the world, whereby all things are regarded as being understood in terms of their constituent parts, which fit together like the cogs in

5

some clockwork mechanism. Thus, every event is determined by initial conditions and is, at least in principle, predictable with some degree of precision. Small changes in initial conditions produce a correspondingly small change in the final state and everything is predictable in accord with essentially linear relationships.

The problem with research based on this view of the world is that it focuses on stability, order, uniformity, equilibrium and linear relationships, and there is no room for the unexpected events that inevitably surprise us and confound researchers, forecasters and planners alike. Nor are the accelerated, positive-feedback-driven change processes that are triggered by such events, and that give rise to an emergent new order, properly taken into account. The preoccupation with stable phenomena has meant that aspects of reality that exemplify tendencies towards instability, disorder, disequilibrium and non-linearity have received less attention and are, in effect assumed away. Our understanding of change has suffered as a consequence.

The inadequacies of conventional approaches in coping with change and transition have given rise to the chaos/complexity perspective in which the predisposition to assume a linear, clockwork world is displaced by concepts which depict a confusing world of non-linearity and surprise, juxtaposed with attributes normally associated with living organisms, such as adaptation, coherence and self-organisation (Capra, 1996; Davies and Gribbin, 1992). Chaos is therefore viewed as 'an order of infinite complexity' (Peat, 1991: 196). For the purposes of this volume, it is sufficient to note the need to balance the previous preoccupation with stable patterns by focusing on change and appreciating the deeper insights that can be derived from an examination of chaos and complexity in tourism.

Case studies of change management in tourism

Case studies are emphasised in this volume partly because this approach provides the most effective means for exploring the relatively new research area of change phenomena in tourism. However, a more compelling reason for adopting a case study approach is that this provides the most effective means of comprehending the inherent complexity of tourism arising from the juxtaposition of stable, dynamic chaos and turbulent systems, and the myriad interactions involved.

Yin (1994) has evaluated the case study method of research. Its essential features are problem definition, research design, data collection, data analysis and reporting. The key problems for the researcher are how to define the case being studied, how to determine the relevant data to be collected, and what to do with the data. The method has parallels with action research in that there is continual interaction between the data being collected, the theoretical issues being studied, and the evolution of management responses (Argyris and Schon, 1978).

In editing this book, we have had in mind the needs of two main groups of readers, who may use case study material in somewhat different ways. Case studies are familiar academic resources on advanced courses, providing students with an opportunity to learn by doing, or by observing and reflecting on real management experiences: 'dealing with cases is very much like working with the actual problems that people encounter in their jobs as managers . . . identifying and clarifying problems . . . analysing quantitative and qualitative data, evaluating alternative courses of action, and then making decisions about what strategies to pursue . . . ' (Lovelock, 1996: 645).

Academic case studies can be differentiated from the managers' experience of the situations they report in several ways. The case study data is pre-packaged, and post hoc: the case study writer has already selected information, and assembled it in a coherent and easily read form which can be studied more or less at leisure. A further distinction is that management problems are dynamic, often calling for some immediate action, with further analysis and major decisions to be taken later. A third, important distinction between academic case studies and the management situations they are based on is that 'participants in case studies aren't responsible for implementing their decisions, nor do they have to live with the consequences . . . ' (Lovelock, 1996: 646).

Case studies provide managers with detailed insights into the practical realities which other managers have already confronted and attempted to resolve. The level of detail provided in the chapter-length case studies in this book indicate the serious attention which a wide range of tourism organisations from around the world have paid to changing conditions. Tourism managers, and those charged with responsibility for policy making in tourist regions, are directly responsible for diagnosing changes affecting their own business and community situations. Most managers share concerns about current operations, trends in their markets, and future developments in the tourism industry. Carriers, destinations, accommodation and other service providers are generally regarded as substitutable by their clients, but they are also mutually dependent for their flows of visitors. Managers are keen to learn from the experiences of their colleagues, both to sharpen the competitive stances which each can adopt in their own circumstances, but also to explore ways of strengthening the long-term future of the tourism industry.

Scales of tourism operation

Another key organising concept for this book is that of scale: tourism organisations can be categorised according to the scale or level at which they operate. At one end of the spectrum is the global scale of operations typified by such organisations as the World Tourist Organisation, whose interests lie primarily in understanding the flow of travellers between countries and the implications for global patterns of tourism of factors such as changing international boundaries for nations and their trading relations, and broad-scale population dynamics. At the other end of this spectrum lie the individual tourism operators running small businesses and dealing directly with individual tourists on a daily basis. In between are large-scale businesses, and local, regional and national companies and destination management and marketing agencies. Some aspects of change in tourism, such as the increasing concern for sustainability, affect all levels of tourism, although in different ways, while others are more closely linked to certain levels or sectors of the tourism system.

This interaction between change and different scales or levels of tourism is reflected in the organisation of the case studies in this book into three sections. The case studies found in the first two sections are generally concerned with organisations responsible for managing tourism at destination, regional, national and international levels, and there is a strong focus on public sector organisations. The case studies in the third section concentrate more on cases involving individual tourism businesses, smaller tourism operations or tourists themselves. In addition to considering different scales of tourism, the case studies collected together in this book provide insights into a range of tourism organisations across many different sectors, in most parts of the globe.

Embracing and managing change in tourism: international case studies

The editors present this book in the belief that it offers more than a collection of interesting and well-argued case studies of tourism management at a time of increasing change in the industry. The case chapters analyse a range of methods of change management adopted by the diverse places featured here, although the issues and the policy responses adopted differ greatly, we draw our readers' attention to a number of themes which recur throughout these case studies:

- multiple stakeholders, that is, groups of people who experience different sets of benefits and problems as a result of local tourist activities, and who also have different abilities to affect the scale and nature of tourism locally;
- the complex nature of tourism, with organisations of different sizes and sophistication, and differing degrees of dependency on local tourism, but together contributing to the visitors' experiences, and the long-term sustainability of that destination;
- the roles of key individuals, including consultants, in recognising, shaping or leading change programmes;
- the adoption of a variety of theoretical approaches in analysing case studies of tourism management in change situations;
- evidence in many of the cases of reflection on, and reconsideration of policy directions, and of the adoption of practical measures for monitoring the effects of policy implementation.

The collection of case studies presented in this book illustrate two main principles: first, the general need to recognise that change is inevitable and, linked to that, the potential benefits of managing change towards a vision which embraces the differing needs of each destination stakeholder group; second, it is clear that there are no universal prescriptions for the effective management of change in tourism since each destination has distinguishing characteristics, and the nature of the problems facing it change over time. Some of the cases deal with incremental, relatively predictable changes, but in other cases the events precipitating change occurred suddenly and have had rapid and dramatic effects, some of which are yet to be fully evaluated.

Each destination therefore has the challenge of identifying the factors causing change locally, and of understanding their dynamics in its own context. Consequently, a policy adopted in one particular situation must not be regarded as a model solution for another destination. Nor indeed would current policy be adequate for dealing with future problems in the same destination. Much more, and perhaps much better research is needed to gain a full understanding of the complex challenges facing tourism managers. The editors hope that *Embracing and Managing Change in Tourism: International Case Studies*, will provide the impetus for tourism managers and researchers to adopt a more systemic approach to analysing the issues, and to embracing change in their policies. As this collection of case studies demonstrates, adaptive change management strategies can be the means for destinations to deal successfully with the complexity and turbulence of the industry and its effects. These are likely to become even more diverse and unpredictable as the scale and forms of the international tourism industry themselves continue change in the third millennium.

References

Argyris, C. and Schon, D. (1978) *Organisational Learning: A Theory of Action Perspective.* Reading, MA: Addison Wesley.

Bertalanffy, L. von (1968) *General Systems Theory*, New York: Brazillier.

Brent Ritchie, J. R. (1992) 'New Realities, New Horizons: Leisure, Tourism and Society in the Third Millennium', *Annual review of Travel* pp. 13–26.

Capra, F. (1996) *The Web of Life: A New Synthesis of Mind and Matter*, London: HarperCollins.

Checkland, P. and Scholes, J. (1992) *Soft Systems Methodology in Action*, Chichester: John Wiley.

Davies, P. and Gribbin, J. (1992) *The Matter Myth: Beyond Chaos and Complexity*, London: Penguin.

de Bono, E. (1987) *CoRT Thinking*, New York: Pergamon Press.

Faulkner, H. W. and Russell, R. (1997) 'Chaos and Complexity in Tourism: In Search of a New Perspective, *Pacific Tourism Review* 1(2): 93–102.

Hall, C. M. (1995) 'In Search of Common Ground: Reflections on Sustainability, Complexity and Process in Tourism Systems – A Discussion between C. Michael Hall and Richard W. Butler', *Journal of Sustainable Tourism* 3(2): 99–105.

Hawkins, D. E. (1993) 'Global Assessment of Tourism Policy: A Process Approach', in Pearce D. G. and R. W. Butler (eds) *Tourism Research: Critiques and Challenges*, London: Routledge, pp. 175–200.

Kirk, D. (1995) 'Hard and Soft Systems: A Common Paradigm for Operations Management?' *International Journal of Contemporary Hospitality Management* 7(5).

Laws, E. (ed.) (1997) *The ATTT Tourism Education Handbook*, London: The Tourism Society.

Langer, E. J. (1989) *Mindfulness*, Reading, MA: Addison-Wesley.

Lovelock, C. H. (1996) *Services Marketing*, 3rd edn, Upper Saddle River, NJ: Prentice-Hall.

McLuhan, M. and Powers, B. R. (1989) *The Global Village: Transformation in World Life and Media in the 21st Century*, New York: Oxford University Press.

Nilson, T. H. (1995) *Chaos Marketing: How to Win in a Turbulent World*, Maidenhead: McGraw-Hill.

Peat, F. D. (1991) *The Philosopher's Stone: Chaos, Synchronicity and the Hidden Order of the World*. New York: Bantam.

Poon, A. (1993) *Tourism, Technology and Competitive Strategies*, Wallingford: CAB International.

Toffler, A. (1970) *Future Shock*, London: Bodley Head.

WTO (World Tourist Organisation) (1992) Annual Statistical Yearbook, Geneva: WTO.

Yin, R. (1994) *Case Study Research: Design and Methods*, 2nd edn. Thousand Oaks, CA: Sage.

Part I

TOURISM MANAGEMENT AND POLICY RESPONSES TO CHANGE

INTRODUCTION
TO PART I

Bill Faulkner

In Chapter 1, attention was drawn to the fact that we live in a changing world and that trends in tourism development can be construed as both a response to and, in some cases, an agent of this change. In Part I we are concerned primarily with the former aspect of tourism's relationship with change. That is, we will focus on cases which highlight how the changes in the broader global environment alluded to in the introduction have affected tourism at the regional, national and local levels, and the public and private sector responses to these impacts.

Several chapters (Smith and Pizam, Dieke, and Akehurst) refer specifically to public sector tourism management and policy responses to such global events as shifts in economic relations associated with trade liberalisation and the emergence of regional trading blocks, the economic modernisation in developing countries and the demise of communism.

In Chapter 2, Smith and Pizam examine the tourism implications of the establishment of the North American Free Trade Association (NAFTA), involving the USA, Canada and Mexico. This is one of many regional associations of countries aimed at developing closer economic relations between members. The European Union (EU) stands out as the most advanced of such associations to the extent that, in this case, closer ties go well beyond purely economic relations to encompass some degree of political union. Examples of other regional blocks include the Association of South East Asian Nations (ASEAN) and the Asian Pacific Economic Co-operation Forum (APEC). With the proliferation of these international linkages, and their tourism implications for individual countries need to be analysed so that appropriate tourism policy and management responses can be identified. The Smith and Pizam chapter takes a step towards doing this by considering the potential impacts of NAFTA from a US perspective.

Dieke also examines how the relationships between several countries within one region affect tourism development. However, he is more concerned with the prospects of, and obstacles to, tourism development in neighbouring countries of tropical Africa, and the potential for developing a co-ordinated approach to the

13

planning and marketing of tourism within the region. In the process, insights are provided into the potential costs and benefits of tourism in developing countries, and the specific tourism management problems experienced by these countries. Issues relating to tourism in developing countries are revisited in Part II, where tourism's role as an instrument of economic development is more specifically examined.

Akehurst's analysis of tourism planning in the Polish city of Kalisz involves a reorientation in the geographical focus of discussion from the national and international level to the more local level of an individual city. In the process, his case study highlights the many facets of tourism development that must be taken into account in formulating a tourism plan at the individual destination level. An equally important theme, however, concerns the tourism planning and development implications of the transition from communism to an open market/ private enterprise regime. Many of the problems encountered in the Kalisz case reflect the difficulty of individuals, government agencies and the institutional framework to adjust to the new order. Within this context, questions concerning the imposition of approaches developed in a different socio-political setting on the local community, and the acceptance and ownership of the plans produced by this community, are central considerations.

Like Akehurst's chapter, the chapter by Le Pelley and Laws on the historic city of Canterbury, England, is concerned with the management of tourism development in an urban situation. However, in this case the changes that require a management response have arisen from an increasing number of visitors attracted by the historic buildings within the walled inner city precinct. This has combined with the growth of Canterbury's resident population to create severe traffic congestion and access problems around the central area. These problems are exacerbated by the walled city feature of Canterbury's older parts. The importance of involving stakeholders in the development of management solutions to such problems is highlighted. In particular, the value of the partnership concept, whereby those affected by tourism and/or who have a stake in its development are co-opted into the management process, is illustrated.

There are two conceptual aspects of the Canterbury case study which warrant emphasis by virtue of their broader tourism management implications. First, the extension of Porter's 'value chain' concept and the reference to the way the walled cities of Europe have formed a coalition to facilitate the sharing of their experiences on planning and management problems through the Walled Towns Friendship Circle illustrates how networking can be mutually beneficial to destinations that might otherwise regard each other as competitors. Second, the 'open/soft systems' model, which is introduced as a framework for managing tourism in walled/historic cities, has potential applications in other similar situations. For instance, it is not hard to visualise extensions of this approach to such settings as major events (where confined spaces within cities become magnets for tourists for the duration of the event) or where a major historic or natural (e.g. a beach) attraction has become less accessible because of the combined effects of urban expansion and the continued growth of tourist numbers.

Some tourist destinations stagnate because, as Butler's (1980) Destination Life Cycle (DLC) model suggests, they fail to adapt to the changing profile of the market and/or their competitiveness is undermined by the degradation of their natural assets and infrastructure. This is illustrated in the case of Coolangatta, an Australian seaside resort described by Russell and Faulkner. Coolangatta provides a classic example of a seaside resort whose evolution reflects the sequence of events described in Butler's model, and where the inertia of the area's history has conspired with the vigorous growth of neighbouring newer destinations to create a tourism 'backwater'. Ironically, however, Coolangatta has reached a turning point in its history where it must decide whether to turn its history into an asset by exploiting the memories of its 1950s and 1960s heyday and appealing to the nostalgia (primarily domestic) market, or develop its infrastructure in the direction that is necessary to capture a share of emerging international markets.

Baum also draws on the Butler DLC model in Chapter 7 to examine the evolution of tourism on a selection of Atlantic island resorts. The unique features of island tourism destinations are explored and it is noted that some cold-water islands represent some of the oldest destinations in the history of mass tourism. The declining fortunes of the older, and historically more accessible, 'cold islands' such as the Isle of Man and the Channel Islands is observed and contrasted with the emergence of more distant 'sun, sand and surf' alternatives. These trends, however, are not simply a reflection of changing market tastes and the inherent stagnating tendencies of destinations described in the DLC. Improvements in transport technology have combined with rising standards of living (disposable incomes) to make distant warm island alternatives more accessible in terms of time and affordable cost. The contrasting stories of the two sets of islands are therefore a reflection of broader changes in the spending power and mobility of relevant markets.

The analysis of the evolution of service provision at ski resorts in the French Alps by Lewis and Green draws specifically on the chaos/complexity perspective alluded to in the general introduction. The authors provide an interesting and potentially very useful extension of the chaos theme by drawing a distinction between entrepreneurs and managers. The former are in chaos/complexity mode owing to their creative response to new opportunities and their efforts to circumvent planners and regulators, while the latter endeavour to maintain a steady state through regulations aimed at controlling change. We therefore have an ongoing tension between generators of change and the managers of change.

The juxtaposition of the rational planning and entrepreneurial approaches in Alpine resorts and tourism generally poses an interesting question regarding the evolution of tourism at a destination: is what we see a rationally planned world descending into chaos or a chaotic world rationalised? There are two other fundamental questions that arise from this conceptualisation of the change process. One concerns the degree to which adaptation is driven primarily by the desire to circumvent regulations and whether or not such an approach can produce positive outcomes in the longer term. The other is related to the extent to which regulation may be instrumental in stifling innovation and, therefore,

15

reducing the competitiveness of the destination in the longer term. Regardless of how one might answer these questions, it is apparent that planning is a flawed concept in a chaotic environment. Unpredictable actions and events often have a tendency to subvert plans and render them dated very quickly. In practice, therefore, embracing and managing change is a matter of being able to accommodate the unforeseen and unintended adverse consequences of the interaction between planners and entrepreneurs.

Reference

Butler, R. (1980) 'The Concept of a Tourist Area Life Cycle: Implications for Management of Resources', *Canadian Geographer* 24(1): 5–12.

2

NAFTA AND TOURISM DEVELOPMENT POLICY IN NORTH AMERICA

Ginger Smith and Abraham Pizam

The North American Free Trade Agreement (NAFTA), involving the USA, Mexico and Canada, is one of the more recent trading blocs to emerge as part of the global trend towards closer economic ties between neighbouring countries. Created in January 1994, NAFTA is the largest free trade area in the world, with a $6.5 trillion market and 370 million people, as compared to the European Community which has a GDP of $3 trillion and a population of 357 million people.

The NAFTA agreement, a document of over 2,000 pages long, represents US, Mexican and Canadian commitment to the free flow of trade in goods and services. Its primary objective is to eliminate barriers to trade in products and services. While the provisions of the USA–Canada Free Trade Agreement (FTA) identified specifically targeted sectors, the strength of the NAFTA is that all services were covered by the agreement unless specifically excluded. Aviation, for example, was excluded from coverage under the NAFTA and operates under bilateral arrangements, which must be negotiated separately. Tourism does not have a special chapter or article in the NAFTA agreement but is covered more generally in sections dealing with trade in services (cross-border transactions), financial and telecommunications investments, and temporary entry.

Regional trading blocks such as NAFTA are expected to have positive tourism implications to the extent that closer economic relations between countries facilitate investment in infrastructure, stimulate intra-regional tourism flows and increase international competitiveness among member countries. The purpose of this chapter is to provide a preliminary assessment of the tourism impacts of NAFTA. This task is compounded by the fact that many of these impacts will be long term and therefore not immediately apparent. Also, isolating NAFTA's impacts from those of other factors is difficult, especially when NAFTA-related changes coincide with other events such as the Mexican economic crisis precipitated by the devaluation of the peso. Notwithstanding these difficulties, however, it is possible to provide a qualified assessment of NAFTA's tourism implications by summarising the impacts that were anticipated prior to its inception and reviewing the debate that has occurred regarding its impacts since 1994. In particular, data on tourist movements between the three countries indicate mixed results to date.

NAFTA's anticipated effects on the tourism industry

Before its implementation, NAFTA was heralded by its supporters as an agreement that would benefit the three countries' tourism industries (Smith, 1994). Specifically, benefits were anticipated in the following areas:

- globalization of North American tourism services
- increase in business and pleasure travel
- facilitation of travel credit finance
- increased tourism investments in Mexico
- increased concern for the environment
- improved commercial relations between USA and Mexico
- improved commercial relations between USA and Canada

Globalisation of North American tourism services

By linking the United States to its first and third trading partners, Canada and Mexico, NAFTA was expected to promote global competitiveness for the tourism industries of the three nations. NAFTA, said it supporters, would allow US, Canadian and Mexican companies the same competitive edge that trade ties have given European and Japanese business concerns. The larger market would help the tourism industry be more competitive in global markets, with Europe, Asia and the rest of the world.

Increase in business and pleasure travel

Travel between the three countries was expected to increase as a result of NAFTA. As more goods and services crossed the border, more travellers of all types would follow, promoting increased use of credit cards, airlines, hotels, restaurants, rental cars, tour buses, cruise lines and other travel-related products and services. Tourism is dependent on economic growth and therefore any NAFTA-generated improvement in the economies of the three countries would automatically result in an increase in discretionary and business-related travel. In addition, since international tourism is classified as an export, and raising exports is one of the single most effective ways to create jobs and stimulate economic growth, any increase in tourism that resulted from the NAFTA agreement would be of significant benefit to the USA, Mexico and Canada.

Facilitation of travel credit finance

Travel credit in the form of credit card usage was to be facilitated by the NAFTA, which provides for national treatment for US banks in Mexico. The transaction volume for US bank and non-bank credit cards – such as Visa, Mastercard, American Express, etc. – was anticipated to increase, along with all the accompanying linkages to frequent traveller programmes and airline, hotel and car rental computerised reservations systems.

Increased tourism investments in Mexico

NAFTA was expected to broaden tourism investment opportunities in Mexico. Not only was the Mexican tourism industry supposed to expand through the influx of American investments in its hotel and foodservice industry, but it was also anticipated to become more attractive to German, French and other tourists who may stay, for example, in Canadian and US-owned properties in Mexico and then continue on northward to visit the two other countries.

It is in the hotel sector that specific language on tourism appears in the NAFTA document under entry procedures and documentation. Hotel managers, accredited through certain educational qualifications, who seek entry to Mexico now, may do so more easily under NAFTA through certain visa processes which speed up entry procedures.

Increased concern for the environment

The tourism industry was anticipated to benefit from NAFTA's historic orientation towards environmental issues. Concepts crucial to tourism planning, management, trade and promotion were to be promoted by the NAFTA and its supplemental agreements. This in turn was anticipated to help facilitate understanding of the common ground between developmental and environmental priorities, and create a better balance between economic development and the preservation of natural and multicultural resources.

To oversee implementation of the environmental supplemental agreement, the three countries agreed to create the North American Commission on the Environment (NAAEC) with a council and secretariat. In addition, there is a joint advisory committee of non-governmental organisations to advise the council on its deliberations. The scope of the Commission is extremely broad. The Commission, through its work programme, may address any environmental or natural resource problem subject to consultations among the parties. This ensures an ongoing, trilateral dialogue focused on improving environmental cooperation throughout North America. Mutual cooperation and transparency of the process, coupled with strong dispute settlement measures, were envisioned to promote improved enforcement of environmental laws.

Besides labour, dispute settlement and import surges, there is an additional agreement that provides for a border-funding supplemental, which may have implications for the tourism industry. The implementing legislation in the NAFTA provided for the establishment of a North American Development Bank (Nadbank) to assist communities in the United States and Mexico in carrying out environmental infrastructure projects. Preservation of the physical, economic and social environment in which tourism is conducted requires on-going investment by the industry in human resource and infrastructure development, and therefore financing such investments would be of great importance to the tourism industry.

Improved commercial relations between USA and Mexico

For Mexico, NAFTA represented an opportunity to stimulate billions of dollars in investment, raise employment and increase the nation's standard of living. As trade barriers between Mexico, the United States and Canada fall under NAFTA, the Mexican tourism economy was to benefit in the following ways:

- expansion of travel infrastructure
- expansion of US and Canadian product and services exports
- increase in local procurement of tourism products and services
- elevation of business standards

It was anticipated that Mexican government infrastructure projects initiated after NAFTA's passage would be open to bids from US and Canadian manufacturers and construction companies. Such opportunities are especially important for small and medium-sized businesses, which comprise a large percentage of the tourism industry. These businesses could not readily overcome high Mexican border barriers that existed before the passage of NAFTA. By lowering costs and dissolving barriers and restrictions on US and Canadian exports, NAFTA will help smaller businesses to invest in or supply the Mexican market even without having to invest in Mexico.

On the basis of the expected NAFTA-induced increase in Mexican travel to US destinations, the United States economy was to benefit in the following ways:

- improved US balance of payments
- support for entry-level employment
- expansion of high-tech employment
- increased purchases of other US goods and services
- increased competitiveness of US businesses

Improved commercial relations between USA and Canada

The United States and Canada have traditionally enjoyed one of the most extensive and beneficial trading relationships. It was anticipated that the liberalising effects of the NAFTA would encourage healthy competition in the tourism industry and strengthen Canadian tourism exports.

Summary

Prior to the NAFTA, there were few major barriers with respect to tourism between the United States, Mexico, and Canada. In essence, the three countries have had close to a NAFTA-like agreement from a tourism perspective. The NAFTA applied finishing touches to tourism relations that were basically positive and already open by guaranteeing market openness between the three countries into perpetuity. With tourism barriers reduced, it was expected that travel organizations would feel the pressure to develop higher quality products, priced more competitively, generating tourism industry growth beneficial to both workers and consumers.

NAFTA's effects: three years after implementation

NAFTA's effects on the US economy have been highly debated. On the one hand the Clinton administration and some state administrations such as Texas, claim that NAFTA's effects have been significant and highly positive. On the opposing side, a number of trade unions and lobbying groups led by an organization called 'Public citizens' claim that NAFTA had numerous serious negative effects on the US economy.

NAFTA's advocates

The Office of the US Trade Representatives (a federal agency that is responsible for developing and coordinating US international trade and direct investment policy) is NAFTA's biggest advocate. USTR claims that:

> during NAFTA's first two years, US goods exports to Mexico and Canada were up by 22% or nearly $31 billion, despite the temporary peso-related decline in goods exports to Mexico in 1995. US goods exports to Canada were up by 26%, while US exports to Mexico were up by 11%.
>
> (US Trade Representative, 1997a)

The US Department of Commerce in recently released data, claims that in 1996, US exports to Mexico surpassed pre-NAFTA levels by $15 billion and reached an all-time record of $57 billion. According to this report, exports to Mexico increased across the board from 1995 in all five categories – food, industrial supplies, capital goods, automotive and consumer goods (US Department of Commerce, International Trade Administration 1997).

The US share of Mexico's goods imports rose from 71 per cent in 1994 to 74 per cent in 1995, and 76 per cent in the first six months of 1996. In 1996, US goods exports to Mexico were 23 per cent higher than in 1995. In the administration's opinion, the fact that 'it took only 18 months for US exports to

Mexico to fully recover from the December 1994 peso crisis instead of the seven years that were needed after the 1982 crisis', attests to the effectiveness and importance of NAFTA (US Trade Representative, 1997a).

US agricultural exports to Mexico and Canada increased by 14.2 per cent in financial year 1996 (October 1995–September 1996). Despite a decline in agricultural exports to Mexico during financial year 1995, agricultural exports to Mexico were up almost 35 per cent for financial year 1996 and reached a record of $5 billion. During the same period, US agricultural imports from Mexico were slightly below the $3.7 billion record for financial year 1995. Agricultural imports from Canada increased in financial year 1996 by 20 per cent to a record of $6.4 billion (US Trade Representative, 1997b).

Jobs supported by US goods exports to NAFTA were up an estimated 311,000 since the NAFTA's implementation. On the other hand, the US Department of Labor estimates that between January 1994 and September 1996 a total of 88,602 US workers were displaced because of increased imports from Canada and Mexico. This number, however, represents only 1 per cent of the total number of involuntary job replacements in the US economy during the above period (US Trade Representative, 1997c).

For those who claim that the NAFTA caused the 1994/95 peso crisis in Mexico and the subsequent 1995 Mexican recession, the USTR response was that 'although the US bilateral goods trade balance with Mexico shifted from a surplus of $1.4 billion in 1994 to a deficit of $15.4 billion in 1995, NAFTA did not cause this trade deficit, nor does a deficit necessarily mean a loss of US jobs' (US Trade Representative, 1997c). The USTR suggests that the crisis was caused by decisions made by the Mexican government on exchange rate policy in the 1980s combined with developments in 1994 in financial markets (US Trade Representative, 1997a).

As previously indicated one of NAFTA's supplemental agreements – The North American Agreement on Environmental Cooperation – created a Commission for Environmental Cooperation (CEC) the mission of which was to 'protect, conserve, and improve the environment through increased cooperation among the parties and increased public participation'. In its first two years of operation, the CEC has begun work on seeking solutions to a number of issues of trilateral significance. The commission is currently focusing on five major themes: environmental conservation; protection of human health and environment; enforcement cooperation and law; trade and economy, and information and public outreach (US Trade Representative, 1997d).

A position similar to that of the USTR and the Department of Commerce was expressed by Texas's Executive Director of Commerce, Brenda F. Arnett. Ms Arnett suggests that since January 1994, NAFTA has created 573,000 jobs in Texas, a 7.5 per cent increase over the job creation figures of pre-NAFTA 1993. Texas's exports to Canada have increased 28.7 per cent in NAFTA's first year and 32.3 per cent in the first three quarters of 1995 (Arnett, 1996).

NAFTA's opponents

As previously stated, NAFTA's opponents include most trade unions, a significant number of congressmen, consumer advocates' groups and other political figures such as Ross Perot. The most active and vocal of these groups is Global Trade Watch, one of the six divisions of the consumer advocate group called Public Citizen. Public Citizen was founded by Ralph Nader in 1971, and describes itself as 'the consumer's eyes and ears in Washington', an organization that 'fights for safer drugs and medical devices, cleaner and safer energy sources, a cleaner environment, and a more open and democratic government'. Global Trade Watch's mission is to:

> educate the American public about the enormous impact of inter-
> national trade and economic globalization on jobs, the environment,
> public health and safety and democratic accountability. Global Trade
> Watch works in defense of consumer health and safety, the environment,
> good jobs and democratic decision-making which are threatened by the
> so-called 'free trade' agenda of the proponents of economic globalization
>   (Public Citizen, 1997a)

As can be seen from the above mission statement, Public Citizen is almost by definition against free trade agreements and/or economic globalisation.

In a report entitled *NAFTA's Broken Promises: Failure to Create US Jobs* (Public Citizen, 1997b), Global Trade Watch, argues that: 'More than 600,000 Americans have lost their jobs due to NAFTA.' The report further claims that out of more than eighty firms that made specific NAFTA job creation and export expansion promises in 1993, 89 per cent had not made any significant steps toward fulfilling their promises of job creation or export expansion. 'Under NAFTA', suggests the report, 'a 1993 $1.7 billion trade surplus with Mexico crashed into a record new NAFTA trade deficit of $16.3 billion in 1996, breaking the previous record $1.5 billion deficit of 1995'. To prove that the 1995 US trade deficit with Mexico was not necessarily due to the peso crisis but from the passing of the NAFTA, the report claims that: 'through the current [Mexican] crisis, Japan and the European Union have maintained trade surpluses with Mexico' (Public Citizen, 1997b).

In another report which was released on 23 December 1996 (Public Citizen, 1996) Public Citizen compiled a total of ninety-six 'facts' to prove that NAFTA has had mostly negative effects on the US economy, and only some minor positive effects on the Mexican economy. The following are examples of claims made in this document:

- Percentage of Americans who say their views toward
  free trade are less favorable than a year ago as the
  result of what they know about NAFTA and GATT   52

- Percent of Mexicans who believe their country has
  had little or no success with NAFTA   67
- Estimated number of US jobs in motor vehicle related
  industries lost due to trade with Mexico in 1995   69,048
- Number of Florida tomato producers in 1991   300
- Number of Florida tomato producers in 1995   75
- Number of workers laid off in a Pocohontas, Arkansas
  Brown Group shoes manufacturer due to 'increased
  imports from Canada' due to NAFTA, according to the
  US Department of Labor   2,400

Last, but not least, Robert A. Blecker, in a 1997 article entitled 'NAFTA and the Peso Collapse Not Just a Coincidence' (Blecker, 1997), claims that:

> the peso had to be devalued in order to implement the Mexican strategy for export-led growth that NAFTA was intended to promote – a strategy that was pushed on Mexico by the US Government and the US corporate interests that stood to profit from this trade agreement. In other words, Mexico had to devalue the peso in order to attract the direct foreign investment and export-oriented manufacturing that the NAFTA agreement was designed to promote.

NAFTA's effects on the tourism industry

Arrivals and departures

As can be seen from Table 2.1, arrivals to the USA from Canada and Mexico have seen an average annual percentage decline of 1.82 per cent and 0.65 per cent respectively for the post-NAFTA years, as compared to an average annual percentage increase of 7.26 per cent and 9.68 per cent for the pre-NAFTA years. During the same periods, US departures to Canada and Mexico have seen an average annual percentage increase of 2.2 per cent and 7.7 per cent respectively during the post NAFTA years, as compared to an average annual decline of 1.7 per cent for Canada and an average annual increase of 4.47 per cent for Mexico. Based on the above data, it is possible to conclude that as compared to the pre-NAFTA years, the post NAFTA years have seen an average annual decline in arrivals to the USA from Canada and Mexico and an average annual increase in the departures to Canada and Mexico. If indeed these changes have been caused by the passing of NAFTA, and not by other events such as the peso devaluation, than it would be plausible to assume that NAFTA has had a positive effect on the tourism industries of Canada and Mexico and a negative effect on the tourism industry of the USA.

To compare the effects that the passing of NAFTA had on the tourism industry of the USA, to the effects that it had on the tourism industries of the other partners, we analysed the traffic data between Canada and Mexico. Table 2.2 lists the departures/arrivals and receipts between Canada and Mexico for the pre- and post-NAFTA years.

As can be seen from the tables, there was no significant difference in the annual average change in arrivals to Canada from Mexico, between the pre- and post-NAFTA periods. The average annual per cent change was almost identical in the two periods (5.19 per cent for post NAFTA and 5.02 per cent for pre-NAFTA). However, as far as receipts are concerned, the data shows a significant decline in the average annual change in the post-NAFTA period (8.8 per cent) as compared to the pre-NAFTA period (11.2 per cent). As to Canadian departures to Mexico, the average annual change for the post-NAFTA period shows a significant increase over the pre-NAFTA period (6.77 per cent vs. −36.53 per cent). The same is true for spending by Canadians going to Mexico. The average annual change for the post-NAFTA period shows a significant increase over the pre-NAFTA period (6.5 per cent vs. 2.2 per cent).

As was the case with the USA, in the case of Canada it is possible to conclude that as compared to the pre-NAFTA years, the post-NAFTA years have seen an average decline in receipts from Mexico, and an average increase in Canadian departures and expenditures in Mexico. Here too, we suggest that if these changes have been *caused* by the passing of NAFTA, and not by the peso crisis, it would be

Table 2.1 US arrivals and departures: Mexico and Canada

	1994 (revised)	1995	1996	1997 (estimate)	Average annual % change Pre-NAFTA (1987–93)	Average annual % change Post-NAFTA (1994–7)
Arrivals (000s)						
Canada	14,974	14,663	15,301	15,884	7.26%	−1.82%
% change	−13.4%	−2.1%	4.4%	3.8%		
Arrivals (000s)						
Mexico	11,321	8,016	8,530	8,960	9.68%	−0.65%
% change	15.2%	−29.2%	6.4%	5.0%		
Departures (000s)						
Canada	12,542	12,933	12,909	13,115	−1.7%	2.2%
% change	4.3%	3.1%	−0.2%	1.6%		
Departures (000s)						
Mexico	15,759	18,771	19,616	20,482	4.47%	7.77%
% change	3.1%	19.1%	4.5%	4.4%		

Source: Tourism Industries, International Trade Administration, Secretaria de Turismo (Mexico), Statistics Canada, Canadian Tourism Research Institute.

Table 2.2 Canadian arrivals and departures: Mexico

	1994	1995	1996	Average annual % change pre-NAFTA (1991–93)	Average annual % change post-NAFTA (1994–96)
Arrivals from Mexico (person-visits)	80,900	62,800	80,500	5.02%	5.19%
% change	9.76%	−22.4%	28.2%		
Receipts from Mexican tourists (C$ millions)	81.5	66.2	79.6	11.2%	8.8%
% change	25%	−18.8%	20.2%		
Departures to Mexico (person-visits)	394,000	406,000	438,000	−5.53%	6.27%
% change	8.5%	3.0%	7.3%		
Spending by Canadian tourists on trips to Mexico (C$ millions)	332.9	344.4	358.4	2.2%	6.50%
% change	12.0%	3.45%	4.06%		

Source: Canadian Tourism Commission (Personal Communication 25 June 1997).

reasonable to assume that, as far as traffic between Canada and Mexico is concerned, NAFTA had a positive effect on the tourism industry of Mexico and a negative effect on the tourism industry of Canada.

Other effects

As previously stated, the passing of NAFTA was anticipated to have numerous positive effects on the tourism industries of the three countries in addition to increasing travel between them. Such effects as increasing the competitiveness on North American tourism services, facilitating travel credit finance, increasing tourism investments in Mexico and improving commercial relations, were all heralded by tourism officials and practitioners as important by-products of NAFTA.

An extensive search conducted in the professional and trade literature, government documents, and the Internet did not identify any significant material related to the existence of evidence about any of the above anticipated effects. However, the few articles published in trade magazines such as *Hotel & Resort Industry*, *Restaurant Business*, *Restaurant & Institutions*, *Lodging Hospitality*, and *Public*

Relations Journal were still discussing the *anticipated* positive effects of NAFTA on the tourism industry. Some trade magazines, such as *Nation's Restaurant News*, *Hotel & Motel Management* and *Successful Meetings*, asked a number of industry officials to express their opinions on the *actual* impacts that the passing of NAFTA has had on their business. From the limited testimony offered, it is possible to deduce that, at least for those companies interviewed, the passing of NAFTA had positive impacts on their business by reducing import tariffs on tourism products, by enabling joint US–Mexican tourism promotion projects, and by spurring the growth of Mexican tourism companies. Examples of such benefits cited included:

- Because Mexican import tariffs on food products such as beef have been reduced significantly, US foodservice operators in Mexico had increased their profitability (Ruggless, 1994).
- As a result of NAFTA, since December of 1994 the state of Texas and its neighbour Mexican state of Nuevo Leon, have jointly promoted the tourism facilities and attractions in both states in a 'Two Nation Vacation' programme that focuses on their shared history and tradition (*Successful Meetings*, 1994).
- Due to NAFTA, Mexico's lodging corporation Grupo Situr was able to swiftly increase its number of properties and move into new business lines such as turnkey hotels. (Austin, 1995).

Conclusion

The question of whether NAFTA has had a positive impact on the US economy and, more specifically, whether it had a beneficial effect on its tourism industry, is difficult to answer at this stage. Because of the peso's devaluation and the resulting Mexican economic crisis, it is not yet possible to determine the true impact that the passing of NAFTA had on the economy of the USA. Some claim that the effect was mostly positive and, as time goes by and Mexico totally recovers from its 1995 crisis, the evidence will show that the USA, not just Mexico, has benefited from NAFTA. Others claim that the US economy will continue to lose jobs and businesses to Mexico for a long time to come.

As to NAFTA's effects on the US tourism industry, at present it is very difficult to evaluate its impacts on the creation of new business opportunities for US hospitality firms and/or their profitability. What is, however, possible to determine is whether traffic between the USA and Mexico, and USA and Canada has increased in the post-NAFTA period as compared to the pre- NAFTA period. The arrivals/departures data show that while Mexico has seen an increase in average annual arrivals from the USA and Canada in the post-NAFTA years, the USA and Canada have seen a decline in average annual arrivals or receipts from Mexico. Unfortunately at this stage it is impossible to determine whether the increase in US and Canadian arrivals to Mexico was due to the lower value of the peso, which made Mexico a less expensive destination to visit, or because

27

of the passing of NAFTA. For the same reason it is also impossible to assess the effects of NAFTA on the average annual decline of Mexican departures/receipts to USA and Canada.

Had the peso crisis not occurred and Mexican departures to USA and Canada in the post- as compared to the pre-NAFTA period declined, while US and Canadian departures to Mexico increased, than it would have been possible to suggest that the passing of NAFTA was of benefit only to the tourism industry of Mexico. However, this not being the case, it can only be suggested that at this point in time it is premature to try to determine the true effects of the passing of NAFTA on the US economy, and, more specifically, on its tourism industry. Only with a period of relative stability in the economies of all three countries will it be possible for researchers and decision-makers to evaluate objectively the real effects of NAFTA on the US economy and its tourism industry.

References

Arnett, B.F. (1996) 'NAFTA Opinion Editorial', State of Texas Mexico Office, released March 8, 1996, retrieved 25 February 1997 from the World Wide Web: http://txdocmx.nafta.net/editorial/

Austin, A.A. (1995) 'Grupo Situr's Goal: 25,000 Rooms', *Hotel & Motel Management*, 20 February: 6–14.

Blecker, R.A. (1997) 'NAFTA and the Peso Collapse Not Just a Coincidence', Washington, DC, Economic Policy Institute, retrieved 4 July 1997 from the World Wide Web: http://epinet.org/bleck.htm

Public Citizen (1996) 'NAFTA Index: Three Years of NAFTA Facts', 23 December 1996, retrieved February 25, 1997 from the World Wide Web: <www.citizen.org.7

Public Citizen (1997a), retrieved 25 February 1997 from the World Wide Web: <www.citizen.org.7

Public Citizen (1997b) 'NAFTA's Broken Promises: Failure to Create US Jobs', retrieved 25 February 1997 from the World Wide Web: <www.citizen.org.7

Ruggless, R. (1994) 'Operators Say Business on Upswing 8 Months after NAFTA', *Nations Restaurant News*, 28 (37) 19 September: 12–14.

Smith, G. (1994) Implications of the North American Free Trade Agreement for the US Tourism Industry', *Tourism Management*, 15 (5): 323–326.

Successful Meetings (1994) 'After NAFTA', *Successful Meetings*, March: 133.

United States Department of Commerce, International Trade Administration (1997) 'NAFTA 1996 Update', Washington DC: US Department of Commerce.

United States Trade Representative (USTR) (1997a) 'NAFTA and the US Economy', retrieved 3 April 1997 from the World Wide Web: http://www.ustr.gov/

United States Trade Representative (USTR) (1997b) 'NAFTA and Agriculture Trade', retrieved 3 April 1997 from the World Wide Web: http://www ustr gov/

United States Trade Representative (USTR) (1997c) 'NAFTA and US Jobs', retrieved 3 April 1997 from the World Wide Web: http://www.ustr.gov/

United States Trade Representative (USTR) (1997d) 'NAFTA and the Environment', retrieved 3 April 1997 from the World Wide Web: http://www.ustr.gov/

3

REGIONAL TOURISM IN AFRICA

Scope and critical issues

Peter U.C. Dieke

In 1980, the Organisation of African Unity (OAU) adopted the Lagos Plan of Action (LPA) and subsequently made the Final Act of Lagos the cornerstone of Africa's development strategy (ECA [UN Economic Commission for Africa], 1992a, 1996a; OAU, 1981; UN, 1986, 1991). The Plan's advocacy of self-reliance and self-sustainability stresses the importance of regional cooperation and integration of national and sectoral plans to turn Africa's development aspirations into actuality, i.e. 'Africa, be your brother's keeper'. The political and economic reasons for this idea are obvious (Dieke, 1995) and include the belief that dependence on external resources alone cannot develop Africa in any significant way; and the need to reverse the economic decline which has afflicted the region since the 1970s – the so-called 'lost decade' – in order to strengthen the capacity of the economies to participate effectively in the evolving global linkages and interdependence. Further, the lessons of the 'scramble for Africa' are well known (Afigbo *et al.*, 1992), not least being the political balkanisation of the continent into arbitrary nation-states. Taken together, the imperatives present Africa with immense challenges: small internal markets, fragile borders, vulnerability to external shocks, narrow economic resource bases, etc. In this sense, regionalism has been perceived as the key to solving these problems and difficulties.

Since the Plan's declaration and from even earlier, African countries have moved in several ways towards the accomplishment of the LPA's goal, as seen in the number of regional economic groupings that have, over the years, been formed in the African continent. Examples of such regional organisations include *Communauté Economique de l'Afrique de l'Ouest* (CEAO), *Conseil de l'Entente* (Entente Council), East African Community (EAC), Economic Community of West African States (ECOWAS) in 1975, Southern African Development Coordination Conference (SADCC) in 1980, and the Preferential Trade Area (PTA) for Eastern, Central and Southern African states in 1981. For reasons of political and other expediencies, some of these bodies do still function (e.g. ECOWAS) while others broke up but have now been relaunched (e.g. EAC), still others in 1996 were to re-emerge with a new focus (e.g. PTA) as the Common Market for Eastern and Southern Africa (COMESA) (see ECA, 1996b). It was expected that these regionally based economic organisations would form the building-block for a Pan African Common Market later enshrined in the Abuja Treaty of 3 June 1991 (OAU, 1991).

The recent prime status accorded to tourism by the United Nations (UN) Economic Commission for Africa (ECA) has added impetus to self-help development strategies in Africa (ECA/WTO, 1984; see also Sako, 1990). In short, the ECA is very concerned about the way tourism is developing in Africa, especially Africa's over-reliance on overseas tourists from major source markets and the implications of this situation for development possibilities. Regional tourism in general is not only desirable, it is necessary if Africa is to participate well in the global 'pleasure periphery'.

Given expectations regarding the potential role of tourism in regional development, among other considerations, ECA's policy objectives for the tourism sector reflect a need for an increase in technical training for African staff (ECA, 1992b), for the creation of African tourist circuits and for regional cooperation to ensure efficient utilisation of the continent's vast tourism resources. Although these goals appear soundly based, they do however mask as much as they reveal pertaining to indigenous African tourism measures. In short, what are the chances of achieving the priority tourism development objectives as envisaged and what are the problems and prospects associated with intra-African tourism? This chapter addresses these two questions and, in particular, relates them to international tourism demand characteristics, but begins with a brief analysis of tourism in the context of developing economies.

Tourism in development – a synthesis

It needs to be admitted here that the two questions posed above and, indeed, the whole debate on tourism and development in less developed countries, are too diverse and complex to be adequately covered in any detail in a chapter of this kind. Considering therefore that the subjects have been examined extensively elsewhere by several authors (for example, ECA/WTO, 1984; ECA, 1986; Teye, 1991; Dieke, 1995 on Africa, and also by de Kadt, 1979; Britton, 1982; Erbes, 1973; Lea, 1988; Jenkins, 1992, 1997; Go and Jenkins, 1997 on developing countries in general), this chapter proposes to describe and characterise, both in quantitative and qualitative terms, the salient aspects of the subject matter together with an assessment of critical issues. To gauge the opinions of these writers it is useful to consider briefly the spectrum of contradictory views about the role of tourism in the development process of the developing world.

First, tourism is perceived by some as a panacea – what Erbes (1973: 1) has labelled 'manna from heaven' for developing countries' fragile economies characterized by a scarcity of development resources such as finance and expertise. These resources are needed to increase the nation's economic surplus, without which the countries would be forced to rely on international aid to support development efforts. Tourism's effect can be felt at two levels, macro or national and micro or local. At the first level, tourism is expected to foster economic growth through foreign exchange earnings and an increase in state revenue and, at a second level, an improvement in the people's well-being in the areas of job creation, revenue/income distribution and balanced regional development. Although tourism causes facilities and services to be provided, there is, however, the possibility of these facilities not necessarily being accessible to local residents if tourism development involves the creation of tourism enclaves, as noted below. In addition, tourism has been criticised for exacerbating the problems of societies: the destruction of social patterns, neo-colonialist relationships of exploitation and dependence, inflationary pressure among others. Both positions, admittedly, have merit.

To take a more balanced view, this chapter explores the issues within a broader framework of international tourism economy, relates the synthesis to African tourism perspectives, and assesses the critical factors which might stimulate and inhibit regional tourism development in Africa. It is expected that these analyses should provide policy guidelines which should be considered before major decisions are taken by African policy makers in this field.

Africa's rank in the world tourism economy: trends and analysis

To appreciate the relative importance of tourism in Africa, it is necessary to situate the discussion in a wider perspective of international tourism and to compare regional variations and performances in order to distil from the analysis a number of critical issues of relevance to the subject matter in focus. On this prospectus, Tables 1–4 show the dispersion and growth rate of international tourism as measured in volume (arrivals) and values (receipts) terms and related to country groupings for the relevant years. These trends are considered at four platforms: global, world regional, regional African, and a synthesis.

Global context

Data available from Tables 3.1 and 3.2 indicate the extent and impact of international tourism, both at international and regional levels. According to a WTO revised estimate (Table 3.1), 561 million tourists travelled world-wide in 1995, a growth rate of 2.8 per cent in comparison with 1994 and 5.5 per cent in 1994 over 1993. It is further estimated that US $381 billion were generated in international tourism receipts in 1995 (10.2 per cent higher than 1994), indicating that international tourism has become a major feature in the world economy. Equally important is the fact that these receipt figures have been expressed in current US dollars and reflect the drop in the value of the US dollars *vis-à-vis* other convertible currencies of key tourist-source countries in Europe and Asia. It

31

Table 3.1 International tourist arrivals and receipts world-wide 1985–95 (millions of arrivals, receipts in thousands of millions of US$ and percentage change)

	1985	*1986*	*1987*	*1988*	*1989*	*1990*	*1991*	*1992*	*1993*	*1994*	*1995*
Arrivals (m)	330	341	367	402	431	459	466	503	518	546	561
% annual growth	3.3	3.3	7.7	9.5	7.3	6.6	1.4	8.0	2.9	5.5	2.8
Receipts (US$ b)	117	142	174	202	218	265	272	309	314	346	381
% annual growth	4.4	21.0	22.6	15.7	8.4	21.2	2.7	13.6	1.8	10.0	10.2

Source: WTO (1996a: 2).

should be noted also that the receipt estimates exclude international transport which is a major source of tourism-induced revenue that amounted to about US $57 billion in 1994 (WTO, 1996: 2). Further to the point regarding the significance of tourism in the global economy, it might also be worth noting that the WTO (n.d.) has estimated that international tourism receipts grew at a faster annual average rate over the 1980s (9.6 per cent) than either merchandise exports (5.5 per cent) or commercial services (7.5 per cent).

In relation to tourist arrivals, while this demand characteristic suggests that international tourism is a high-volume industry, the increase in 1995 was rather sluggish when compared with the data recorded in 1994. Part of the reason for this situation can be explained by the depressed nature of the economies of the major tourist-generating countries which were, for the relevant period, rocked by high unemployment and recession. Nevertheless, global tourism activity has continued to grow (Table 3.1). In fact between 1985 and 1995 international tourist arrivals and receipts have respectively grown at the average rate of 5.5 per cent and 12.5 per cent per year. Based on past and current trends, and given international tourism's inherent resilience to bounce back after a recession or a political upheaval (e.g. the Gulf War in 1991; see WTO 1991), there is no doubt that international visitor arrivals and receipts will continue to expand.

Thus the WTO (1994: 4-6) has projected to the year 2000 and beyond an average annual percentage increase in global tourism arrivals of 3.8 per cent between 1995 and 2000, and 4.4 per cent and between 1990 and 2010. Furthermore, it is suggested (WTO, 1996b: 3) that in 1995 tourism receipts accounted for one-third of the value of world trade in the service sector. In other words, tourism receipts represented 8 per cent of total world exports of goods and 30 per cent of the total world exports of services. One other key factor underlying the growth and development of international tourism includes the growth rate for air traffic world-wide with over 360 million passengers carried on international air services in 1995, an increase of 5 per cent over the preceding year (WTO, 1996b), with consequent economic implications for specific regions.

Regional context

At the regional level, Table 3.2 presents international tourist flows, receipts and overall market share for selected years. The key trends might be usefully described as follows. The situation in 1995 suggests that the Middle East was the fastest growing region (12.6 per cent for arrivals and 32.1 per cent for tourism receipts), followed by South Asia and East Asia and the Pacific region. The Americas showed a substantial growth of tourist arrivals in 1995 of 3.3 per cent while international tourism receipts for the whole continent stagnated at 0.7 per cent above the 1994 level. Europe continued to be the most visited region of the world in 1995, with close to two-thirds of international tourist arrivals at a growth rate of 2.3 per cent, one point below the world average. In fact, Europe and the Americas maintained their position of leadership in international tourism in 1995, together accounting for almost 80 per cent of tourist arrivals and 71 per cent of receipts. Africa witnessed a slight improvement in the growth rate of tourist arrivals but the trend in tourism receipts in 1995 was an increase of 3.4 per cent from 1994 down from 7.1 per cent previously. Conversely, tourism arrivals increased from 1.0 per cent to 2.4 per cent over the same period.

Brief comments on the above trends might be appropriate, particularly relating to short-haul versus long-haul international travel. Much of intra-European travel (i.e. travel within and between Western European countries) is relatively short haul (and therefore more comparable with a lot of domestic travel in large countries like Australia and the United States) (see Edwards, 1987; Go, 1997; Theobald, 1994; Williams and Shaw, 1988). This distinction is important because it helps to explain the dominance of Europe, in volume and value terms, in the international tourism market place. Second, as noted elsewhere (Dieke, 1997), it highlights the relevance of intra-regional tourism flows in Africa, among other African source markets. The latter issue is considered further below.

Regional African context

A close examination of Table 3.3 indicates the origin of the main tourist flows in Africa in 1995. As can be seen, there were three major source markets, in order of importance: residents of African countries (i.e intra-regional tourism); Europeans; visitors from the Middle East, from the Americas and from Asia. It is clear that intra-regional flows are by far the dominant tourist flows constituting over 44 per cent of the total in 1995. Intra-regional flows witnessed a significant increase of about 162 per cent in this year over 1985, representing an average annual growth of 10.1 per cent since 1985. Europe constitutes the second largest tourist-generating market for Africa. In 1995, for instance, tourist flow into the continent from Europe stood at approximately 6 million, compared with above 4 million a decade earlier, an annual average growth of 3.5 per cent. Finally, arrivals from the Middle East have significantly increased between 1985 and 1995. Flows from the Americas, mainly the United States, appear to have stagnated in the past ten

Table 3.2 International tourist arrivals and receipts in selected country groupings, 1990, 1994–95

(a)

Country groupings	Tourist arrivals (in millions and % annual change)			Tourist receipts (in US$ billions and % annual change)			Market shares (%)					
	1990	1994	1995	1990	1994	1995	1990 Arr.	Rec	1994 Arr.	Rec.	1995 Arr.	Rec.
World	459.2 (6.6)	545.9 (5.5)	561.0 (2.8)	264.7 (21.2)	345.5 (10.0)	380.7 (10.2)	100.0		100.0		100.0	
Africa	15.1 (9.2)	18.3 (1.0)	18.7 (2.4)	5.2 (15.0)	6.4 (7.1)	6.7 (3.4)	3.3	(2.0)	3.3	(1.9)	3.3	(1.7)
Americas	93.6 (7.5)	107.0 (3.3)	110.6 (3.3)	69.5 (16.4)	95.7 (6.1)	96.4 (0.7)	20.4	(26.2)	19.6	(27.7)	19.7	(25.3)
East Asia/Pacific	53.1 (14.6)	76.9 (10.6)	83.0 (7.9)	38.8 (13.7)	61.7 (18.3)	71.9 (16.2)	11.6	(14.7)	14.1	(17.9)	14.8	(18.9)
Europe	286.7 (5.0)	329.8 (5.1)	333.3 (1.1)	144.0 (27.9)	173.2 (9.6)	195.3 (12.8)	62.4	(54.4)	60.4	(50.1)	59.4	(51.3)
Middle East	7.6 (−1.2)	9.9 (9.9)	11.1 (12.6)	5.1 (−4.3)	5.1 (6.3)	6.7 (32.1)	1.7	(1.9)	1.8	(1.5)	2.0	(1.8)
South Asia	3.2 (4.1)	3.9 (10.9)	4.3 (8.2)	2.1 (2.8)	3.2 (13.4)	3.7 (16.6)	0.7	(0.8)	0.7	(0.9)	0.8	(1.0)

Table 3.2 continued

(b)
Country groupings (continued)

Country groupings	Tourist arrivals (in millions)			Tourist receipts (in US$ millions)			% change		Market shares (%)	
	1990	1994	1995	1990	1994	1995	94/95	90/95	1990	1995
Developed countries	284.2	314.7	321.4	190.0	230.2	242.1	5.2	5.0	71.6	65.1
Developing countries	128.2	157.3	170.1	70.3	102.1	113.3	11.0	10.0	26.6	30.5
Central/East Europe	46.7	74.3	75.6	4.9	14.4	16.2	12.9	27.4	1.8	4.4
OECD	303.0	331.4	340.3	195.8	237.1	249.0	5.0	4.9	74.0	67.0
European Union	205.9	225.0	230.3	121.3	139.1	151.1	8.6	4.5	45.8	40.7
NAFTA	72.0	78.6	81.5	54.1	73.0	71.5	-2.1	5.7	20.4	19.2
ASEAN	21.0	25.6	28.0	14.0	23.1	25.7	11.2	12.9	5.3	6.9
Mediterranean countries	155.0	169.0	173.0	75.3	92.2	103.2	12.0	6.5	28.4	27.8

Sources: WTO (1996a: 6–9, (1996b): 18–19).

Table 3.3 Tourist arrivals in Africa according to main source markets, 1985 and 1995 (thousands of arrivals, growth percentage and market share)

Source markets	Tourist arrivals in Africa (000s)		Average annual growth rate (%)	Market share (%)	
	1985	1995	1985–1995	1985	1995
Africa	3,164	8,280	10.1	32.6	44.3
Europe	4,107	5,766	3.5	42.3	30.8
Middle East	194	652	12.9	2.0	3.5
Americas	430	475	1.0	4.4	2.5
Asia/Pacific	163	259	4.8	1.7	1.4
Other countries	1,653	3,275	7.1	17.0	17.5
Total Africa	9,710	18,707	6.8	100.0	100.0

Source: WTO (1996a: 95).

Table 3.4 Tourism trends by sub-regions 1990–95

Sub-regions	Tourist arrivals (000s)	% change over	% of total Africa		Tourism receipts (US$ m)	% change over	% of total Africa	
	1995	1994	1990	1995	1995	1994	1990	1995
Eastern	4,058	10.2	18.9	21.7	1,563	8.5	21.5	23.5
Middle	269	−2.5	2.3	1.4	115	−5.0	2.0	1.7
Northern	7,240	−11.0	55.7	38.7	2,536	−3.2	44.0	38.1
Southern	5,584	18.8	13.5	29.9	1,866	10.5	21.5	28.1
Western	1,556	6.4	9.6	8.3	574	1.1	11.0	8.5
Total Africa	18,707	2.4	100.0	100.0	6,654	3.4	100.0	100.0

Countries of the sub-regions

Eastern: Burundi, The Comoros, Djibouti, Eritrea, Ethiopia, Kenya, Madagascar, Malawi, Mauritius, Mozambique, Réunion, Rwanda, Seychelles, Somalia, Uganda, United Republic of Tanzania, Zambia, Zimbabwe.

Middle: Angola, Cameroon, Central Africa Republic, Chad, Congo, Equatorial Guinea, Gabon, Sao Tomé and Principé, Zaïre (now Democratic Republic of Congo).

Northern: Algeria, Morocco, Sudan, Tunisia.

Southern: Botswana, Lesotho, Namibia, South Africa, Swaziland.

Western: Benin, Burkina Faso, Cape Verde, Côte d'Ivoire, The Gambia, Ghana, Guinea, Guinea Bissau, Liberia, Mali, Muaritania, Niger, Nigeria, Sénégal, Sierra Leone, Togo.

Source: WTO (1996a: 25).

years. The Asian market has exhibited rather strong growth, reflecting the low volume base.

Table 3.4 indicates the sub-regional pattern. North African countries had the largest share of both arrivals (38.7 per cent) and receipts (38.1 per cent) in 1995, followed by Southern and Eastern Africa in descending order. Western and Middle sub-regions are the poor cousins of the continent's tourist industry. This Northern dominance is explained not only by the sub-region's proximity to the major European markets but, more importantly, its long-standing economic and other ties with these areas. There was also the suggestion back in 1972 that the sub-region is 'simply a natural extension of European resorts, in the path of the inevitable southern push towards the sun and, initially at least, towards less crowded beaches' (Hutchinson, 1972: 45). It is further argued, with respect to many developing countries that 'where foreign enterprises were present in a country's tourist industry they would be the most successful' (Britton, 1982: 340). This explains why the countries of Eastern and Southern Africa are, in tourism terms, significant, as the case study of Kenya shows: 'pioneer facilities developments were in place because Kenya had a vigorous expatriate community which sought to advance foreign commercial interests, including tourism' (Dieke, 1993: 13). On the basis of the statistics (Tables 3.1–3.4) we can draw a number of conclusions.

A synthesis

These trends in demand and therefore receipts and market share, can be summarised in the following way:

- All regions have expanded their tourism sectors following the advent of jumbo jets and the emergence of the giant hotel chains. These developments have enhanced the extent of tourist travel and the number of destinations, and the competition among them which increased noticeably.
- While the expansion has benefited all regions, this has been on an unequal basis and thus highlights a heavy geographical concentration of both tourist arrivals and receipts.
- In absolute terms, the industrialised nations of North America and Western Europe and Japan, have absorbed about 80 per cent in both counts.
- There has been a global redistribution in the foci of tourism activity associated with the emergence of the 'tiger economies' of the Asia-Pacific Rim. Economic growth in these countries has fueled both growth in demand and the awareness of new tourist destinations.
- Although the quantitative importance of tourism varies greatly from one region to another and may appear very small in developing regions when compared to developed regions, tourism's significance to the local economy of developing regions should not be underestimated.
- Finally, tourism in Africa had a modest revival of growth in 1995, although

this did not necessarily lead to a growth of net benefits. However, Africa accounts for a mere 2 per cent of world tourist arrivals and receipts – a figure that has not changed much, even though the continent accounts for over 12 per cent of the world's population (see ECA, 1991: 5).

In practical terms, three questions arise from this synthesis: What is holding African tourism back? What are the constraints on African tourism success in the 1990s and beyond? What can be done to remove the constraints? These questions are addressed in the following sections but the main emphasis will be on finding possible answers to the third question through an examination of intra-regional tourism.

Towards regional tourism in Africa

To exemplify the need to develop 'African intra-regional tourism', perhaps a starting point is to define the concepts of 'regionalism' and 'regional tourism', and to consider why Africa needs regional tourism. This section also evaluates measures taken to achieve that aim, and then discusses some of the broad policy issues arising from the analysis.

Concepts of 'regionalism' and 'regional tourism'

'Regionalism' means simply cooperation among African countries to integrate their markets in order to engender faster economic growth and enhanced living standards – the cornerstone of the Abuja Treaty establishing the African Economic Community (OAU, 1991). In terms of tourism development, regionalism implies that each African country will cooperate at all levels in the development of 'their attractions, capital infrastructure, natural and human resources to serve the needs of the domestic and international (inter-regional and intra-regional) tourism sectors' (Teye, 1991: 288). This definition has connotations at two levels, government and individual.

At the *government level*, there is a need to draw on local resources in the construction and running of a tourism industry and to create facilities adapted to a particular market segment – leisure, business, visiting friends and relatives (VFR). The second implication of the regionalism concept is that, at the *individual level*, there is the need, for instance, for leisure travel to be directed towards Africa, utilising local natural and social resources. In both instances it can be argued that developing intra-African tourism carries with it the corollary that a tourist product using indigenous African features and more local resources is central to its success. Nevertheless, the obvious challenge is to promote an original, authentic product that takes account both of Africa's needs and the demand generated by promotion.

From the point of view of tourism statistics, what is fundamentally left undefined so far is what is 'regional tourism' in an African sense? As used here, the term is taken to mean simply the movement of tourists resident within the Africa region. 'Tourists', as defined by the WTO relate to genuine visitors whose major travel motivations are for pleasure, business, cultural, religion, visiting friends and relatives, meetings or conferences. Meanwhile the 'African region', again as generally defined by the WTO, does not include Egypt, which is regarded as part of the Middle East. Thus tourists from Egypt are completely excluded. While tourists may be nationals of African and non-African countries, their key to inclusion is that they are domiciled in Africa.

Support for regional tourism

There are many reasons for seeking to develop African intra-regional tourism. On a general level, the argument that international tourism has the potential to contribute to economic development has been noted above. Regional tourism may have specific features which are of significance to warrant Africa's support for it as part of their development strategies.

First, the Niamey Conference on 'Intra-African Cooperation in Tourism' (ECA, 1986) expressed serious concern over Africa's over-dependence on foreign sources as its main tourist 'trigger markets' and the economic and other consequences for the continent associated with the situation. Given the previously observed asymmetrical patterns of international tourism market share, the general demand fluctuations and the issue of 'leakages', the conference felt that Africa should concentrate on domestic and intra-regional tourism. This strategy would eliminate such distortions and would result in two obvious outcomes: increasing the profitability of tourism infrastructure and enhancing overall intra-Africa trade.

Second, demand, be it present or potential, for intra-African tourism does exist (see Table 3.3). On the basis of an analysis of road transport statistics, one regional adviser in the transport and communications unit of the Economic Commission for Africa (ECA), shows that Africans travel a lot even in the face of current unstable economic and political environment (personal communication with the author). This view seems to apply more to business tourism than any other type of tourism and may well suggest that one of the most important aspects of travel in Africa is related to 'business purposes'. The significance of all this is that, in the words of the regional adviser, 'weekly markets and trade have always animated the economic, social, cultural and political life of Africa'. Growth in future demand for regional tourism in Africa can be realised given improvements in education levels, a renaissance of cultural awareness, increasing urbanisation and increasing population growth (with about 50 per cent of youth being under the age of 20) which Africa has witnessed in recent years. It is possible here to make a few comments on demand determinants.

While it might be true that the main travel motivations for Africans are business related, it is doubtful whether these alone are sufficient to trigger tourist movements of the type and scale being suggested in this chapter. There are, of course, other potentially significant components of travel such as politics (e.g. official missions), visiting friends and relatives, holiday and recreation, etc. The scale and impact of these travel categories will vary from country to country. Second, the demand would appear to highlight not only those factors which motivate travel in Africa, e.g. improvements in education levels and urbanisation but also structural factors, e.g. changes in population structure. Of equal, or perhaps greater importance is the fact that the demand variables identified thus far are silent on the economic considerations which facilitate travel, e.g. sufficient disposable income, adequate means of transport and relative prices. These are issues for further debate.

As has been argued elsewhere (Adejuwon, 1986) leisure travel is not integral to the lifestyle of an average African. Demand is constrained partly by Africans' low income per capita and partly as a consequence of extended family commitments and obligations. The effect of all this on Africa might be to depress demand for leisure travel, while simultaneously stimulating the VFR market segment. The latter assessment might raise a question as to whether the VFR has particular significance for the intra-regional tourism economy when it is considered that this class of tourists tends to stay with friends or in forms of accommodation other than hotels.

A third reason to support regional tourism pertains to the wide range of physical, historical and cultural resources that contribute to the tourist product. Although these may vary greatly they will have an added advantage of meeting the specific interests of specific cultural and ethnic groups. For example, it will be easier to attract visitors to a location or site with a historical monument where its existence has particular relevance in motivating those groups.

Fourth, from a marketing angle, regional tourism cooperation has merit. An increasing number of foreign tourists who visit Africa are now opting for tour circuits as opposed to resort holidays. This will enable them to see as much of the continent as possible and visit many places and so enrich their holiday experience. For Africa, such a prospect can only be realised if several countries combine to offer more attractive products and programmes, renew and diversify each region's tourism supply, and thus contribute to the differentiation of the local product. This policy will prove more attractive and more easily understood by foreigners than national products. A regional concept will harness to greater effect marketing and promotion resources, while a more coordinated approach to negotiations with tour operators can be developed. Additionally, a united Africa would be better able to counter, through education programmes targeted at the travel trade in the tourist-generating countries, the negative impressions and images associated with African destinations. At the same time, by acting in unison, the continent might strengthen its bargaining position in order to win more concessions from transnational companies.

Finally, as alluded to above, other considerations include the need for Africa to overcome the disadvantages of small-sized economies or markets, to consolidate the political independence of African countries and to strengthen its overall position *vis-à-vis* other competitors. Some other spin-offs might be the opening up of those countries with fewer tourism resources and those located away from major tourist routes. The mutual interest which regional tourism will generate may provide a catalyst for the development and profitability of national or regional airlines. To support these policies, African countries have taken several measures, with varying degrees of commitment and success.

Regional tourism support programmes

African countries recognise that they cannot move forward with regard to these prospects unless they take concerted action. The existing structures in the continent can help here: at the political level, there is the OAU and at the economic, social and cultural levels there are the ECA, and the regional and sub-regional organisations. The key here is to forge close cooperation among these bodies in order to pool their human and financial resources, and to ensure the implementation and monitoring of the programmes. The recent formation of the Regional Tourism Organisation of Southern Africa (RETOSA) in April 1996 is a logical outcome of this new thinking. Cooperation with such international organisations as ILO/WTO, members of the UN system and the European Union (EU) will avoid duplication of work. With all this in mind, African countries have proceeded to establish cooperation in tourism on the basis of two structural principles, which are inter-governmental and sectoral-technical in nature.

There are three levels of inter-governmental cooperation. The first is the Conference of African Ministers of Tourism, a consultative body which meets every two years. The fundamental objective of this body is to coordinate and evaluate, through *ad hoc* mechanisms set up for the purpose, Africa-wide tourism initiatives and to seek their implementation. Some specific examples of the Conference's areas of cooperation are in the harmonisation of tourist legislation and promotion, defence of the interests of African operators, etc. Second, there are Tourist Promotion Committees at regional and sub-regional levels, whose consultative role and responsibilities involve the creation and development, promotion and marketing of Africa's tourist product. Membership of the various Promotion Committees is drawn from both the public and private sectors, home and abroad. Finally, the Inter-Agency Coordination platform coordinates the action of regional, sub-regional and international institutions (e.g. UN system) operating in Africa. It also develops strategies for tourism and mobilises the funds required to implement the programmes selected.

The second structural principle, sectoral cooperation, is implemented through three professional associations: the African Hotel Association (AHA); the African Tourist Training Centre Association (ATTCA); and the African Travel and Tour Operations Association (ATTOA). These are, in the main, technical bodies whose memberships are continent-wide and whose objectives are wide-ranging and reflect the interests of their members in the development of tourism in Africa.

These support measures should perhaps be better viewed against the background of the general problems afflicting Africa which are themselves barriers to development of the tourism sector. Therefore, the programmes should be applauded as a good start to the extent that they demonstrate that at last African policy makers and others interested in Africa's future are living up to expectations as they attempt to articulate a concerted response. That said, these support programmes are of course not an end in themselves. They are no more than facilitating mechanisms and their success or failure will depend on how rhetoric dovetails with action and, in particular, how quickly targets are achieved. Without doubt, many obstacles – some internal, others external – need to be considered and perhaps overcome if regional tourism cooperation in Africa is to materialise.

Critical issues

The desire of African leaders to pursue intra-regional tourism must be accompanied by a similar desire and commitment from the African populace. However, these two-fold desires, no matter how well-intentioned, do not necessarily work as these intentions must be accompanied by situational factors which transform desire into actuality. It is important that the issues are not seen only in a narrow, African focus, but also in a wider sense of business relationships and policies particularly relating to international tourism; to underscore the interdependence of tourism and its constituent parts, a multi-sectoral and multi-faceted industry, especially if it is accepted that in tourism we live in a 'global village'.

First, an explanation of why tourism flourishes in some developing societies and languishes in others must be sought. Africa is perhaps a classic case of the latter case, with disastrous consequences for intra-regional tourism in the continent. Most arguments have turned either to aspects of the culture of society, or to certain economic realities. Adejuwon (1986), for instance, has argued that indigenous Africans are not leisure-minded. This implies that the values of African society do profoundly influence holiday-taking behaviour. It further means that the traditions of African family life prevent Africans from participating in recreational tourism. In this sense, it is not leisure but visiting friends and relatives (VFR) tourism that predominates in much of Africa. Since intra-regional tourists frequently lodge with friends or family, the implication must be that there is a corresponding reduction in the need for hotel accommodation. Thus the VFR market segment is not contributing to the tourist economy despite the argument to the contrary (see *Journal of Tourism Studies*, 1995).

There is evidence to suggest (see Ifemesia, 1982) that the obligations of kinship drain the resources of a potential African tourist, who is also overwhelmed by other imperative demands to rob him of the opportunity to 'get away from it all'! This situation is further compounded by the low income levels which reflect economic conditions in Africa. Given this reality, and particularly Africa's perception of holiday-taking, the prospects for intra-African tourism might be limited. The question must be, how does one change this perception to foster Africa participation in holiday tourism?

Second, for regional tourism to survive, it is essential that Africa maintain internal cohesion and avoid political instability and political conflicts between countries. While the different levels of tourism development in many countries in Africa provide many potential opportunities for the more developed to provide assistance to those newly entering the field, it seems natural to suppose that there is an element of competition. This is a natural consequence of the commercial realities of tourism, but in Africa it may be reinforced by differences in political outlook. The demise of the East African Community (EAC) in the mid-1960s was indicative of how separatist forces could potentially derail or threaten regional cooperation in tourism. The main criticisms and concerns centred on Tanzania's accusation that Kenya was engaged in a 'dirty tricks' campaign over Kenya's diversion of tourists to its own advantage (see Mascarenhas, 1971). The situation had two effects; first, to break up the EAC and, second, to force the two countries to close their common borders for a very long time. In retrospect, it can be argued that the main lessons learned from this episode are three-fold; the need for sovereign states in Africa to make political concessions; to show great states-manship; and to set aside national interests in order to serve continent-and sub-national-wide development goals.

Third, it is important to acknowledge the place of foreign interest groups in the tourism sector – be they in tour operations, airlines or hotels – in developing Africa's tourism resources and also in stimulating demand. From the point of view of economies of scale advantages, international corporations have extensive market connections, established reputations, and the financial muscle and recognised expertise to enable them to make or break a tourism destination. The evidence confirms that the most 'successful' tourism developments have been realised because of the vertical relationships which many African countries have forged with foreign tourism enterprises. One might suspect that these companies will be upset by Africa's current attempt at regionalism in tourism activities as such initiatives could alter the status quo in some way. Such a development has the prospect of being counter-productive considering the competitive environ-ment that characterises international tourism and all the consequences that flow from it.

One possible outcome of this situation could be destination substitution, stemming from poor and inadequate levels of service provided in the host destination. With a growing awareness of consumerism and the recent intro-duction of the European Union (EU) Directive on Package Travel, travel companies in the EU countries are obliged to prevent such poor service levels, to enrich tourist holiday experience and thus avert possible legal actions. The implications of all this is that tour operators are becoming increasingly selective about their choice of agents with whom to do business. Something about which little is known is what goes into these negotiations, the sensitive issues, the criteria for decision-making, the behind the scenes interplay of forces that shape public policy (X.X., 1983). The nature of the subject matter may give one grounds to speculate about corporate praetorianism in international tourism as transnational companies might move their capital elsewhere if they felt threatened – a development that would be a major setback to Africa's tourism programmes.

Fourth, another important situational factor for intra-regional tourism cooperation is the nature that regional tourism might take as there are no known studies on the subject. Similarly it is unclear what the characteristics of African tourists are, or what their likes and dislikes are. It is evident that international tourists require particular standards of facilities, foodstuffs, services and as Cohen (1972) suggests, the international (mass) tourist travels in 'an ecological bubble' of a familiar environment which reminds him or her of home. Several questions are pertinent. Can the same be said of African tourists or would their profile differ from that of other international tourists in any major way? How compatible are the two market segments?

As a general proposition it is obvious that tourists from major tourist-generating countries are at the top end of the spending spectrum, which the majority of African budgets cannot match. While this statement may be true, supporters of African cause might conversely argue that what is good for the European visitor is equally suitable for the African tourist. This observation may, of course, prompt a further question: Is it possible that, with economic development and 'modern-isation' in Africa, the African market's tastes and preferences will increasingly resemble those of Western countries? Whatever the answer to this question, the challenge of course is to adapt tourism product, especially accommodations, to the needs and financial capabilities of the African market. Second, it is inconceivable to expect those African countries currently serving major international tourist countries suddenly to sever this relationship in favour of intra-African tourists, or for countries at an early stage of tourism development to design a tourist product which excludes Western tourists. If, as noted, the idea of a 'global village' has substance, the truth is that Africa will in the foreseeable future continue to depend on Western countries as its main source of tourists.

Fifth, one is reminded that most African economies are characterised by 'soft' currencies which have no exchange value beyond the national frontiers of issue, and are thus not used in international trading transactions. Given this situation, and also the shortage of foreign exchange, promoting intra-African tourism would prove difficult without a financial framework to permit a free flow of capital. It can be argued that perhaps one possibility is to resuscitate the moribund currency used in the countries of East African Community before the union broke up; another alternative might be to adopt the CFA franc that has for years served the former French West African states. In both cases, however, caution must be exercised because they call for considerable political courage. More important, perhaps, is the understanding that support for these regional currencies is not shared by many African 'leaders of thought'. Critics would therefore associate the prescriptions with the 'new colonialism' ethic and tourism must bear its own share of the blame for perpetuating this culture.

As Chief Emeka Anyaoku, the present Secretary-General of the Common-wealth, claimed in a 1992 television documentary:

> the old regional [colonial] arrangements were intended to meet the needs and demands of a different order which inevitably tied Africa's economic life to the needs of the governing European metropolis. Independence created its own needs which had to be met in new ways. ... Colonialism in Africa was essentially a process that began with carving up and dividing the Continent at the 1884/1885 Berlin Conference. The consequence of horse-trading by the European colonial powers paid little regard to Africa's old national or ethnic loyalties. ... We must never forget that whatever its apparent practical benefits, colonial rule, by the nature of human history and experience, had to come to an end. That is why, in spite of all the coups and civil conflicts, nowhere in Africa have people sought or will ever seek the return of colonial rule.

Such words of wisdom are instructive and indeed the tourism literature is replete with this characterisation of tourism as an agent of colonialism (see Britton, 1982). This is a subject that is reserved for another occasion.

Finally, other inhibiting factors worthy of consideration can be summarised as follows: the unreliable nature of African air services in terms of route connections, the expense involved and service quality. Transport bottle-necks slow down development and economic activities as a whole; so does the low level of free movement from one country to another. The current global changes in the geo-political landscape (e.g. the integration of markets in the old communist bloc countries, the 1992 single European Agreement) are a cause for concern. The emerging democracies might prove to be boom tourism destinations and Edgell (1990) has suggested, there might be a trend for more and more tourists to want to holiday in their own 'backyard', leaving Africa further marginalised. These problems notwithstanding, we wish to conclude the chapter by formulating a few generalisations about intra-regional tourism cooperation in Africa.

Conclusion

Developing intra-regional tourism in Africa is not an easy task and will require all the energies and talents the tourism leaders can muster. In practice, this objective depends strongly on the following factors:

- the development of tourism between adjoining countries of the same sub-region (the Regional Tourism Organisation of Southern Africa or RETOSA, formed in April 1996, provides an excellent example) and, in the longer term, between different sub-regions (West and North Africa, for instance);

- simplifying frontier formalities for adjoining countries to facilitate movements between them;
- development of business and professional congresses, trade fairs, conferences and seminars at regional and sub-regional level as a means of exchanging experience and promoting trade;
- promotion of special air fares (excursions, etc.) using national or sub-regional airlines within a given sub-region and between the main hubs;
- creation of accommodation facilities commensurate with the needs and means of the African market;
- public awareness campaigns to highlight the role of tourism in the development process and encourage participation;
- reducing tariffs to national and intra-regional customs;
- devising and promoting all-inclusive trips and packages (weekends, one week) especially for economic, cultural and sports events of regional or sub-regional scope;
- harmonising development plans and legislation, as well as standardising the components of tourism products.

In general terms, regional tourism should take into account the special characteristics of each country, the nature of the tourism infrastructure, human resources for employment, and the economic, cultural and social environment. While much progress has been made in some of the above matters, others (e.g. the will to make political concessions, political instability, etc.) may become obstacles to coordinated action among tourism authorities. Clearly problems vary considerably from sub-region to sub-region and it is difficult to reconcile the interests and aims, for example, of North and West Africa, Central Africa or Southern Africa. In Africa, as elsewhere in developed and developing countries, regionalism should precede internationalism and there is every assurance of its viability. The examples of the North American Free Trade Association (NAFTA), the European Union (EU), the Association of South East Asian Nations (ASEAN) and others confirm this notion.

References

Adejuwon, Franklin J. (1986) 'Trends of tourist demand in Africa', *Tourist Review* 41, 1: 20–24.

Anyaoku, Emeka (1992) 'The image of Africa', London Independent Television Network (ITN) Channel 4 Documentary, 8 April.

Britton, Stephen G. (1992) 'The political economy of tourism in the third world', *Annals of Tourism Research* 9, 3: 331–58.

Cohen, Erik (1972) 'Towards a sociology of international tourism', *Social Research* 39, 1: 164–82.

Dieke, Peter U.C. (1993) 'Cross-national comparison of tourism development: lessons from Kenya and The Gambia', *Journal of Tourism Studies* 4, 1: 2–18.

Dieke, Peter U.C. (1995) 'Tourism and structural adjustment programmes in the African economy', *Tourism Economics* 1, 1: 71–93.

Dieke, Peter U.C. (1997) 'Overview on tourism in Africa', paper presented at the Workshop on Promotion of Tourism in Africa held on 3 April at the National Teachers' Institute, Kaduna, Nigeria, to mark the 7th All-Africa Trade and Tourism Fair in Nigeria 29 March–6 April.

ECA (UN Economic Commission for Africa) (1986) *Development of tourism in Africa: the intra-African tourism promotion of tourist product*, Addis Ababa, Ethiopia: ECA (Transcom/68).

ECA (1991) *African alternative framework to structural adjustment programmes for socio-economic recovery and transformation – a popular version*, Addis Ababa, Ethiopia: ECA.

ECA (1992a) *The ECA in the 1990s: policy and management framework for facing Africa's development challenges*, Addis Ababa, Ethiopia: ECA (E/ECA/CM.18/4).

ECA (1992b) *Directory of vocational training facilities in tourism in Africa*, Addis Ababa, Ethiopia: ECA (Transcom/91/500).

ECA (1996a) *Serving Africa better: Strategic directions for the ECA*, Addis Ababa, Ethiopia: ECA (E/ECA/CM.22/2).

ECA (1996b) 'COMESA: a successful venture in regional integration', First Session of the conference of African ministers responsible for trade, regional cooperation, integration and tourism held in Addis Ababa, Ethiopia, 14–16 February (ECA 96/133).

ECA/WTO (World Tourism Organisation) (1984) 'Intra-Africa tourism cooperation', joint presentation to the Regional Conference on Intra-African Tourism Cooperation jointly organised by ECA/WTO held in Niamey, Niger, 2–5 October.

Edgell, David L. (1990) *International tourism policy*, New York: Van Nostrand.

Edwards, A. (1987) *Choosing holiday destinations: The impact of exchange rates and inflation*, Special Report No. 1109. London: Economist Intelligence Unit.

Erbes, Robert (1973) *International tourism and the economy of developing countries*, Paris: OECD.

Go, Frank M. (1997) 'Asian and Australasian dimensions of global tourism development', in Frank M. Go and Carson L. Jenkins (eds.) *Tourism and economic development in Asia and Australasia*, London: Cassell, pp. 3–34.

Go, Frank M. and Carson L. Jenkins (eds) (1997) *Tourism and economic development in Asia and Australasia*, London: Cassell.

Hutchinson, Alan (1972) 'Tourism in Africa: Africa's tourism leaps ahead', *African Development* November, pp. 41–50.

Ifemesia, Chieka (1982) *Traditional humane living among the Igbo: an historical perspective*, Enugu, Nigeria: Fourth Dimension.

Jenkins, Carson L. (1992) 'Tourism in Third World development – fact or fiction?', inaugural lecture given at The Scottish Hotel School, University of Strathclyde, Glasgow, UK, 30 April.

Jenkins, Carson L. (1997) 'Impacts of the development of international tourism in the Asian region,' in Frank M. Go and Carson L. Jenkins (eds.) *Tourism and economic development in Asia and Australasia*, London: Cassell, pp. 48–64.

Journal of Tourism Studies (1995) The VFR market, Special issue, *Journal of Tourism Studies* 6, 2.

de Kadt, Emanuel (1979) *Tourism: passport to development?* New York: Oxford University Press.

Lea, John (1988) *Tourism and development in the Third World*, London: Routledge.

Mascarenhas, Ophelia C. (1971) 'Tourism in East Africa – a bibliographical essay', *Current Bibliography on African Affairs* 4, 5: 315–26.

OAU (Organisation of African Unity) (1981) *Lagos plan of action for the economic development of Africa: 1980–2000*, Geneva: International Institute for Labour Studies.

OAU (1991) *Treaty establishing the African Economic Community*, Abuja, Nigeria: OAU.

Sako, Filifing (1990) 'Tourism in Africa: an expanding industry', *The Courier* 122: 69–72.

Teye, Victor B. (1991) 'Prospects for regional tourism cooperation in Africa', in S. Medlik (ed.) *Managing Tourism*, Oxford: Butterworth-Heinemann, pp. 286–96.

Theobald, William F. (ed.) 1994) *Global tourism – the next decade*, Oxford: Butterworth-Heinemann.

UN (United Nations) (1986) *Africa's submission to the special session of the UN general assembly on Africa's economic and social crisis*, New York: UN General Assembly.

UN (1991) *New agenda for the development of Africa in the 1990s*, New York: UN General Assembly.

Williams, Allan M. and Gareth Shaw (1988) *Tourism and economic development: Western European experiences*, London: Belhaven Press.

WTO (World Tourism Organisation) (1991) *Impact of the Gulf crisis on international tourism*, WTO Special Report, May, Madrid: WTO.

WTO (1994) *Global tourism forecasts to the year 2000 and beyond: Africa Vol. 2*, Madrid: WTO.

WTO (1996a) *1995 International tourism overview*, Madrid: WTO.

WTO (1996b) *Tourism market trends – Africa 1995*, Madrid: WTO.

WTO (n.d.) 'On tourism trends to the year 2000', Madrid: WTO.

X.X. (1983) 'Multinationals at work: an insider assessment', in *The yearbook of world affairs*, London: The London Institute of World Affairs, pp. 168–86.

4

TOURISM PLANNING AND IMPLEMENTATION IN THE POLISH CITY OF KALISZ

Gary Akehurst

This chapter describes a project funded by the Foreign and Commonwealth Office of the United Kingdom Government (as part of its Local Authority Technical Links Scheme), Southampton City Council and Kalisz City Council, the overall aim of which is to provide assistance to Kalisz in Poland in the preparation and implementation of a City Development Plan. Work on the project commenced in September 1993 and continues at the present time. Stage 1 started in September 1993 and was completed in April 1995, followed by a presentation of an urban development plan to the City Council of Kalisz. Following intensive discussions with city officers and elected representatives, four areas of particular concern were identified: land use and planning, traffic management, retail policy and tourism development. Stage 2 involves the detailed implementation of the Stage 1 tourism plan and is currently under way. A key element of the project throughout its implementation has been the transfer of UK urban planning and local government experience to elected representatives and city officers in Poland, in the knowledge that this transfer should be of permanent benefit to the development of the City of Kalisz as an important regional centre and community. It is difficult, and indeed presumptuous to suppose that western tourism planning concepts can be transferred to a country recently emerged from communist control. This transfer can lead to tensions, misunderstandings and inevitably questions are raised as to the appropriateness of often culturally alien planning concepts that are being proposed. The Kalisz project, part of which is reported here, has been described by the British Ambassador to Poland as 'one of the best British projects currently being undertaken in Poland'.

This case describes some of the practical planning considerations and implementation problems of a tourism development plan in the City of Kalisz, being undertaken by the author with Jeff Walters, Head of Economic Development at the City of Southampton Council, over a three-year period. Getting agreement on what tourism development should be encouraged has not been easy and there is a sizeable number of elected city representatives who remain to be convinced that tourism activities should be developed over and above manufacturing activities, which are the main job- and wealth-generators in the city. This case provides an insight into who decides 'on the nature, scale and speed of development for tourism' and 'to what extent . . . the concerns of the various groups involved, such as visitors, residents, investors and employees [are] recognised and responded to' (Laws, 1995: 4). What is clear however, is that most of the literature on tourism planning and development, with the notable exception of Inskeep (1991), the World Tourism Organisation's National and Regional Tourism Planning (1994) developed with Edward Inskeep, and Laws (1995), offers little practical assistance in the development of actual tourism plans and their implementation, whether at city, regional or national level. As tourism plans developed within consultancy projects rarely get published within the public domain, the rigour of the methodology and structuring of these tourism plans are seldom openly discussed. This is understandable because few consultancy firms would wish to divulge their fee-generating 'secrets'. In this author's experience, many of these consultant-generated tourism plans are not essentially difficult to prepare although they are often shrouded in a spurious mystery. Furthermore, they often lack a rigorous structure and the involvement of the planner in the implementation of their plans. This case is offered as a modest way of moving forward our understanding of the practicalities of tourism planning based on a live case rather than a theoretical model.

The development of a tourism plan (as a subset of a city plan encompassing town planning, traffic management, tourism and retailing) is outlined and how this plan is being implemented is then described. The tourism plan establishes what needs to be done to achieve the city's objectives and what tourism products should be developed, when, where and how. It also establishes who will co-ordinate, promote, monitor and evaluate the proposed actions. In this respect, three sub-plans have been developed: an organisation, marketing and promotion plan; a tourism product plan; and a quality plan. This has meant starting from scratch a tourism section within the City Council's Department of Economic Initiatives (to promote standards, formulate policy, monitor implementation and advise City Councillors), a proposed commercial Tourism Investment Promotion and Marketing Agency (as a one-stop shop for practical assistance for investors) and a proposed Tourism Development Agency to develop the infrastructure at a small number of chosen sites.

Background

Poland

Poland is working hard to expand its tourism industry. In 1994 fewer than 200,000 British visitors went to Poland, and most of these were ex-patriate Poles returning to their roots and business people participating in the emerging capitalist markets. David Wickers writing in *The Sunday Times* newspaper on 23 July 1995 has said:

> Here is a country with all the right 'un' words, such as untouristy, underrated and unspoilt – the sort of attributes usually only found a long way from home – yet the country is only a two-hour flight away. Poland is also one of the meatiest countries in Europe, with an amazing history that invites you to sink your teeth in deeply – but then bites back hard, wounding you with raw emotions. Poland has been pillaged, annexed, crushed, partitioned and obliterated. For more than 100 years it was wiped clean off the map of Europe. During the second world war Polish Jews were exterminated on a scale that defines the very notion of evil. Even peacetime was a painful chapter, the surviving Poles being systematically uprooted by Stalin from their homelands in order to eradicate any sense of identity. The most surprising fact about Poland is that there is anything left to see, let alone enjoy.
>
> Although better known abroad for the damage heaped on the environment by the foul breath of its factories, Poland has some remarkable natural assets. There are 21 national parks and more than 100 protected areas. It is a land of bears and bison, wolves and wild boar, lynx and elk. A quarter of the country is forest. There are 9,000 lakes. And everywhere there are signs of a growing commitment to conservation. Could you imagine British Telecom allowing thousands of storks to build their nests, as big as dog baskets, on the top of telegraph poles? Come to Poland.
>
> The domesticated countryside is a Bruegel canvas come to life. The summer fields are family affairs, toddlers to grannies scything grass, making hay, ploughing with chestnut-brown horses whose manes toss in the wind. Rural Poland is as much a vision of a different time as place, stirring some deep sentiment for the way we were.
>
> The road is not all roses. There is an awful greyness to Polish towns, usually defined by copses of red and white striped factory chimneys. There are too many flats with cracked and crumbling masonry like a collection of Ronan Points, multi-storey disasters waiting to happen. The glorious exceptions are the born-again cities, such as inner Gdansk and Warsaw, reconstructed inch by inch after their wartime razings, or the fabulous Krakow, which escaped without a bruise.

Poland has made considerable progress towards stabilising and reforming its economy over the last five years. It is attempting to encourage foreign investment following some thirteen years when the country was in default on its communist-era debt. It was the first former Warsaw Pact state to embark on 'fast-track economic reforms' with restrictions lifted on foreign trade and currency movements. Economic stabilisation, western debt relief and export-led growth has led to a debut in the Eurobond market. Associate membership of the European Union may be achieved by 2002. In 1994 Gross Domestic Product grew by 5 per cent, labour productivity increased by 17 per cent and exports rose by 22 per cent. Inflation, however, remains at over 30 per cent. (*Financial Times*, 5 June 1995).

The Polish tourism industry is going through a period of transformation with the privatisation of travel agencies, tour operators and hotels previously operated by ORBIS, the former state tourist organisation which controlled all inward and outward travel and tourism. Tourist arrivals appear to be increasing, with a rise from 3.8 million in 1991 to 18.8 million in 1994 (according to the Institute of Tourism in Warsaw, based on World Tourism Organisation data) and foreign currency tourist receipts grew from US $1 billion in 1991 to US $6.2 billion in 1994 (Jermanowski, 1994: 10; Hunt, 1992). While tourism data is difficult to obtain and in many instances unreliable, broad upward trends can nevertheless be identified.

During the communist period, and since, the key tourism areas appear to have been the Baltic Coast (the most popular domestic tourism destination), the Mazurian Lake District in the north-east, the Western Carpathian Mountains (including the Tatra Mountains) and the Sudeten Mountains in southern Poland (Dawson, cited in Hall 1991: 196–9). Unfortunately, the City of Kalisz is not in one of these main tourist areas.

It would be a mistake to believe that the concepts of the market, marketing research and communications are alien to post-communist Polish society. Indeed, the Polish people involved in this planning project have fully lived up to the name sometimes given to the Poles: 'the Italians of Northern Europe'. Polish people are natural traders with an eye for value for money, but they recognise the merits of marketing. While the advantages of marketing may be instinctively understood and readily practised, nevertheless, the very mention of the word 'planning' is met with understandable hostility – planning is associated with the communist era. The project team quickly became adept at describing 'planning' processes in ways which were non-threatening. It took some time to persuade people that free markets need elements of planning in order to deliver societal wealth and obtain desired goals. In this sense, 'planning' is a facilitator not a constrainer.

Kalisz

As indicated in Figure 4.1, Kalisz is located in the valley of the Prosna River on the eastern verge of the Calisian heights, some 200 km south-west of Warsaw. It is Greater Poland's second biggest city and capital of the province since 1975. It covers an area of 55 sq. km and has a population of 107,000. Claudius Ptolemy, the Alexandrian cartographer, mentioned the city in the second century. The fastest growth took place in the fifteenth and sixteenth centuries, although numerous wars and epidemics in the seventeenth and eighteenth centuries hampered its development. During the First World War, around 80 per cent of the old city was destroyed, but during the 1920s and 1930s the remarkable old city

Figure 4.1 Location of Kalisz in Poland

was completely reconstructed. In September 1939, the city was occupied by the Germans and there followed a period of relentless repression. Following the war, industry was developed. Kalisz lies in a rich agricultural region, but the main industries within the city limits include textiles, food processing, aircraft components and pianos.

The *Interim Report* written by the project team (Walters *et al.*, 1994) provides a picture of Kalisz as a city of great potential, set in the heart of a productive and skilled area of Poland. This vision of the potential of the city was set alongside an analysis of its strengths and weaknesses in the early 1990s, and in the transitional stage towards the development of a market-based city economy. The report went on to set out a strategic overview of the priorities for the city in the next five years. The City Council accepted this interim report virtually in its entirety, although it was clear that there was a need to phase in developments over a long time span because of funding difficulties.

The *Interim Report* identified key tourism issues facing Kalisz over the next decade. Tourism is clearly not a panacea, but there are considerable rewards from adopting an effective and well-focused tourism strategy through the jobs and income generated in the local economy. Kalisz has tourist potential because of three factors:

- its claim to be Poland's oldest city;
- its wealth of buildings of architectural interest, set in a pleasant environment; and

- it is a major cultural centre, historically associated with famous actors, musicians and artists.

The city is, however, poorly placed in relation to motorways, express railways and airports, and the level of support services to sustain tourism is low. Tourism infrastructure and amenities (such as hotels and cafes) are poor and will need development but funds for tourism development are limited.

The *Interim Report* went on to recommend that urgent attention should be given to basic infrastructure:

- tourism must be packaged, with the city giving attention to establishing a tourism unit within the City Council's management structure, which is quite separate from the existing Culture Department;
- wherever possible, buildings of historical interest must be opened to the public and noticeboards (in Polish, German and English) erected to allow more interpretation;
- hotel and guest house accommodation at affordable prices must be provided;
- a one-stop Tourist Information Bureau should be established at a central location;
- taxis must be regulated and adhere to minimum standards.

To the credit of the local authorities, and with little fuss or discussion with the project members, several recommendations were taken up. A Tourist Information Bureau was opened in October 1994 with specifically recruited staff and a new computer accommodation service, and work started on the construction of a visitor interpretation centre (built in a traditional wooden Polish style) at a site of Roman remains. The provision of affordable low-cost accommodation was, and still remains, a major restraint but in the next two years at least one new hotel will have been constructed.

The majority of tourists that Kalisz can realistically hope to attract are likely to be domestic weekenders. Overseas visitors are likely to be restricted to those of Polish origin, but a small number of specialist tourists may be willing to visit because of the historical and architectural qualities of the city. Car-borne visits are likely from Germany and the Czech Republic. With promotion, Kalisz may be able to attract business people on visits to other towns and cities in the region but this may not be easy.

Possible tourism products include:

- reconstruction of the Roman past and the development of associated tours;
- interpretation and tours of medieval and Napoleonic architecture;
- a country park linking city centre parkland, the river and the lake at Golochow, and development of cycling and walking trails;
- development of existing city events such as the music festivals, Chopin

concerts and competitions, jazz/piano festival, cycling Grand Prix and supermarathon, plus new festivals;
- organised excursions, for example, to the chateau at Golochow;
- religious tourism, given the proximity to Czechtohowa (home to Poland's most sacred religious centre); and
- conferences.

There are encouraging signs. The people of Kalisz may have been 'held back' by the central planning regime of communism but they are learning new market skills very quickly. This natural intelligence, combined with a hospitable and hard-working nature, is complemented by an excellent school and university system. Once the organisations described below have been set up, and marketing, training and evaluation systems are in place, Kalisz will become a recognised tourist centre, primarily for domestic visitors. The possibility therefore exists to promote Kalisz as 'the birthplace of Poland'.

A tourism blueprint for action – a development scenario

The *Final Report* written by the project team (Walters *et al.*, 1995), built on the work undertaken for the Interim Report and has established *a tourism blueprint for action*: This blueprint addressed such questions as:

- *What* needs to be done to achieve the city's objectives?
- *What* tourism products should be developed?
- *When*, *where* and *how* should these tourism products be developed given potential international and domestic demand?
- Who will *co-ordinate*, *promote*, *control*, *monitor* and *evaluate* the actions proposed for tourism development?

Linked to these fundamental questions are development time-scales, that is, short-term (up to two years), medium-term (two to five years) and long-term (over five years). Keeping things clear and focused was important, not least because reports and meetings had to be translated into Polish. It is felt that something sometimes gets lost in the translation process, perhaps some of the subtleties and underlying meanings, and sometimes the project members have found their recommendations have been quietly ignored. So be it; this is a transfer of knowledge process and the people of Kalisz and their elected representatives need to take ownership of *their* tourism plan, because they will be living with its consequences in the years ahead. It may well be that we are assuming too much of Polish people in this time of transition from communism to a free market system, and that there is a fine line to be drawn between assistance and patronisation.

Within the tourism action plan three essential sub-plans have been prepared:

- An *organisation, marketing and promotion plan* (to encourage investors, develop market research, promote the city, provide information services and an efficient accommodation booking service);
- A *tourism product plan* (amenities, attractions and infrastructure);
- A *quality plan* (to ensure environmentally friendly tourism and protection, and quality services for tourists).

These plans required careful consultation with city representatives and this will continue at regular intervals in the future with the people of Kalisz. It needs their support, sense of ownership and willingness to make it work. Outside experts cannot impose their thoughts and plans on the city and, rightly, it is doubted that the people of Kalisz would allow this. These considerations also raise issues of whether western tourism planning concepts and procedures are appropriate within a developing economy. There is no easy answer to this, and indeed the only answer must surely be to listen carefully to what objectives are being sought and devise plans which are realistic and sympathetic to both people and environment in the attainment of those objectives.

The objectives identified for the Kalisz city tourism plan are:

- to promote Kalisz as an attractive destination for tourists;
- to develop a sustainable, environmentally sound tourism industry;
- to generate income and employment opportunities for the people of Kalisz;
- to encourage and facilitate profitable investment in tourism, which leads to the refurbishment and enhancement of the city's infrastructure;
- to enhance the cultural and educational life of the city.

Tourism needs to be developed and promoted to establish a very clear and distinctive image for Kalisz, which exploits to the fullest extent the tourism potential, and which differentiates Kalisz from other Polish cities.

In the short term, action centres on organisation, promotion and small-scale product development requiring only modest levels of investment. This recognises the realities of investment in Poland and the lack of investment funds. Accordingly, emphasis is being placed on such development as:

- special interest trips in relatively small groups of around 15 people, including architectural and heritage appreciation, trekking, cycling and water sports;
- chauffeured car tours with local guides;
- bus and mini bus tours with local guides;
- private sector investment in bed and breakfast accommodation and camping facilities;
- letting of private residents' rooms to visitors.

In the medium-term (that is, between four to eight years) it was recommended that tourism product developments should continue the modest investments as previously outlined, but investment in hotels should be a priority. There are already signs that such investment is beginning – a German company is starting work shortly on the building of a hotel in the city. However, long-term developments will depend on economic stability and the success of previous tourism ventures.

Organisation, marketing and promotion plan

There was an urgent need to define clearly the functions and responsibilities of all parties (both public and private sector) involved in tourism within the city limits. There was a confusion as to who would do what and when, and above all, who would co-ordinate often disparate activities which were likely to be undertaken in parallel rather than in any strict sequential time path. The project team recommended that the functions of tourism marketing and promotion and investment promotion be devolved to a new body – a tourism investment promotion and marketing agency, answerable to a new department of tourism within the City Council. Following detailed analysis and audit, it was felt that what was required is:

- the effective organisation of tourism within the City Council;
- attention to the requirements of investors;
- manpower planning and training;
- effective marketing and promotion of tourism within Kalisz.

Guidelines have been carefully developed by the project team to establish the broad actions required in the short, medium and long-terms with an indication of priorities.

The City Council will need to ensure that:

- tourism policy formulation and implementation is democratically debated and controlled with reference to city, government and regional objectives;
- potential investors are encouraged and assisted;
- standards of services to tourists are promoted and maintained, including the classification and registration of accommodation;
- information and research services, such as data collection and analysis are provided for decision-makers in the City Council and its agencies and departments;
- manpower for the tourism industry is planned for and trained to high standards.

A clear organisational structure is obviously required which provides reliable and efficient services to tourists, operators, potential investors and City Council agencies. Each organisation must be clearly defined, with clear lines of accountability and well-defined areas of responsibility. Co-ordination of these diverse functions, by a new department of tourism which is accountable to elected representatives of the city, was felt to be absolutely essential. However, the City Council decided that in the medium term, rather than establish a new council department (with all the consequent costs involved), a tourism officer would be appointed and stationed within the Department of Economic Initiatives. This tourism officer was charged with undertaking the preparatory work for a draft tourism plan prior to coming to the UK for specific training at Portsmouth University, Southampton Institute and the City of Southampton Economic Development Division, in tourism plan development, tourism assets audits and tourist attractions evaluation. Consequently, in April 1996, the tourism officer came to the UK with the Head of the City of Kalisz Economic Initiatives Department for a one-week stay. This post of tourism officer required a person with good tourism experience. Lack of experience can be overcome if there is a willingness to listen, consult widely and ask for help when unsure of what to do next.

In addition to the establishment of an organisational structure, and following the UK visit of the tourism officer and the Head of the City of Kalisz Economic Initiatives Department, a draft tourism plan was prepared by the City Department of Economic Initiatives and submitted for discussion within the council. This plan was discussed first in a specially constituted Kalisz Tourism Forum, consisting of elected representatives and members of tourism organisations, and chaired by the City Vice-President, with an agreed constitution and working arrangements. This Tourism Forum has proved to be the most durable of the actions taken so far and is working well. It has given the various stakeholders in tourism development a real voice in shaping that development. The establishment of a tourism investment, promotion and marketing agency has not materialised at this stage. This is not surprising at this time in post-communist Poland where there is a relatively cautious attitude to risk and a lack of investment funds. However, the main reason for the delay may lie in the relative lack of Polish individuals willing and financially able to take the risk in developing such an agency.

In theory, this draft tourism plan, as prepared by the City Council, should have outlined proposed developments in the short, medium and long term and should be reviewed annually, with a major review every three years to examine the attainment of set objectives and the organisation of tourism in terms of efficiency, effectiveness and other performance indicators. In reality, and at the time of writing, things are in a state of flux – a first draft plan was frankly less than adequate. Perhaps something had been missed in the translation from Polish to English or perhaps the project team had not communicated their thoughts sufficiently clearly and in depth. Whatever the reason, a tourism plan structure was faxed to Kalisz Council with a recommendation to use it in preparing a new draft tourism plan (see Appendix for the suggested structure to be used in a revised tourism plan). At the current time (1997), the City of Kalisz has recruited the services of a Polish tourism specialist from Poznan to drive forward the tourism plan on a day-to-day basis and the tourism officer has left the employ of the city council.

Three new organisations are proposed and, subject to City Council approval, will be established in the medium term in order to develop tourism in Kalisz:

- *A City Council department of tourism* which advises City Councillors, formulates policy, monitors and evaluates implementation, promotes standards and promotes the city. Much of this co-ordination work will be undertaken in partnership with other council departments, the Wojewodstwo (regional government) and the private sector.

- A commercial *Tourism Investment Promotion and Marketing Agency* (Kalisz Marketing), reporting to the proposed department of tourism, which will:

    - act as a 'one-stop shop' for investors by providing practical assistance to investors, including identification of suitable opportunities;
    - vigorously promote the city to potential investors and tourists;
    - maintain, in collaboration with the proposed City Council department of tourism and departments of economic initiatives and planning, a comprehensive list of possible investment sites;
    - advise potential investors on planning, investment and tax laws (including laws concerning possible repatriation of profits to foreign countries and foreign corporations owning land and property) and possible fiscal incentives;
    - evaluate the merits of proposed projects and assist (if necessary) in obtaining finance;
    - act as an intermediary between developers, city, regional and central government and funding sources to secure project implementation;
    - prepare, publish and distribute an investors guide which gives clear, concise information about the practicalities of investing in Kalisz.

    It is anticipated that this agency would, after initial start-up funding, be self-funding, raising its operating revenues from fees charged to investors for services and consultancy fees.

- A *Tourism Development Agency* possibly at regional level (or indeed at city level) which develops the infrastructure initially at a small number of carefully selected sites, and then sells or leases plots of land for tourism development. Careful consideration will need to be given to a range of fiscal incentives and possible tax holidays in order to encourage investment. One model of a development agency is that in the United Kingdom, where certain run down areas of the London docklands have been successfully developed. Such an agency should be established by either central Government or the Regional Council. Certainly, the Regional Council in partnership with the City Council should have a guiding strategic role working with the emerging private sector, and where foreign investment has a role to play.

In the context of Kalisz's history, a Tourism Development Agency is a rather radical proposal but it is felt that this is essential. A prime requirement for success will be to keep government control to a minimum but at the same time ensure that the unique and appealing characteristics of Kalisz are maintained. A Development Agency is a free market facilitator which cuts through bureaucratic planning procedures to achieve objectives subject to fundamental safeguards of law and environment.

Manpower planning and training is a particular priority. At present there is an ample supply of manpower in the region, but if tourism is to develop there is an urgent need for the training of hospitality and tourism managers, particularly hotel and restaurant managers, and operatives such as waiters, chefs, housekeepers and receptionists. Training is required at all levels from City Council department to the local hotel.

Current training needs are to be reviewed and on-the-job and off-the-job training will be undertaken in collaboration with local educational institutions, such as Poznan University, the Institute of Tourism in Poznan, the local gastronomy school, and foreign tourism companies.

In the short-term, high priority is being given to:

- establishment of an on-the-job training scheme for managers, supervisors and guides under the control of the new department of tourism, using local and foreign instructors;
- establishment of a training section in the Department of Economic Initiatives, which will promote tourism training in collaboration with institutions and companies, seek funding and sponsorship of the on-the-job training schemes, and advise the City Council on future training policies and manpower requirements;
- the training of several tourism department staff in the United Kingdom (in terms of tourist attraction evaluation, financing, promotion and market research);
- securing commercial sponsorship for on-the-job training through investment and tax incentive schemes.

After three years, the Department of Economic Initiatives (or, if set up, the new Department of Tourism) will need to consider setting up teams of field instructors to take training into tourism businesses, collaborate with Poznan University on the provision of certificated off-the-job training schemes and develop overseas exchange schemes with vocational training centres elsewhere in Europe. The financing of tourism training could, if government regulations allow, be based on a restaurant and hotel bill training levy of 2 per cent.

Marketing and promotion activities will be vital to success. Achievement of the City Council's tourism objectives will depend, in part, on the image of Kalisz tourism which is projected not only within Poland but across the world. Some of the factors which attract tourists to an area may include:

- scenery differences;
- different cultural traditions;
- heritage and architecture;
- cost advantages to the tourist;
- climate differences compared with home countries;
- quality of transport links; and
- the availability of suitable holidays including packaged holidays.

While the City Council should be able to capitalise on the strengths of Kalisz (as the oldest city with a rich historical tradition) sustained, radical and creative marketing and promotion activities will undoubtedly be required to overcome possible weaknesses. It will also be vital to systematically and regularly canvass visitor views about, and experiences of the facilities and attractions in Kalisz, and for such feedback to be communicated to responsible partners and stakeholders.

Short-term actions include the essential building of consumer and investor awareness of tourism opportunities in Kalisz. This requires:

- The establishment of the Tourism Investment Promotion and Marketing Agency (provisionally named 'Kalisz Marketing') as mentioned above, with the clear brief of working with the Department of Economic Initiatives and, if established, the new Department of Tourism to promote tourism;
- Within the tourism plan (as mentioned above), target markets should be established, focusing primarily on:
  - domestic Polish visitors, visiting friends and relatives or places of religious significance;
  - ethnic Poles overseas, especially Americans;
  - Europeans and Pacific Rim residents, over 45 years, seeking high quality holidays in a clean and safe environment, rich in history and culture;
  - young people aged 18–35 years seeking activity and specialist holidays such as trekking, rafting, canoeing, etc.;
  - residents in the nearby countries of Germany and the Czech Republic.
- The Department of Economic Initiatives or Department of Tourism in partnership with Kalisz Marketing should undertake systematic and regular marketing research to determine for each month of the year:
  - number of trips and nights spent;
  - prime objects of trips (holiday, business, visiting friends and relatives [VFR], study, other);
  - trip length (1–3 nights, 4+ nights);
  - mode of transport used;
  - origin of visitors;
  - age of visitors;
  - average spend of visitors per day and by spend category; and
  - percentage of repeat visitors.

61

The tourism plan should forecast tourist arrivals, mode of travel, and other market research data. In addition, there will need to be a regular accommodation survey collecting data such as number of bedrooms available, average length of stay in accommodation, average room and bed occupancy, average achieved room rates, numbers of staff and possible recruitment difficulties.

Other marketing efforts will include:

- a programme of press visits, identifying and inviting key journalists;
- promotion to the travel trade, including preparation of a multilingual travel guide;
- preparation of a target list of Polish towns and regions and foreign country towns, to circulate promotional material;
- preparation of a student travel initiative, including poster campaign in Polish university cities;
- ensuring promotional literature is regularly sent to Polish embassies abroad and tourist information centres in Warsaw.

This is not a full list but is indicative of promotional activities which can usefully be carried out.

In the *medium-term* the city will need to concentrate resources on markets which offer the greatest potential returns. Kalisz Marketing will need to actively promote the growth of the city's tourism industry, with marketing research and promotional material which is reliable, consistent and attractive. Public relations in the form of newspaper and magazine articles is a cheap and effective way of building awareness in potential markets. Over time, Kalisz Marketing will need to actively promote 'flagship' tourism developments which set the tone, scale and sense of direction. Kalisz Marketing will need to go on to establish good working relationships with tour operators, coach operators and other tourism businesses in order to establish profitable products and services.

In the medium-term, consumer and investor awareness will have been increased by the activities of the Department of Economic Initiatives or the new Department of Tourism and Kalisz Marketing. Marketing strategies will need modifying where necessary in the light of experience and following prolonged and systematic collection of market data.

In the *long-term*, the markets for tourism in Kalisz may have matured beyond the tour operator phase, into ones where direct selling to independent tourists becomes important. Long-term actions include the continued promotion of the city by Kalisz Marketing, identifying further priorities and market gaps and close attention being given to the development and promotion of domestic tourism, as disposable incomes begin to increase significantly. The marketing and promotion functions of Kalisz Marketing will be reviewed in the light of experience.

Tourism products plan

A tourism products plan needs to consider carefully accommodation, products, attractions and facilities. Short-term actions are highlighted in Table 4.1 (Accommodation, Attractions and Facilities) and Table 4.2 (Tourist Information Services). Medium-term actions are shown in Table 4.3. The medium-term is defined as being from two to five years.

Following successful implementation of essentially small-scale tourism developments in the short term (including refurbishment of existing hotel stock), a larger-scale development of tourism products may be possible. This presupposes a relatively stable economy together with an inward flow of investment funds into Poland. In the long term (that is, over five years) product developments will depend on a number of factors including the pace of development in preceding years, investment funding and growth in the economy.

A Tourism Quality Plan

It is vital that the city develops tourism which is sympathetic to the environment, and which enhances the life of the people of the city. A quality plan is being developed which establishes guidelines for encouraging environmentally friendly tourism, protects the countryside and enhances the life of the city and ensures quality services to tourists, such that the city promises only that which can successfully be delivered.

The city will need to establish very clear physical planning guidelines and environmental control guidelines. Consideration of these issues is really outside the scope of this chapter but to be able to control and monitor tourism development, and co-ordinate tourism development with other types of construction, planning must take place at a national and regional or city level. Building construction needs co-ordination with the detailed topography of an area in order to minimise environmental impacts. Matters to be considered in a tourism quality plan are listed in Table 4.4.

Conclusions

This case describes one distinctive approach to the development of tourism within a city. It should be readily acknowledged that there are many different ways of promoting the development of urban tourism. The methodology developed by the project team recognises that there are several discrete stages in the development process, founded on audit, analysis, dialogue, commitment, stakeholder and visitor feedback, overall strategy, plans for action, target setting, and regular monitoring and evaluation of results. Above all, the local community through elected representatives must agree and take ownership of their own plan for development.

Table 4.1 Tourism products plan – accommodation, products, attractions and facilities

Accommodation

- Reviewing existing hotel and guesthouse accommodation stock and encouraging private businesses to refurbish to acceptable standards (as defined by market research)
- Kalisz Marketing to review and update existing site portfolio and circulate details to commercial attachés, investment agencies and major hotel groups
- City Council to consider a Hotel Development Incentives Scheme in order to increase stock of acceptable accommodation
- The Department of Economic Initiatives or the new Department of Tourism and Kalisz Marketing to encourage the provision of private rooms for visitors by preparing a comprehensive list of private city residents willing to let rooms
- The Department of Tourism, in consultation with the City Council, to agree standards, classification and registration scheme with code of conduct for the operation of accommodation
- The Department of Tourism to agree a standard sign for registered accommodation and encourage its display in a prominent place
- Private businesses with Kalisz Marketing assistance encouraged to develop Bed and Breakfast accommodation with donor aid
- Establish an efficient self-funding accommodation booking system based on the new Tourist Information Centre, established in October 1994

Tourism Products, Attractions and Facilities

- Kalisz Marketing to set up a comprehensive inventory of existing and potential attractions
- The Department of Tourism to appraise state of daytime and evening entertainments, festivals and concerts
- The Department of Tourism in partnership with Kalisz Marketing, to generate a priority system for improvement of individual sites at different levels of expenditure. Implementation will depend on resource availability but priority should be given to improvements likely to generate income
- Private businesses with Kalisz Marketing to develop key trekking, walking, cycling and horse-riding routes and centres
- Kalisz Marketing to identify centres for special interest holidays (archaeology, architecture, etc.) particularly archaeological and heritage trips focusing on Roman remains (target market – older persons over 40 primarily from Western Europe);
- Kalisz Marketing to encourage private chauffeured car tours with guides, beginning with one-day tours with official licensed cars, drivers and guides
- Identify and develop with private firms a limited number of camping and caravanning centres
- The private sector encouraged to develop mini-bus tours with guides, and with the active support of Kalisz Marketing and donor aid
- Kalisz Marketing to assist private businesses to establish restaurants and cafes, particularly in the city centre
- The Department of Tourism to establish a guides, taxis and chauffeured car driver registration and licensing scheme
- Kalisz Marketing to review ways of extending the tourism season
- Examine the feasibility of establishing small handicraft and souvenir production centres, which can be visited

Table 4.2 Tourism products plan – tourist information services

Tourist Information Services

Tourists of whatever nationality need good information services – where to stay, at what price, where to eat and what to see. A number of actions are being carried out, which include:

- a review of existing tourist publications and their distribution arrangements;
- a review of existing hotel bedroom literature, and agreement on production of new publicity material and distribution;
- assurance that the new Tourist Information Centre (now located in premises near the City Hall), has agreed annual funding, management, responsibilities, accountabilities and operational plans;
- production of simple tourist guides in Polish and foreign languages (German and English);
- preparation and publication of a simple foreign language 'What's on' guide, which is designed to be self-funding;
- agreement on the number and siting of information boards;
- agreement on standard design and contents of information boards, adding foreign language captions where necessary;
- review of signs at individual attractions, replacing or supplementing where necessary;
- to ensure adequate directional signs throughout the city;
- to ensure that there are knowledgeable and helpful wardens at attractions;
- preparation of an independent restaurants guide;
- preparation of a simple shops guide, particularly craft, souvenir and antiques shops;
- to encourage shops to display attractive tourist products;
- to ensure that all restaurant menus are translated into at least English

Table 4.3 Tourism products – medium-term actions

- Continue to identify and develop with private firms a limited number of camping and caravanning centres
- Kalisz Marketing to assist in the development of cultural centres with music and folklore festivals
- The private sector encouraged to develop and present conferences, seminars and business meetings
- Private companies encouraged to develop upmarket coach tours, using Kalisz as the main centre
- Kalisz Marketing with the tourism department to encourage the development of additional hotels of various standards (luxury, middle range and budget) on the periphery of the old city centre
- Limited development by the private sector, with Kalisz Marketing encouragement, of self-catering accommodation just outside the city
- Private sector development of sports facilities (including golf) at the lake at Golochow with perhaps multi-activity centres based around low infrastructure requirements

Table 4.4 Tourism quality plan – issues to be considered

- Desirable development densities in designated zones, given availability of building land, the supply of services such as water and power and the level of existing infrastructure
- Avoidance of ribbon development and control of creeping development
- Using the natural topography with its opportunities and constraints
- For quality tourist developments, a maximum building density for hotels and apartments (including ancillary facilities, car parking and landscaping) of 100 tourist beds per hectare of land should be adhered to, together with a maximum height for buildings depending on topography
- All development should be based on local vernacular (styles, materials, colours and features) and designed to create as little change to the natural environment as possible. Maximum benefit is gained by exploiting local views, sunsets and natural topography in building design
- Building height should not exceed that of surrounding buildings and be in sympathy with the overall town
- Tight planning control is necessary to prevent indiscriminate industrial development spoiling tourism attractions
- Clustering of small tourist developments is more sympathetic to the environment and local economy
- The city should offer mixed developments, geared towards providing interest and services for both tourists and residents

This is clearly not an exhaustive list but is indicative of those matters which deserve especial consideration.

This development process requires a very clear action plan, as to what needs to be done, when, where and how, and who will co-ordinate, promote, monitor and evaluate actions. A tourism development plan needs three essential sub-plans, using three broad planning time-scales of short, medium and long term:

- an organisation, marketing and promotion plan (paying particular attention to investment encouragement and support, and market research);
- a tourism product plan (amenities, attractions and infrastructure);
- a tourism quality plan (environmentally friendly tourism and quality services for tourists).

So far, the most successful development has been the establishment of a Tourism Forum, which has led to intensive debates about what kind of tourism development is needed in this part of Poland.

This project is different because those who have prepared an urban development plan are now charged with the responsibility of implementing their tourism development proposals. This gives an opportunity for tourism planning consultants and tourism educators to 'practise what is preached'. The project team believe that this approach can be successfully applied elsewhere, in both developed and developing countries, and are now working with other clients to put this methodology into practice.

Appendix

Recommended Structure for Kalisz Tourism Plan

1 **Foreward by City President** (commending plan to the people of Kalisz).
2 **Executive summary** (main recommendations in bullet points or listed paragraphs).
3 **Terms of Reference** (carefully formulated to achieve desired results and outputs).
4 **Overall tourism development objectives and policies** (what needs to be done; natural, cultural, economic and tourism resources conserved for continuous use in the future – i.e. sustainable development plus income generation, employment enhancement, integration into overall city development plans and policies, etc.).
5 **Existing tourism markets and forecasts**

- numbers of current visitor arrivals, visitors profiles (length of stay, transport mode, etc.), patterns of movements;
- accommodation currently used by each market segment with trends if possible;
- *projected* visitor arrivals – short term, medium term, long term.

6 **Current tourism policies, planning and organisational structure** (currently, who does what and how; limitations to current organisational structures and processes – setting the scene for the development actions).
7 **Audit or inventory of existing and planned tourism resources**

- summary of existing and planned tourism sites, attractions, facilities, accommodation, infrastructure, etc. *with most details in appendices*;
- *map or plan* showing location of principal tourism attractions, hotels, etc.

8 **Recommended development actions** (with clear time-scales, cost estimates and identification of *who* is responsible and accountable for *what*; in addition, the development actions should be *divided into sections* – product developments, organisational/institutional proposals, manpower, quality and environmental guidelines, investor promotion and support, etc.; clearly shown as short-term: 1–2 years; medium-term: 2–5 years; long-term: +5 years):

- possible alternative development scenarios;
- organisational structures (from review of current structures with recommendations for changes – city council, public and private sectors – roles of various organisations and agencies including Voievodship, PTTK and regional development agencies);
- manpower planning, education and training (projected manpower requirements and development of competencies and skills);

- marketing strategies, promotion programmes and market research;
- development of accommodation required (B & B, hotels, camp sites, etc. with projected accommodation needs in short, medium and long term broken down by market segment);
- development of a fully integrated accommodation reservation system linked to other regional and national systems;
- development of clearly defined tourism products with clearly identified market segments;
- quality, physical and environmental guidelines (ways to control tourism developments, lessen impacts, conserve natural environment, etc.);
- plan for actions in seeking financial capital for defined developments (tourist attractions, facilities, infrastructure and services; the mechanisms for attracting capital investment especially development of an investor's guide to Kalisz.

9 Monitoring, control and evaluation of progress
This requires:

- very clear lines of responsibility and accountability, including Tourism Forum arrangements;
- clear and quantified performance targets with control functions to correct variances;
- feedback from stakeholders and visitors.

10 Appendices
- checklist for evaluation of tourist attractions and services;
- inventory of current and planned tourist attractions, facilities, etc.;
- detailed projections of visitor arrivals, manpower, etc. (if prepared);
- list of all agencies and organisations involved in tourism.

Acknowledgements

I place on record the appreciation of the close working partnership between the project team and the officers and members of Kalisz City Council, which has made this project possible. I pay particular tribute to the contribution of Wojciech Bachor, President of Kalisz City Council, Zigismund Kasmierczak, former Economic Development Officer, City of Kalisz, Iwana Duda, Head of the Economic Initiatives Department, City of Kalisz, Marius Tomasiewski and Pavel Wozny our translators and James Beadle, Local Government International Bureau, London.

I hope that the work reported here will assist further with the progress of the City of Kalisz and the realisation of its ambitions. I commend the drive, vision and entrepreneurialism of the City of Kalisz and its people, to anyone considering locating a business in the region of Kalisz.

References

Financial Times (1995) 'Poland's Progress', 5 June.

Hall, D.R. (1991) *Tourism and Economic Development in Eastern Europe and Soviet Union*, London: Belhaven.

Hunt, J. (1992) *Poland*, EIU International Tourism reports 199, No. 2, London: Economist Intelligence Unit.

Inskeep, E. (1991) *Tourism Planning: An Integrated and Sustainable Development Approach*, New York: Van Nostrand Reinhold.

Jermanowski, C. (1994) *Tourism Development Objectives*, Warsaw: State Sports and Tourism Administration.

Laws, E. (1995) *Tourist Destination Management*, London: Routledge.

Walters, J., Sanger, A. and Whitehead, A. (1994) *A Development Plan for Kalisz, Interim Report*, Southampton City Council, Overseas Development Administration, Kalisz City Council and Public Policy Research Centre, Southampton, April.

Walters, J., Sanger, A., Akehurst, G. and Whitehead, A. (1995) *A Development Plan for Kalisz, Final Report*, Southampton City Council, Overseas Development Administration, Kalisz City Council and Public Policy Research Centre, Southampton, March.

Wickers, D. (1995) 'In Pole Position', *The Sunday Times*, 23 July: 5–6.

World Tourism Organisation (1994) *National and Regional Tourism Planning. Methodologies and Case Studies*, London: Routledge.

5

A STAKEHOLDER-BENEFITS APPROACH TO TOURISM MANAGEMENT IN A HISTORIC CITY CENTRE

The Canterbury City Centre Initiative

Barbara Le Pelley and Eric Laws

The urban context to tourism

Cities are in transition, under a variety of pressures including demographic trends; changing preferences for city centre or urban living and working locations; and transportation, retailing and tourism developments (Jacobs, 1972; Leontidou, 1990; Soane, 1993). The conglomeration of large numbers of people in a small geographic area, typical of city conditions, imposes significant management and policy problems, not least in terms of the maintenance of their health and economic welfare (Ashton, 1992). Furthermore, cities also rank amongst the cultural treasures of civilisation, they provide entertainment and artistic outlets for the surrounding population, and function as administrative and shopping centres. These attributes, combined with their nodal location in area transport networks (Page, 1995), make many cities and towns magnets for tourists, particularly when they are marketed appropriately (Kotler *et al.*, 1993; Law, 1993). The clustering or honeypotting of tourist resources and tourism complexes, results in the creation of districts in which tourism is the dominant activity (Ashworth and Voogd, 1990; Jansen-Verbeke, 1988).

A number of issues arise for policy makers seeking a balance of benefits for a city's key stakeholders, notably the tourists themselves, residents, sub-regional service users and the business community. Tourists have a tendency to gravitate towards major sites within a city, and this concentration may result in the disadvantages of saturation and congestion. Crowds of tourists can seem intimidating, create problems of noise and litter, and overuse can damage historic structures. It has also been noted that residents experience the effects of tourism to differing degrees depending on where they reside, their occupations, and their length of residency (Davis *et al.*, 1988; Sheldon and Var, 1984), but the general

70

consequence is that the local population may feel crowded out and come to resent the repeated invasions (Davies *et al.*, 1988; Law, 1993). The appropriate priorities and policy directions are contingent on the characteristics of individual cities (Buckley and Witt, 1989; Law, 1992; Ross, 1992). These issues and the responses to them are further exacerbated by the general conservation and interpretation needs of historic cities (Rumble, 1989; Godfrey *et al.*, 1994), and investigated in a number of English historic cities by the English Historic Towns Forum (1990–96), and Ashton and Tunbridge (1990), as indicated by studies of Canterbury (Le Pelley, 1994; Le Pelley and Laws, 1995).

Tourists and other city space users

Ashworth and Tunbridge (1990) have drawn attention to the contrasting needs of four different users of city spaces, including heritage tourists from the surrounding region; recreating residents who choose to spend part of their leisure time in the city; non-recreating visitors on family or business visits; and non-recreating residents using the city spaces in the normal course of their lives.

Although each of these four groups has a different pattern of behaviour, and makes somewhat different spatial use of the city, tourists share many basic needs with habitual city space users. These needs include parking, restaurants, cafes, toilets and information services, but although tourists increase the pressure on these facilities beyond what would be required to provide services to the inhabitants, any expansion of public facilities has to be funded by the local tax payers.

Additionally, tourists expect specific types of facilities, notably their interests extend beyond the primary attractions to embrace secondary elements of the tourism system including catering and souvenir shopping. Consequently, a tourist city experiences pressures to provide different retailing facilities from those required by residents of a similar sized city without a significant tourist sector. Shop keepers respond to the signals of tourist spending by altering their stock lines in favour of faster moving or higher margin lines. This contributes to crowding out the staples of retailing on which local residents depend. Furthermore, the tourist oriented retailers' success tends to increase rental values, further undermining the viability of traditional city centre retailers at a time when, in many areas, they are already under pressure from out of town retail developments. The overall effect may reduce the utility to residents of the city's shopping facilities and, if unchecked, lead to a decline of the city centre as an effective and pleasurable place to shop in the face of a decrease in the availability of resident-oriented goods on offer. This decline in attractiveness as a local shopping centre is further exacerbated when accompanied by physical overcrowding of the city's streets by tourists.

Tourists are temporary visitors who often wish to wander, stopping to gaze at varied features of the city scape and its heritage, but in so doing they may intrude into spaces which residents regard as 'private'. Consequently they can slow the progress of residents and commuters using the streets for their everyday business, contributing to tensions between these groups (Doxey, 1975; MacCannell, 1976; Urry, 1990). Many tourists are conducted around their destination city in groups (whose size is usually determined by the capacity of the coaches in which they travel), and whether independent or in groups, they tend to cluster around specific attractions.

The challenges facing Europe's walled cities

Historic walled cities are by nature compact – for example the walled central core of Canterbury is less than half a mile in diameter, consequently, the different patterns of space usage overlap significantly, resulting in congestion and conflict. Figure 5.1 shows both the increase in pedestrian use at selected sites in Canterbury between 1975 and 1993 (although it should be noted that the survey methods were enhanced in 1993), and the categories of people passing each point in 1993. A sample of people passing each location were interviewed, and the segmented bar chart shows how the distribution of tourists, residents, shoppers and commuters differed around the city.

The city walls, the historic street patterns and buildings are themselves a significant constraint on the potential for change. Even in cities such as Bruges where the walls have mainly disappeared, the historic street layout and presence of such features as a moat imposes a controlling factor on the economic life of the city. The urban fabric of a historic city is therefore both its primary attraction, in that this is what appeals to tourists in the first place, and also its Achilles' heel, in that its traditional structure was not designed to accommodate the pressures which result from modern tourism.

Thus, tourism exacerbates many of the existing historic city problems whilst adding additional issues to the agenda for analysis and policy making. The presence of tourists (and their cars and coaches) adds to the pressures on narrow streets, it further complicates the flow of people around them because tourists are unfamiliar with the city, and because their interests differ from those who use it routinely. Many of Europe's walled historic towns have now come together in mutual support groups such as the Walled Towns Friendship Circle in order to share experience on the planning and management problems they all face. This chapter now discusses Canterbury's strategic approaches to achieving a specific aim of the Walled Towns Friendship Circle (1991):

> To encourage an increase in tourism whilst considering responses to the challenges posed by modern traffic conditions; pedestrian precincts; delivery services to homes and businesses; building and road maintenance; grants and assistance for preservation work and the responsibility and burdens of conservation.

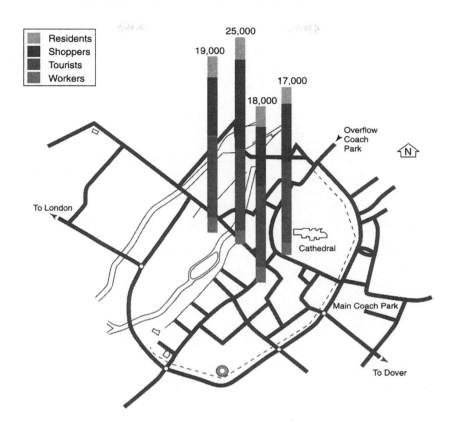

Growth in Canterbury city centre – pedestrian flows and concentrations

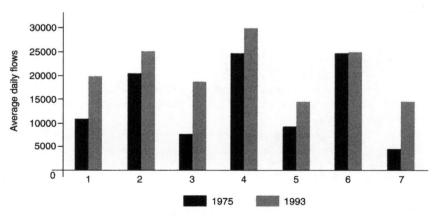

Figure 5.1 Space usage in Canterbury 1993
Source: Kent County Council Transportation Study, 1975; *Canterbury Tourism and Shopping Survey*, 1993.

The challenges facing Canterbury

Canterbury is one of the most important historic cities in Britain, and it exhibits many of the tourism management problems which are familiar to those involved in the management of walled and other historic cities. Canterbury City Council, and its predecessors, have been actively involved in tourism since pilgrims were attracted in large numbers immediately after the murder of St Thomas Becket on 29 December 1170, and the city's 'touristic' importance was celebrated in Chaucer's *Canterbury Tales*, first published in 1380.

Canterbury City is home to 36,621 residents (OPCS, 1991), and contains more than 1,500 listed buildings. The importance to the nation of conserving Canterbury was recognised in 1968 when the entire walled city, together with the medieval suburbs was designated as a Conservation Area of Outstanding National Importance. More recently, the cathedral, St Augustine's Abbey and St Martin's Church have been designated as a World Heritage Site by ICOMOS, the International Commission on Monuments or Sites. Within the city there are twenty-four Ancient Monuments ranging from underground remains to above ground buildings. Canterbury is one of only five towns in Britain to have been designated as an Area of Archaeological Importance.

Visitor pressure

Canterbury benefits and suffers from its location near to both London and the main Channel crossings linking Britain with the continent of Europe (the Ports of Dover, Folkestone, Ramsgate and Sheerness, and the Channel Tunnel which became fully operational in 1995). The Trans-European Transport Networks, supplemented by the Channel Tunnel Rail Link and TGV to Paris and Brussels have considerably increased mobility for citizens in the heartland of Europe and are creating significant additional pressure on historic and cultural tourism centres, Figure 5.2 indicates the location of Canterbury. This visitor pressure is a source of potential conflict with the inhabitants of Canterbury, where visitor numbers have escalated rapidly. The city now receives in excess of 2 million day-visitors annually, a visibly high proportion being groups of school children from the near continent, and over 500,000 staying visits. The rate of increase has leapt from an average increase of 50,000 per year over the previous two decades to an estimated 100,000 annual increase in 1995, primarily in the form of groups making day-visits by coach.

A recent comparative study of the raw ratios of visitors to residents in Europe's main historic tourism cities illustrates the unusually serious nature of congestion within Canterbury (Table 5.1). Although the figures are averages, and relate to various years within the early 1990s, and do not take into account seasonal and spatial issues, particularly clustering, it is clear that, Canterbury's ratio of 55:1 (2,250,000 tourists to 41,000 inhabitants in 1993) significantly exceeds cities with double the tourist numbers (the number of inhabitants on which this ratio is based

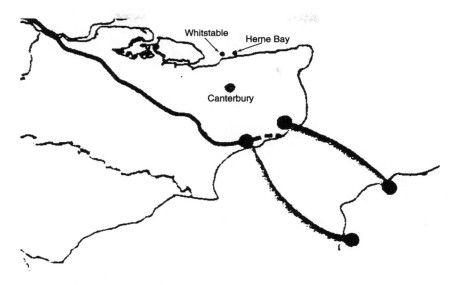

Figure 5.2 The location of Canterbury

Table 5.1 Tourist/resident ratio in historic cities

Cities	Year	Tourists	Residents	Ratio
Amsterdam	1992	7,651,000	1,300,000	6:1
Aix-en-Provence	1991	1,000,000	124,000	8:1
Florence	1991	4,000,000	408,000	10:1
Oxford	1991	1,500,000	130,000	12:1
Bruges	1990	2,740,000	117,000	23:1
Salzburg	1992	5,415,000	150,000	36:1
Canterbury	**1993**	**2,250,000**	**41,000**	**55:1**
Canterbury	*1975*	*1,000,000*	*37,000*	*27:1*
Venice	1992	8,627,000	80,000	108:1

Sources: Le Pelley (1994), Giotti (1994).
Note: Data for Canterbury also provides a comparison of 1975 with 1993, indicating that the tourist/resident ratio has almost doubled in the period.

is inconsistent with census figures, but is a reasonable approximation). The ratio calculated for Canterbury contrasts with Florence (10:1)and Salzburg (36:1). Venice, itself regarded as one of Europe's greatest heritage cities, and under threat from many pressures, experienced a tourist to resident ratio of 108:1 in 1992 (Giotti, 1994). Taking into account clustering, the majority of Canterbury's tourists occupy 10% of the area within the city walls, their main impact is on the 3,500 residents who actually live within the walled historic city core, and on residents from Canterbury's catchment area who are now beginning to shop elsewhere.

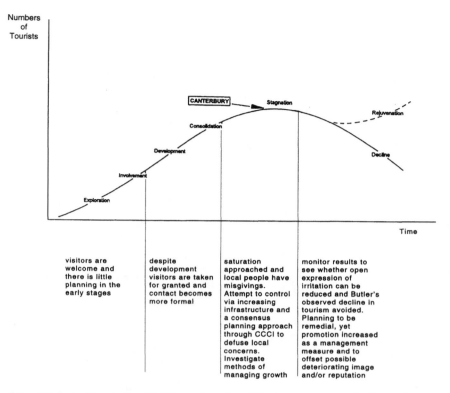

Figure 5.3 An irridex-destination life-cycle model of Canterbury as a tourist destination
Source: PATHS proposal, Le Pelley and Laws (1995).

The consequences of the market situation outlined above can be understood from the combined perspectives of the tourist destination life cycle (Butler, 1980), and the causation theory of visitor–resident irritants (Doxey, 1975). Figure 5.3 indicates that Canterbury is approaching the zenith of its life cycle, where the presence of visitors is beginning to drive local people away, and in response, an urgent revision of the city's visitor management strategies is being undertaken.

Policy responses in Canterbury

For many years, the Council has considered it important to provide a good experience for all visitors in the expectation that they will encourage friends and relatives to come. Strategic guidance for tourism planning has been set out in the city's *Local Plan*, the urban plan for the city. Tactical approaches have included selective promotion for Canterbury as a staying visitor destination, rather than as a day-visit destination, visitor surveys, and the development and implementation of programmes for visitor management and visitor amenity improvements,

Table 5.2 Impact of street improvement works and stakeholder outcomes in Canterbury

Street works	Pedestrians per day		Comments
Pedestrianisation and repaving	1975 1993	11,500 20,000	The West Gate Furthest location from tourist coach park. Since repaving post-1985, new businesses have opened. Street is at pedestrian capacity much of day.
Pedestrianisation and repaving between 1975 and 1985	1975 1993	22,000 25,000	Main entrance to museum in the High Street Few tourists recorded here in 1975. Equal numbers recorded to those outside the cathedral by 1985. Nearby street has not been repaved and has few pedestrians, even though traffic has been excluded.
Pedestrianisation, repaving, new visitor information centre, shopping development, visitor attraction between 1975 and 1993	1975 1993	7,000 17,500	St Margaret's Street Major increase in tourists recorded. High quality restoration of important historic buildings has created new visitor attraction in this street.
Repaving and pedestrianisation 1975–1985. New shopping centre 1993	1975 1993	25,000 30,000	The Parade (central shopping area) Street now at pedestrian capacity.
No change – pedestrian area 1975–1993	1975 1993	10,000 15,000	Burgate, near the cathedral Visitor numbers increased but pressure at this point has not increased at same rate. Management measures have deflected the pressure.
No change – pedestrian area 1975–1993	1975 1993	25,000 25,000	St George's Street Modern part of the town. Attracts more residents than tourists. Businesses have added extra floorspace.
No change – traffic still uses street	1975 1993	6,000 10,500	Palace Street Increase in tourists due to location of new coach park. Traders clamouring to be included in pedestrian area.
Old Coach Park New Coach Park	1991 1996	412,950 925,000	New coach park location has changed emphasis of visitor penetration. Management measures have been introduced to reduce impact of visitor flows.

particularly to the streets. Table 5.2 lists the main projects undertaken to improve streets for the benefit of tourists and other stakeholders in the city between 1975 and 1993.

However, by the late 1980s, it had become apparent that the local authority's strategic *Local Plan* and short-term tactical actions were no longer sufficient. A more systematic approach to the long-term management of Canterbury's tourism industry was needed, involving a wider set of actions than could be undertaken by the local authority alone and involving other key interests in the town. This change in emphasis was made more urgent by a number of impending developments. The catalyst for the present strategic approach was changes in the South East of Britain resulting from the construction of the Channel Tunnel (progressively opened to services during 1994 and 1995), and the accompanying series of realignments and improvements to the area's roads. Improved access held the promise of more visitors in the area, but also implied the threat to Canterbury's tourism industry of a range of competing destinations in Kent, Nord Pas de Calais and further afield.

At the same time, Canterbury's status as the region's retailing centre was being challenged by out of town shopping centres, which offered easy parking and less crowded conditions, located near to other towns in the area. This was perceived as a major threat by local business, since Canterbury is the sub-regional centre for East Kent, with 12 million multi-purpose visitors annually, 2.5 million of whom are tourists, mainly on day-visits (Grant *et al.*, 1996).

Visitor management experience and new initiatives in Canterbury

The changes threatening Canterbury were the context to an impetus from the Chamber of Commerce and the Dean and Chapter of the cathedral in 1991, who were concerned about the escalating negative attitudes in the town. The City Council's Planning Department was approached for advice on finding a solution. Experience in nearby Herne Bay suggested that the City Council would be reluctant to become involved in any city management scheme without the commitment of private sector partners. The Planning Department therefore recommended a feasibility study which would also clarify and agree the aims and objectives of the scheme. A steering group was set up to conduct a SWOT analysis and a comparative study of other schemes. This led to production of a consultation document suggesting a tailor-made approach for Canterbury, a provisional work plan and an indicative budget. The consultation document was widely circulated, and stimulated the local College of Higher Education to become a partner. The college saw involvement as providing its tourism management students and lecturers with practical experience of destination planning. It also took responsibility for statistical monitoring of the planned initiative, and appointed a Research Fellow with experience of Main Street initiatives in the USA.

Canterbury's planning strategy has long espoused a long-term commitment to the city's architectural and cultural heritage. The City Local Plan (1981) had recognised the pressure which large numbers of tourists placed on the attributes of the city which attracted them in the first place, and this concern was echoed in the 1994 *Canterbury District Draft Local Plan*. These documents form the statutory basis for the city's visitor strategy, but its implementation requires the co-ordination of many local Council functions especially Planning, Conservation, Highways and Environmental Health, with those undertaken by Kent County Council as the strategic highway authority. A sensitive approach is required to the key problem in Canterbury, the physical constraints imposed by its narrow medieval street patterns, and its buildings. Furthermore, many private sector organisations are directly or indirectly involved in the quality of visitors' experiences, and dependent on visitor-generated business to varying degrees.

The importance of partnerships

Destinations are complex groupings of primary attractions and secondary services, and it is increasingly recognised that their sustainability depends on forming effective and long-lasting relationships or alliances with suppliers and other organisations which contribute to clients' experiences (McKenna, 1993; Gummesson, 1994; Laws, 1995). Adopting Porter's (1987) insight that an industry can be analysed by the contribution of its members to the value chain, the management philosophy becomes one of collaboration, even between organisations which compete within the market (Laws, 1996).

The Canterbury City Centre Initiative (CCCI) was established in October 1994 as a legal entity, limited by guarantee, with funding partners drawn from leading local and regional public and private sector bodies. Core annual funding was provided by some eighteen partners, £20,000 for agreed visitor management projects, £55,000 for the Visitor Manager and the Research Fellow, and expertise, gifts in kind and personnel time to the value of about £50,000 annually for three years was also secured (Grant *et al.*, 1996).

The CCCI aimed to develop a sustainable management strategy for tourists and shoppers by involving the tourist trade and residents in its evolution. The Canterbury City Centre Initiative has four main objectives:

- to identify and respond to the needs of residents and visitors
- to stimulate the prosperity of the business community
- to add value to the experience of visiting Canterbury
- to preserve the character of the city

Figure 5.4 indicates the framework of the Canterbury City Centre Initiative. The CCCI's management structure consists of a management board of directors drawn from the partners and two steering committees – Strategy and Monitoring. There are five single-issue task force groups consisting of Board members and

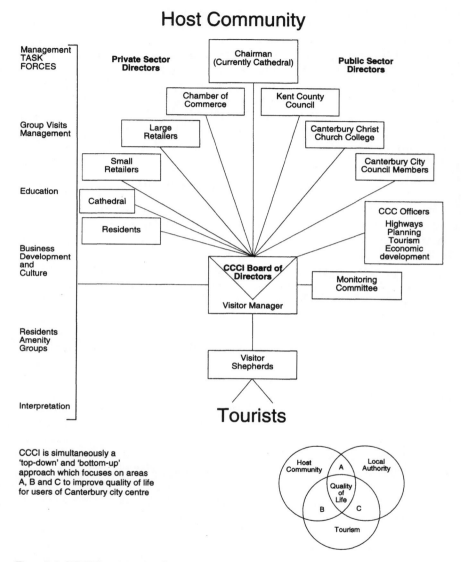

Figure 5.4 CCCI management framework

others with specialist expertise and two specialists, a specific Canterbury Visitor Manager, funded by Canterbury City Council and a Research Fellow to oversee the monitoring, based in the local College.

The City Visitor Manager

This role was recognised as being central to the success of CCCI, and one which would require a sensitive approach to reconciling the many interests in Canterbury. The appointment was as a City Council employee, but also afforded independence in the town centre, encouraging cooperation between groups who were not experienced in working together. The Visitor Manager forms the link between all partners and undertakes projects on their behalf to an agreed action plan, while developing awareness throughout the city about CCCI. An example of her work is the appointment of several 'Visitor Shepherds' located at the new coach park, the waymarking of visitor trails and the provision of maps to distribute mass tourist groups onto four designated routes between the new coach park (B in Figure 5.2) and the town centre. The Visitor Shepherd project was set up in 1995 as a three-month pilot, funded by the private sector. Their success was such that they are to operate year round, supported by additional public sector funds, with their remit extended into the city centre. A major spin-off benefit from the CCCI has been increased recognition by the City Council of the advantages which can be achieved by coordinating its own activities. A joint member-officer group has conducted an investigation of coach parking in the city, and is now considering whether to extend its remit to visitor management issues in general. An inter-departmental team of officers known as the 'Visitor Policy Group' has been formed to coordinate departmental input. Within the town itself, there is a developing feeling of consensus about the actions which need to be undertaken.

Involvement in CCCI

The strength of the CCCI is that it draws together all those who have an interest in managing Canterbury's visitor infrastructure and provides a forum for communication between them about their own work. The partners continue to implement their respective management tasks, and jointly fund the CCCI to carry out tasks which no one else is doing, or those which need a coordinated approach. Initially, it was envisaged that the Management Committee would undertake projects, but it was recognised that these were too complex. Instead, management problems were split into individual tasks, and assigned to task forces, of which there were five at the time of writing: group visits management, business culture and development, residents, education and interpretation. The five management task forces were set up to achieve solutions to problems through brainstorming, to identify those who have the responsibility and power to deal with them, and to persuade them to act. The task force concept has the advantage of reducing pressure on individual members of the CCCI, and enables a wide range of interest groups to participate (by invitation) in, and influence the CCCI's activities. The City Visitor Manager attends all task force meetings.

The visitors' experience of Canterbury

An increasingly dominant theme in the research and management literatures is concern to achieve customer satisfaction (Gronroos, 1990; Zeithmal *et al.*, 1990; Ryan, 1996). This is held to be a major determinant of repeat visits, while failure threatens to undermine the viability of tourism in one area as future visitors divert to alternative destinations (Laws, 1996; Chadee and Mattson, 1996). Furthermore, dissatisfied visitors exhibit complaining behaviour, and in expressing their disappointment or frustration, they can cause stress for staff and others, including residents, with whom they come into contact (Laws, 1991).

The first difficulty which visitors to Canterbury experience is in parking, or in reaching the historic city from outlying coach and car parks, followed by the general problem of locating the points of tourist interest within the city. The city's PARC (Park and Ride in Canterbury) Plan, being undertaken by the City and County Councils jointly, sets out the strategy for parking for cars and coaches, and pedestrian flows in the city. The aim is to keep long-term business parking out of the city centre by promoting a Park and Ride scheme. Central off-street parking space has been earmarked for shoppers and tourists, but priced to discourage long stays, complemented by a voucher scheme and on-street parking bays to provide for residents' needs (Roberts and Parker, 1992).

Considerable efforts are being made to improve Visitor Welcome; the CCCI is organising a number of staff training courses including the Tourist Board's 'Welcome Host' and 'Welcome International' courses. Foreign language training is also being made available for contact staff through subsidised tailor-made in-house courses.

Coach groups

Coaches are an increasingly dominant means of bringing visitors to Canterbury, particularly from the continent of Europe. In 1995, some 65 per cent of coaches came to Canterbury from mainland Europe, specifically, France, 36 per cent, Belgium, 14 per cent and Germany, 11 per cent (Coach Working Party, 1997). As Table 5.3 indicates, total coach arrivals in 1995 exceeded 18,000.

Around 400 to 500 coach groups arriving weekly is not uncommon, and recent price wars between the cross-Channel ferry companies, and between them and the Channel Tunnel, have increased the pressure. Canterbury's original coach park (location A in Figure 5.1) at Longport, with just 40 spaces, was unable to cope with the demand and presented a poor first impression of Canterbury, compounded by a dangerous road crossing on which several tourists had been injured or killed.

As a result of forecasts during the early 1990s of increasing coach group visits, an urgent decision was taken to move the coach park during the winter of

1994/95, ready for the main 1995 season. The only available large site, capable of holding in excess of 100 coaches was at Kingsmead market site on the east of the city centre (location B on Figure 5.1). Logistically, this is not the ideal location as the main coach arrivals to the city come from the west, although there is a steady recorded increase in coaches arriving from the ferry terminal at Ramsgate. The Kingsmead coach park was seen as a limited-life solution whilst a more permanent arrangement is investigated. Nevertheless, it is being used as an opportunity to test the efficacy of various visitor management techniques, and to see whether good quality facilities would attract coach drivers to the new coach park: it is hoped that they will no longer drop passengers illegally in the streets. Annual surveys are undertaken of driver and passenger reactions to the new coach park; effectiveness of the four tourist trails from the coach park to the town centre are also being evaluated, as is the contribution made by the Shepherds to the success of Canterbury's visitor management strategies. After 18 months of operation, it can be concluded that the visitor management methods are effective and that the facilities provided for drivers encourage them to use the coach park. Further refinements are planned as a result of the monitoring of journeys. Additional surveys are being programmed – the information from them is proving to be one of the most effective ways of developing the visitor management strategy.

The problems associated with coach-based visits result from congestion caused by the vehicles and by the passengers moving into Canterbury and around it in groups. These problems are compounded by the arrival of many coaches at the same time of the day, due to their common starting points in London or in the nearby European centres. The purposes of the PATHS initiative (Pro-active Approaches to Tourism Hosting Strategies) is to develop a methodology for managing tour coach visits to Canterbury through collaboration with tour and coach operators in Belgium, France and the United Kingdom and also with other historic towns which are facing similar problems. Already, two significant steps have been completed. A major market research exercise was carried out into the incoming tour group market from Belgium and France. The 'Allons-Y' survey was a collaborative exercise between the BTA Paris Office, South East England Tourist Board, Kent County Council and CCCI. The research was jointly funded

Table 5.3 Coach park usage 1991–1996

Year	Coaches
1991	8,259
1992	12,157
1993	13,636
1994	14,087
1995	18,211
1996	18,500

Source: Report of Coach Working Party (1997)

by the partners and by contributions of £250 each from thirty attractions throughout South East England. Over 8,000 questionnaires were mailed to tour operators and their clients. A 10 per cent response rate provided invaluable information on market characteristics for four main client groups, school parties (primary and secondary), senior citizens, *comités d'entreprises* and clubs and societies, in addition to the requirements of tour operators. The 'Allons-Y' results are being used to develop client-friendly tour packages amongst the subscribers to the research.

The second major step has been collaboration with other English historic towns through the English Historic Towns Forum, the Association of Leading Visitor Attractions, the English Tourist Board, the British Incoming Tour Operators Association and the Confederation of Passenger Transport in producing an agreed Code of Practice for visits by coaches to historic towns. The purpose of the Code of Practice is to show how positive cooperation between local authorities, coach and tour operators, guides and the attractions or facilities being visited can be turned into effective management by everyone making an effort to understand the issues being raised by this form of tourism, and to acknowledge each other's concerns and aspirations.

Table 5.4 Analysis of the Kingsmead Coach Park

Advantages

+ Owned by City Council.
+ Already in operation.
+ Can accommodate present numbers of coaches.
+ Located away from residential properties.
+ Offers a pleasant walking route into the centre.
+ Site is easily signed and operated.
+ No planning objections (although permanent use would have to be notified to the Secretary of State, as it would be a departure from the *District Plan*).

Disadvantages

− Access is difficult from the main approaches roads, and adds to congestion on the ring road, contributing to traffic spill into residential areas.
− The Kingsmead site does not provide an attractive introduction to the city.
− The distance from the centre is too far for elderly visitors, for whom a shuttle service has been provided at a nominal fee.
− The site is remote from other attractions, notably museums. This conflicts with the wider visitor management strategy, to develop more attractions.
− There is conflict at the entrance between passengers and traffic.
− High use of the riverside walk into the city centre detracts from its previously quiet environment.
− The site is operating near to capacity during the peak season (although it could be extended if a Depot is relocated).
− Use as a coach park compromises the future development of the area for other uses.

Source: Coach Parking In Canterbury, 1996.

The objective is to involve coach tour operators and their drivers in accepting the concept and benefits to be achieved by working with the host communities, and a move towards general acceptance of the idea that it is not unreasonable for tourists to expect to walk to reach the centre of any of Europe's historic towns.

A study was undertaken of alternatives for coach parking in Canterbury as the present situation is acknowledged to be less than ideal. The study included detailed reviews of the experience of comparable cities such as Windsor and Salzburg. Some ten sites around Canterbury were evaluated and the working party reported that 'it is clear that sums in the region of a minimum of £3 to £3.5 million would have to be allocated to the provision of a new coach park. The two options with increased revenue costs in the region of £0.5 million per year are not affordable under the current government's financial controls and cannot be supported by the City Treasurer and Deputy Chief Executive' (Coach Party Working Report, 1997).

Another option is to upgrade the existing, interim, coach park. The working party put an estimated cost of £100,000 on improvements needed 'covering such works as improved ticketing arrangement, construction of picnic area together with landscaping, river banks and upgrading to the existing toilets. These works would be necessary if the coach park is to remain on this location' (Coach Party Working Report, 1997). Table 5.4 lists the advantages and drawbacks of continuing to use the interim coach park, and given this analysis, the working party recommended that the Council commits itself to the relocation of the coach park by the year 2000 on an 'in principle' basis.

Pedestrianisation

Tourists are unfamiliar with the cities they visit, and they tend to congregate in specific areas. In Canterbury, as indicated in Figure 5.1, these are particularly the Cathedral precincts and the main shopping street, the High Street. This was historically the main A2 road linking London and Dover, but was pedestrianised when a bypass was constructed in the 1960s. This improvement has encouraged tourists to spread further afield within the city, rather than limit their visit to the area around the cathedral, but more pedestrianisation and additional initiatives are planned to encourage tourists to experience different areas of the city. Experience in many other towns in Britain indicates that local traders have objected to the exclusion of vehicles from shopping streets during the day. In one of its consultation documents, Canterbury City Council justified pedestrianisation in the following way: 'The Council has (also) been implementing a policy of pedestrianisation and general street refurbishment. This is not a luxury item. It makes sound economic sense to keep the town centre in good order as this is what attracts shoppers to Canterbury.'

Managing visitor flows to the cathedral

The main attraction in Canterbury is the cathedral. This too experiences severe pressures from visitors, and in response the Dean and Chapter have developed visitor management policies appropriate to a church. The first recorded visitor management task was to ensure that the medieval pilgrims approached St Thomas Becket's tomb on their knees! The modern concern is to ease the flow of visitors around the main points of interest in the cathedral. The religious and architectural importance of the cathedral attracts many school parties, both from Britain and abroad, and these groups were identified as a major cause of complaint by others who used the cathedral. A scheme, 'Operation Shepherd' was developed with a number of tactics to control the entrance of groups. In the first year, 1992, only 13 per cent of groups booked their visit to the cathedral but in 1993, 45 per cent of school groups pre-booked. In June 1995, an entrance charge of £2.00 was introduced for all tourists visiting Canterbury cathedral and its precincts, although local people can continue to enter free with the benefit of a pass. Close monitoring of this new policy approach is being undertaken to assess whether visitor numbers to the cathedral and the city have changed significantly. Preliminary results indicate that an initial drop in cathedral visits is continuing but that visits to the city in general have increased. The cathedral continues to be a significant reason for visiting Canterbury.

Expanding the city's tourist appeal

To increase the range of attractions in Canterbury, the City Council has financially supported and encouraged private projects such as 'Canterbury Tales', housed in a renovated church, and portraying the experience of Chaucer's pilgrims on their journey to Canterbury. The City Council also operates four Visitors' Bureaux, two in Canterbury, the others located in Herne Bay and in Whitstable, both nearby seaside resorts within its administrative area (see Figure 5.2) in partnership with the Chamber of Trade.

The Council has also invested in a museum explaining Canterbury's heritage more formally, and in a Roman museum. Through collaboration with English Heritage and other site owners, a successful lottery bid has enabled the creation of a Visitor Centre, renovations and improved interpretation of that part of the World Heritage Site at St Augustine's Abbey and St Martin's Church. The Queen Bertha trail and a year-long programme of events in 1997 mark the 1,400th anniversary of St Augustine's arrival in England.

The Council has a stringent conservation policy on shop fronts and the use of building materials, partly in response to unsympathetic reconstruction after war-time bombing. A theatre, improved sports facilities and new walking and cycle tracks, together with new gardens and a street market, have further enhanced the appeal of the city to its residents.

Analysis of the Canterbury City Centre Initiative

An analytical framework is needed within which the many aspects of managing tourism in walled and historic cities can be investigated. The interdependecy of the elements which together make up tourist destinations can best be understood from the perspective of a soft, open, systems model, (Mill and Morrison, 1985; Leiper, 1990; Checkland and Scholes, 1990, Laws, 1995). The 'systems' aspect of this type of model has the advantage of focusing attention on all the major inputs needed to provide tourism services, and on the outcomes of tourism processes for all groups with interests in the destination. The 'soft' feature of the approach is concerned with the interactions of tourists and residents in tourist destination areas. The model is 'open' because it recognises the legislative, cultural and technological contexts for tourism processes. A further aspect highlighted by this analytical framework is the consequences of tourism for the area's environment. On a theoretical level, systems theory provides a way of focusing the insights from many social sciences on destination processes and their consequences. Krippendorf (1987) has pointed out that the relationships between ecological and other effects and the phenomenon of mass tourism 'cannot be identified if they are viewed from a narrow disciplinary angle'.

Systems theory argues that the efficiency of the destination's operations will be affected by changes to any of the elements of which it is composed. For effective management, three aspects need to be clearly understood:

- the effects on outputs of any change to its inputs,
- the ways in which its internal subsystems and processes are linked,
- how the subsystems and processes are controlled.

The range of tourists' needs during their stay in any destination is met by a variety of organisations, many are private sector, profit-oriented companies, but this diversity leads to the need for coordinated destination management. Typically, this role is undertaken by various governmental bodies, at national, regional or local scales. An experienced tourism administrator has pointed out that 'Social and economic policy makers in general have not regarded tourism as important, and so state bodies responsible for tourism activity are often ill equipped to deal with its development' (Likorish, 1991). He criticised the poor co-ordination between government departments and between the major trading sectors, arguing that tourism is usually accorded relatively low priority, and limited resources, resulting in poor co-ordination and piecemeal programmes. This is further exacerbated by the method of public sector funding which is on an annual basis, inevitably leading to short-term planning.

The European Union Green Paper 'The Role of the Community in the Field of Tourism' (EU, 1995) sheds further light on this problem. It discusses the public property which forms the key attraction for tourists (natural heritage, cultural heritage, etc.) as a 'free good' in economic terms, i.e. one which is not taken into account in evaluating the impact of tourism and its costs. In Canterbury, this phenomenon has been recognised as a major conceptual problem and has been addressed by setting up the CCCI to adopt a coordinated and systemic approach, over a three-year horizon, to the problems posed by tourism.

Application of a soft, open systems framework to Canterbury

All tourist destination systems consist of elements (or sub-systems) in the form of natural or primary attractions such as climate, supported by secondary features such as hotels (Jansen-Verbeke,1988). Figure 5.5 shows that, in Canterbury, the primary elements are the historic city itself, and particular attractions especially the Cathedral. The secondary elements include hotels, guest houses and the range of attractions, shopping and catering in the city centre. Additional elements are the information services available to visitors, catering, and car and coach parking. The inputs into the destination system are the expectations and attitudes of its tourists and residents, the skills of people employed in its tourism-related

INPUTS	CANTERBURY DESTINATION SYSTEM	OUTCOMES	
	Present		Future
Tourists' expectations	PRIMARY ELEMENTS	IMPACTS Economic Community Environment Ecology	The
	Cathedral		
	Historic city centre		Canterbury
Entrepreneurial creativity			City Centre
	SECONDARY ELEMENTS		
Employee skills			Initiative
	Hotels		
Investors capital	Catering	STAKEHOLDERS OUTCOMES (selected)	and
Local Authority planning	Retailing		
	Attractions	Tourists Residents	PATHS
Residents' expectations and attitudes	Information services	Employees Retailers	Projects
	Parking	Coach operators Local authority	
	Infrastructure		
	EXTERNAL INFLUENCES Transport developments Competition Tastes Legislation Currency exchange rates		

Figure 5.5 Soft, open systems model of tourism in Canterbury

sectors, public sector resources, investors' resources, and managerial and technical skills including, particularly, recognition in the local planning process of the need for a positive framework for tourism management leading to a continuity of approach.

The systems model incorporates the significance of changing competitive conditions, and improvements to the infrastructure or transport network as external factors. The method focuses attention on the outcomes of the system's functioning for particular stakeholder groups during a given time period. Evaluation of the outcomes against the costs of inputs and policy objectives provides the basis of feedback, thus introducing a future-time dimension to the model.

The foregoing implies that the system just happens by itself, but it does not. There has to be a vision, and an instigator of action. The CCCI is attempting to change the outcomes for the benefits of the stakeholders by taking deliberate actions.

Stakeholder analysis of the Canterbury initiatives

Stakeholder theory recognises that there are plural interests, and particular groups have varying degrees of involvement in tourism. They are affected by it in differing ways, and they enjoy differing abilities to influence future developments. The main groups in Canterbury are its visitors, residents involved in tourism, and others who are not, the organisations which speak for the cultural life of the city and its historic fabric, local businesses, and other organisations which organise aspects of tourists' experiences in the city, but are themselves based elsewhere, and the residents of East Kent who regard Canterbury as their sub-regional service centre. The quality of a tourism system's operations can be assessed by examining the outcomes for each stakeholder group (Laws, 1995). Table 5.5 presents a summary of responses to various problems and the improvements made for selected stakeholders in Canterbury's tourism system.

Monitoring the initiatives

Control over the quality and consistency of a system's functioning and outputs depends on an effective feedback channel between the monitoring and decision-making sub-systems of the destination. This can be achieved by market research to understand the expectations of visitors and coach operators, and techniques to assess the impacts of tourism on the quartet of destination concerns, its society, economy, environment and ecology (Laws, 1995). The Canterbury City Centre Initiative includes independent research on its effectiveness conducted by a researcher based in a local College, and began with a base study to establish the nature and extent of tourism in Canterbury.

Table 5.5 Recent improvements for Canterbury's stakeholders

Stakeholders	Indicative tourism problems	Responses
Residents	Congestion	Attract tourists into less visited areas
Tourists	Congestion Limited range of attractions Lack of information	New walks Signing Maps Pedestrianisation New attractions
Coach operators	Restricted and undeveloped parking	New coach park Shepherds Good facilities for drivers
Cathedral	Crowds detract from experiences of visitors and worshippers	Shepherding scheme Entrance charges
Environment	Air pollution Litter Inappropriate changes to historic buildings	Air quality monitoring at selected sites Park and Ride to reduce cars in city New location for coach park Special planning controls
Retailers	Presence of tourists encourages street traders in competition with local shops	Increased town centre management activities and extra policing Appointment of 'City Warden'
Local Authority	Diversion of funds from local issues to pay for visitor management	Seek financial and practical help from wider tourism interests through involvement in CCCI

Further development of the Canterbury City Centre Initiative

The first phase of the CCCI is scheduled to terminate in October 1997. The CCCI was conceived as an ongoing, evolving mechanism for improvement for residents, visitors, business and the city environment. The partners have expressed their wish to continue and extend the initiative. The following will be given priority:

- interpretation of the city for residents and visitors
- promotion of the benefits of tourism to residents
- improving the level of care to visitors
- enabling businesses and other city users to maximise the benefits of tourism overall
- improving the quality of the city and the quality of life for those who live and work in Canterbury

The CCCI has no executive powers, its main role is to persuade others to cooperate to achieve its objectives. For the second phase, which embraces wider coordination and management remits, CCCI will need both extra resources and more authority. As part of the enhanced scope envisaged for CCCI, the existing role of Visitor Manager is to be expanded to City Centre Manager. The recommendation, to the Council policy committee, to support the future strategy of the CCCI, to fund the post of City Centre Manager and to provide 50 per cent funding for CCCI for a three-year period from October 1997, was approved by the City Council in January 1997.

In formulating its ideas for the second phase, the CCCI canvassed the opinions of local businesses prior to approaching the City Council for further funding. The Chartered Institute of Bankers commented: 'the conflicting interests of "Canterbury" in its widest definition can and should be reconciled and I am sure that CCCI can contribute to this'. St Peter's Association (a residents' amenity society) stated that 'the work the Initiative carries out is essential to coordinate and provide a balanced policy to safeguard the interests of those who live, work and shop in our City'. The Dean and Chapter of Canterbury said: 'We strongly feel that the CCCI should proceed into a second phase . . . a partnership between all agencies concerned with tourism . . . can only be good for both the quality of the tourists' experience of Canterbury and also for safeguarding the interests of the City and its cultural and commercial life.' Stagecoach East Kent (the local bus company) expressed its commitment 'to working with the community and other business interests . . . The CCCI has enabled the company to become actively involved . . . '

The key aims identified in the CCCI second phase discussion document include business and employment development, safeguarding the natural and man-made environments, and marketing to promote awareness of the area for investors. The economic objectives for tourism include developing an agreed action plan, achieving a sustainable balance, offering a quality experience and maximising the employment potential.

Discussion and conclusion

This chapter has reviewed some of the issues confronting historic walled cities which are also tourist destinations, and it has examined how one major destination, Canterbury, has recognised the particular difficulties and oppor-tunities facing it, and is responding to them through the Canterbury City Centre Initiative. The first phase of the CCCI has been successful in demonstrating its ability to achieve partnership between the many interest groups in the city, and has enhanced the experiences of visitors and residents. Recognising that much remains to be done, a second phase is planned, in which the CCCI recognises the need for a comprehensive interpretation strategy in Canterbury.

It is becoming increasingly important for destinations to control the effects of tourism and implement policies to gain the most benefit and to minimise the harmful consequences which result from tourist activity. In particular, the consequences for other stakeholders, including local residents and businesses, need to be made explicit if tourism is to be sustainable as the basis for continuing economic and cultural development in historic walled cities. This process begins with a clear analysis of the specific issues and characteristics in a particular destination as a basis for a coherent, systematic view of the various policies to resolve them. Implementing policies requires resources, and a range of skills to carry them out. Experience in Canterbury highlights the fact that sustained success depends on the long-term commitment of a key figure who can stimulate, coordinate and maintain the vision through the efforts of a range of contributing organisations with sometimes conflicting priorities.

The European Union's Green Paper 'The Role of the Union in the Field of Tourism' points to the social and economic importance of tourism (Section 3a, 1995). However, Canterbury, in common with the majority of small towns which also happen to be major tourist destinations, does not itself have the resources to fully implement effective tourism management policies. Such communities will increasingly turn to their national governments and the European Union to help to fund their efforts to mitigate the negative impacts of tourism at the local level. This is a responsibility which all governments and the European Union should be willing to accept if they are serious about the significance of tourism, and if they wish to compete effectively with alternative destinations around the world while preserving the essential characteristics of the many small historic cities which are fundamental elements in the evolution of Europe's cultural heritage.

References

Special sources

Canterbury City District Plan, Report on Choices and Strategy, Canterbury City Council, 1981.
Canterbury District Draft Local Plan, Canterbury City Council, 1994.
Canterbury Tourism and Shopping Survey, Canterbury City Council, 1993.
Coach Working Party Report, Canterbury City Council, 1997.
Economic Development Strategy, Canterbury City Council, 1996/97.
Kent County Council Transportation Study, 1975.
'Minutes of the Policy Committee', Canterbury City Council, September 1996.
Tourism Profile, Canterbury City Council, 1994.
Tourism: A Serious Business, Canterbury City Council, 1996.

General references

Ashton, J. (1992) *Healthy Cities*, Open University Press, Buckingham.
Ashworth, G. J. and Tunridge, J. (1990) *The Tourist Historic City*, Belhaven, London.
Ashworth, G. J. and Voogd (1990) *Selling the City: Marketing Approaches to Public Sector Urban Planning*, Wiley, Chichester.

Buckley P. and Witt, F. (1989) 'Tourism in Difficult Areas', *Tourism Management* June: 138–53.

Butler, R. (1980) 'The Concept of a Tourist Area Life Cylce: Implications for Management of Resources', *Canadian Geographer* 24, 1: 5–12.

Chadee, D. and Mattson, J. (1996) 'An Empirical Assessment of Customer Satisfaction in Tourism', *Service Industries Journal* 16, 3: 305–20.

Checkland, P. and Scholes, J. (1990) *Soft Systems Methodology in Action*, John Wiley & Sons, Chichester.

Davis, D., Allen, J. and Cosenza, R. (1988) 'Segmenting Local Residents by their Attitudes Interests and Opinions Towards Tourism', *Journal of Travel Research* Fall: 2–8.

Doxey, G. U. (1975) 'A Causation Theory of Visitor–Resident Irritants, Methodology and Research Inferences', *Travel and Tourism Research Association Proceedings*, pp. 195–8, Salt Lake City.

European Union 'The Role of the Union in the Field of Tourism', EU, Brussels.

Giotti, G. (1994) *Cities of Art: Problems and Perspectives*, CISET, University of Venice.

Godfrey K. B., Goodey, B. and Glasson, (1994) 'Tourism Management in Europe's Historic Cities', in P. E. Murphy (ed.), *Quality Management in Urban Tourism, Balancing Business and Environment*, University of Victoria, p. 192–202, Canada.

Grant, M., Human, B. and Le Pelley, B. (1996) 'Canterbury City Centre Initiative, Visitor Destination Management in Practice', *BTA Insights*, July: C1–13.

Gronroos, C. (1990) *Service Management and Marketing: Managing the Moments of Truth in Service Competition*. Lexington Books, New York.

Gummesson, E. (1994) 'Making Relationship Marketing Operational', *International Journal of Service Industry Management* 5, 5: 5–20.

Jacobs, J. (1972) *The Economy of Cities*, Pelican, Harmondsworth.

Jansen-Verbeke, M. (1988) *Leisure, Recreation and Tourism in Inner Cities*, Netherlands, Geographical Studies, Amsterdam.

Kotler, P., Haider, D. H. and Rein, I. (1993) *Marketing Places*, The Free Press, New York.

Krippendor, J. (1987) *The Holiday Makers*, Heinemann, London.

Law, C. M. (1992) 'Urban Tourism and its Contribution to Economic Regeneration', *Urban Studies* 29, 3: 599–618.

Law, C. M. (1993) *Urban Tourism: Attracting Visitors to Large Cities*, Mansell, London.

Laws, E. (1991) *Tourism Marketing, Service and Quality Management Perspectives*, Stanley Thornes, Cheltenham.

Laws, E. (1995) *Tourist Destination Management, Issues, Analysis and Policies*, Routledge, London.

Laws, E. (1996) *The Inclusive Holiday Industry, Relationships, Responsibility and Customer Satisfaction*, Thomson International Business Press, London.

Le Pelley, B. (1994) 'Canterbury: Managing Tourism and Heritage', in P. E. Murphy (ed.) *Quality Management in Urban Tourism, Balancing Business and Environment*, University of Victoria.

Le Pelley, B. and Laws, E. (1995) 'PATHS towards PEACE: A Stakeholder Benefits Approach to City Centre Management in Canterbury', in M. Robertson (ed.) *Proceedings of the Urban Environment Conference*, South Bank University, London, pp. 76–85.

Leontidou, L. (1990) *The Mediterranean City in Transition*, Oxford University Press, Oxford.

Leiper, N. (1990) *Tourism Systems*, Massey University Press, New Zealand.

Lickorish, L. (1991) *Developing Tourism Destinations*, Longman, Harlow.

MacCannell, D. (1976) *The Tourist: A New Theory of the Leisure Class*, Macmillan, London.

McKenna, R. (1993) *Relationship Marketing: Successful Strategies for the Age of the Customer*, Addison Wesley, Reading, MA.

Mill, R. C. and Morrison, A. M. (1983) *The Tourism System*, Prentice Hall, Englewood Cliffs, NJ.

Page, S. (1995) *Urban Tourism*, Routledge, London.

Porter, M. (1987) 'From Competitive Advantage to Corporate Strategy', *Harvard Business Review* 3: 43–59.

Roberts, A. G. and Parker, A. G. (1992) 'The Parc Plan, Canterbury: The Introduction of a Balanced Transport Policy', *Proceedings of the Institute of Civil Engineers and Municipal Engineers*, Paper 9907, June: 101–10.

Ross, G. (1992) 'Resident Perceptions of the Impact of Tourism on an Australian City', *Journal of Tourism Research* 20, 3: 7–10.

Rumble, P. (1989) 'Interpreting the Built Environment', in D. Uzzell (ed.) *Heritage Interpretation*, vol. 1, Belhaven Press, London, pp. 24–32.

Ryan, C. (1995) *Researching Tourist Satisfaction: Issues, Concepts, Problems*, Routledge, London.

Sheldon, P. and Var, T. (1984) 'Resident Attitudes to Tourism in North Wales', *Tourism Management* March: 40–49.

Soane, J. (1993) *Fashionable Resort Regions: Their Evolution and Transformation*, CAB International, Wallingford.

Urry, J. (1990) *The Tourist Gaze*. Sage Publications, London.

Walled Towns Friendship Circle (1991) Brochure, Tenby.

Zeithmal, V., Parasuraman, A. and Berry, L. *Delivering Quality Service, Balancing Customer Perceptions and Expectations*, The Free Press/Macmillan, New York.

RELIVING THE DESTINATION LIFE CYCLE IN COOLANGATTA

An historical perspective on the rise, decline and rejuvenation of an Australian seaside resort

Roslyn Russell and Bill Faulkner

As shown in Figure 6.1, Coolangatta is Queensland's most southerly coastal resort. Located 110 kilometres south of the state capital, Brisbane, and at the southern extremity of the Gold Coast, it stands adjacent to the New South Wales town Tweed Heads on the Tweed River. Together they have become known as the Twin Towns although the state border, a barrier now invisible, separates them.

The story of tourism in Coolangatta began over a hundred years ago. From the early beginnings, when there existed only one hotel used by passing travellers, Coolangatta became one of Queensland's most popular resorts. Indeed, for a substantial part of the area's history over the last fifty years, the local economy has been almost singularly tourism-dependent, with fine surfing beaches as its main natural attraction. In recent years, however, Coolangatta has become a dated seaside town clinging to memories of the past, and in this respect its history has exhibited many of the characteristics of Butler's (1980) destination life cycle model, albeit somewhat blurred.

In this chapter we utilise Butler's model in conjunction with an historical analysis approach to explore the dynamics of tourism development at Coolangatta and provide insights into issues affecting the rejuvenation of this area as a tourist destination. While the destination life cycle model serves as a framework for structuring our analysis, the historical perspective highlights the relevance of memories to the consideration of rejuvenation strategies for destinations such as Coolangatta, which has a rich tourism history. Far from being an unproductive and self-indulgent glorification of the past, the history of an area is 'the storehouse of experience through which people develop a sense of their social identity and their future prospects' (Tosh, 1991: 1). Thus, in the Coolangatta instance, the shared history of a community is both a potential asset and a constraint so far as the rejuvenation process is concerned. It is an asset to the

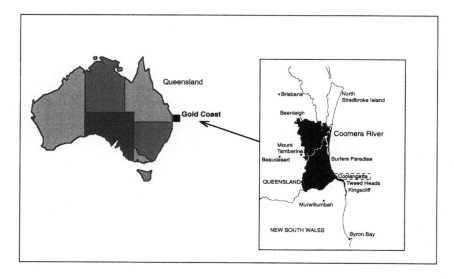

Figure 6.1 Location of Coolangatta

extent that nostalgia provides a potentially powerful tool for repositioning Coolangatta in the tourism market place and, at the community level, it may foster the common sense of purpose and agreed set of goals that is so essential for developing and implementing a rejuvenation strategy. It is a constraint in the sense that historical inertia has the capacity to restrict options, and tourism development strategies at any time must be consistent with prevailing community beliefs and genres.

In the analysis that follows, we begin with some background on the destination life cycle model and then apply this framework to the analysis of the evolution of tourism at Coolangatta. Implications for the consideration of approaches to the rejuvenation of tourism at this destination are then considered.

Butler's destination life cycle model

In a statement which was influential in the analysis of destination histories, Stanley Plog suggested that :'We can visualise a destination moving across a spectrum, however gradually or slowly, but far too often inexorably toward the potential of its own demise' (Plog in Butler, 1980: 6). This scenario is based on the observation that, as a destination evolves, it becomes more accessible to a wider market, facilities for servicing tourists expand and, eventually, it attracts visitors who are both different and more numerous. This process, he argues, results in the area becoming saturated and the unique features that initially made it so attractive to visit are gradually destroyed. According to Plog, therefore, destinations contain the 'seeds of their own destruction', with obsolescence and decline being the all too frequent outcome of their encounter with tourism.

Butler's model draws upon the product life cycle (PLC) concept originating from the discipline of marketing to describe the evolution of tourist destinations. The PLC is based on the proposition that a product's sales follow an asymptotic S-shaped curve, reflecting its passage through different stages of market performance – introduction, growth, maturity and decline. Butler envisages tourist destinations experiencing a series of developmental stages, beginning with exploration and progressing to involvement, development, consolidation and culminating in stagnation and decline when the carrying capacity has been exceeded and/or the destination's assets are degraded. However, Butler extends Plog's argument by recognising the potential of a turning point along the curve, whereby decline and demise could be averted by rejuvenation and vitalisation. A summary of the characteristics of each stage of Butler's model are presented in Table 6.1.

Although cases described in the literature do not lend themselves to crude classification of instances where Butler's assertions are proved or disproved, it is nevertheless possible to distinguish between generally conforming and non-conforming situations, as we have in Table 6.2. This summary reveals that, while there are many deviations from the predicted sequence of changes, there are also many conforming cases. If anything, however, each study highlights the need for some flexibility in the interpretation of patterns by recognising the uniqueness of each destination's development pattern and the variety of factors that modify the sequence of changes, or obstruct or accelerate progress through the cycle. Variations involving the omission of stages can be affected by such factors as natural disasters, government intervention, pre-existent local wealth, physical limitations or the scale of operation.

Given the degree of support for the model revealed in Table 6.2, Lundgren's (1984 in Weaver, 1990: 9) observation that: 'Butler put into the realistic cyclical context of reality that everyone knew about and clearly recognised, but had never formulated into an overall theory' seems justified. Even the harshest critics of the model do not reject it outright and generally concede its usefulness as a heuristic device, while at the same time cautioning against the misconceptions inevitably involved if the model is interpreted too literally (Cooper, 1994). Thus, in his review of the application of the life cycle model, Cooper concludes that 'acceptance of the concept appears to be strengthening, particularly as an organising framework for destination development' (1994: 345).

Most applications of the tourist destination life cycle model to date have concentrated on using it in three different ways: first, as a framework for examining and analysing the evolution of a destination; second, as an aid to marketing and planning; and, third, as a forecasting tool (Cooper, 1994: 341). There is little doubt that the life cycle concept is applicable to the first two purposes, at least to the extent that it is useful for identifying influences and constraints and the bearing of these on the directions destinations have taken. But we need to be wary about pushing the model beyond its intended purpose. As Cooper and Jackson (1989: 381) have observed: 'Butler's original conceptualisation of the

Table 6.1 Summary of characteristics of Butler's model (based on Butler, 1980)

Exploration stage	• Small numbers of tourists (allocentrics, explorers) • No specific tourist facilities • High level of contact with locals • Low impact on region
Involvement stage	• Residents begin to provide tourist facilities • High level of contact between tourists and locals • Advertising to attract tourists is initiated • Tourist seasons emerge • Pressure is put on governments and public agencies to improve transport and other facilities for tourists
Development stage	• The tourist market is well-defined • Heavy advertising in tourist-generating areas • Local involvement declines as they lose control of development • Locally provided facilities will disappear. External organisations will replace the locally provided facilities with larger, more elaborate and modern facilities
Consolidation stage	• Rate of increase in numbers of tourists will decline although total numbers will still increase • Total visitor numbers exceed the number of permanent residents • Marketing and advertising will be extensive • Efforts will be made to extend the tourist seasons • Economy of the region is now dependent on tourism • Opposition and feelings of discontent are evident among the host community
Stagnation stage	• Peak numbers of visitors will have been reached • Capacity levels for many variables will have been reached • The area will have a well-established image but will no longer be fashionable • Heavy reliance on repeat visitation and business-related visitors • Natural and 'authentic' attractions have been superseded by imported 'artificial' facilities • Existing properties experience frequent change in ownership • Type of tourist will most likely be the mass tourist or psychocentric.
Decline stage	• Unable to compete with newer attractions • Experiences a declining market • Increase in day trippers or weekend visits • Property turnover will be high • Tourist facilities will be replaced by non-tourist related structures. Local involvement increases again
Rejuvenation	• Combined efforts from government and private bodies • May concentrate on attracting a specific interest group rather than allocentrics again • Artificial attractions such as theme parks may renew interest in the area

Table 6.2 Case studies summary

Cases most clearly demonstrating substantial conformity to Butler's model	• Thai beach resorts (Cohen, 1982) • European coastal islands (Pearce, 1989) • Sauble Beach, Ontario (Strapp, 1988) • Grand Island, Louisiana (Meyer-Arendt, 1985) • Atlantic City (Stansfield, 1978) • Minorca (Williams, 1993) • Lancaster County (Hovinen, 1981) • Isle of Man (Cooper and Jackson, 1989) • Hahndorf, Bright, Cairns (Ross, 1994) • Papua-New Guinea, Fiji, Vanuatu (Douglas, 1994) • Grand Cayman, Antigua (Weaver, 1990)
Cases most clearly demonstrating apparent deviation from Butler's model	• Pacific islands (Choy, 1992) • Thai beach resorts (Cohen, 1982) • Lancaster County (Hovinen, 1981) • Hahndorf, Bright (Ross, 1994) • Grand Island, Louisiana (Meyer-Arendt, 1985)

tourist cycle did not envisage its use as a prescriptive tool and many problems emerge when the cycle is used in this way.' The model is no more a crystal ball than is general history and, in this sense, its value as a forecasting tool is limited.

Given the complexity of these background factors, it is not surprising that the Butler model has been deficient as a predictive tool to date. It has been used to highlight the transition from one stage to another in hindsight, rather than to forecast the future. Thus, its application requires the methodology of the historian and its principle product, according to Cooper and Jackson (1989: 384), is the 'light cast upon both the decision makers and the process of decision making'.

One of the main strengths of Butler's model revealed by the above discussion, therefore, lies in its utility as a framework for historical analysis. Through such analysis, we can identify the many variables and 'turning points' that have influenced the evolution of tourism in a region. It follows that, with the better understanding of the underlying dynamics of tourism development at a destination this analysis can provide, we will have a better foundation for both generating and considering future scenarios, considering future options and developing appropriate tourism marketing and planning strategies. In particular, through this perspective, we can focus on turning points in the destination's evolution that might signal the onset of stagnation and decline, and which make Plog's spectre of inevitable decline a potential reality unless some remedial action is taken.

The History of Tourism at Coolangatta

The evolution of tourism at Coolangatta is described schematically in Figure 6.2, where phases of tourism development are identified in terms of Butler's model and key events affecting the progress of tourism development are indicated. As there is no consistent data source for visitor numbers over the whole of the period concerned, it has been necessary to interpolate from fragmentary historical data in the construction of this plot. While the figure indicates that the growth in visitor numbers corresponds broadly with Butler's S curve and the sequence is initiated by an exploration phase and proceeds towards stagnation, changes in the intervening phases are not strictly in accord with the model. The utility of the model as a heuristic device for destination planning is nevertheless illustrated through the way it highlights how Coolangatta deviates from patterns of development that are common elsewhere. This point is elaborated through the description of each phase that follows.

Drifters, explorers and early involvement (pre-1900s)

The story of tourism in Coolangatta began over a hundred years ago. In the late 1800s to the turn of the century, the area was largely inhabited by timber-getters, who harvested the fine Red Cedar timber from the surrounding rainforests, and farmers primarily engaged in dairy production. Urban development at Coolangatta at this stage was largely geared to servicing this rural population and transport linkages from Brisbane, 110 kilometres to the north, were tenuous. A combination of railway and coach transport had to be used, the lack of bridges between Coolangatta and the railhead at Southport meant that several large rivers had to be forded, while the easiest route for part of the trip involved the use of the beaches as a thoroughfare. Under these circumstances, only the most intrepid traveller resembling Cohen's (1972) explorers and drifters traversed the area and, accordingly, tourism development had reached the exploration stage. Local involvement in the industry began as early as 1884 when local character, Robert Harrison, opened the Commercial Hotel on land fronting onto Coolangatta beach (Sullivan, 1982). However, the servicing of visitors was a secondary function of the hotel as it had only two guest rooms.

Individual mass tourism, the first wave (incipient development, 1900–30)

With the extension of the railway to Coolangatta in 1903, the flow of visitors increased and local involvement intensified through the provision of a range of accommodation options. Mass tourism was therefore thrust upon Coolangatta while it was yet in its infancy and, while the railway may not have been built to service a tourist centre, it was certainly instrumental in creating its character.

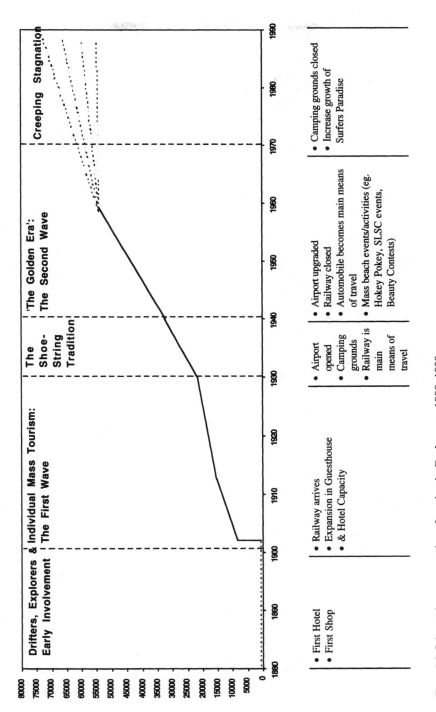

Figure 6.2 Schematic representation of tourism in Coolangatta 1880–1990

As few accommodation or other facilities existed, visitors improvised their own. Coolangatta rapidly became the venue for family camping. Holiday-makers were very independent mass tourists retaining much of the venturesomeness of pioneers and content to live primitively.

Patty Fagan, a flamboyant Irishman, had discovered Greenmount (naming it after his Irish birthplace) while constructing the railway station, sheds and residences on the Nerang–Tweed Heads line. He built and was the proprietor of the Greenmount Guesthouse, which opened in 1908. He and those who followed to establish the Stella Maris, the Beach House and the St Leonards' guest houses that were synonymous with Coolangatta's heyday, were visionaries, albeit on a small scale. The guest houses were small two-storey buildings and none were capable initially of catering for more than forty guests. Some, however, were capable of stretching their guest lists by sleeping an overflow of those prepared to rough it on verandahs. They offered communal eating at a single table and shared access to such niceties as hot and cold running water and flushing toilets. The builders of these early forms of tourist accommodation arrived with the first waves of visitors. They quickly identified with the community. They reflected both the involvement, and to a lesser extent, the development stages of Butler's paradigm. They were entrepreneurs who established low-key operations sympathetic to the local social environment. Operations continued to be low-cost and low-key even when the volume of business increased. There was no sell-out to corporate ownership, no impersonalisation of host/guest relationships and no significant export of profits.

By 1913 there had been considerable growth. There were three hotels, eight boarding houses, a real estate agent, a fruiterer, butcher, hairdresser and baker (Longhurst, 1994). The permanent population was only 300 by 1914 but this was 'one of the highest population concentrations on the Gold Coast strip' (Vader and Lang 1980: 25). By 1935 it had risen to a mere 1,884 (*Border Star* [B.St. 31.1.35]. So few could host so many during peak holiday periods only because visitor demands on local resources were minimal since their holiday life-styles were simple and the accommodation for most was portable. The bulk of the visitors were not guest house clientele. For the few visitors choosing to use accommodation houses, they could expect basic accommodation for 8 shillings (80 cents) a day or 2 guineas (approx Aus. $4.20) a week with meals included. These visitors were generally a less-than-rich market which initially comprised families and honeymooners and, later, young singles (Sullivan, 1982).

The first rail excursion from Brisbane on 24 October 1903 brought 1,000 visitors (Longhurst, 1994: 141). Early excursions were often organised on a group basis and, in this sense they resembled those that carried English industrial workers from the Midlands to their seaside picnics. For example, 'An excursion was run from Ipswich, two train loads arrived well filled. The excursionists comprised the employees of the Railway Brass shop with their families, friends and The Railway Band' (*Tweed Daily* [TD] 2.3.15). It was opportune that availability of mass transport coincided with the emergence of surfing as a popular sport. At a time when Coolangatta's beaches were regarded by many as the best surfing beaches and reputedly the safest, two popular innovations – rail travel and surfing – converged.

It was during the earlier part of this period (1900–14) that Coolangatta's status as the 'Cinderella' of the Gold Coast first became evident. This status was apparent in two respects. First, the accommodation establishments were more downmarket than the grander establishments further north around Southport and the tourist clientele attracted to the area as more distinctively working class. Second, Coolangatta was then the southern extremity of the Nerang Shire Council and was consistently given a low priority in the allocation of funds for civic improvements over this period and it was the Council's refusal to be responsible for sanitary arrangements for campers in the area that eventually triggered the formation of the Coolangatta Shire Council in 1914. As one local historian observed, 'few wars of independence have been fought over sanitation' (Longhurst, 1994: 145).

Providing changing-sheds for the surges of visitors was an added burden for the impoverished new council. A cluster of change booths at the foot of Greenmount catered for guest house visitors but, until the surf clubs could finance the first public dressing sheds by public subscription, artful use had to be made of the natural shrubbery. The year of 1914 was a time of war, economic nervousness and priority adjustments. Even for established councils it was not an appropriate time for improving the infrastructure. At this stage, the limited size and ratepaying capacity of the Coolangatta community meant that it could not support the public investment necessary to accommodate periodic visitor influxes.

The shoestring tradition (1930s)

Coolangatta became the most popular of a string of camping areas along the coast, partly because it was where the railway came closest to the coast. The campers were mainly blue-collar workers, many of whom came from the Ipswich area. At the time, Ipswich was the site of the railway workshops that were possibly the largest industrial plant in the state. Workers and their families travelled free during their Christmas holidays and at other times enjoyed concessional privileges requiring them to pay only one quarter normal fares. For others less favoured, the fares were substantial. Adult return fare from South Brisbane would cost 6 shillings, about a sixth of a weekly wage in the 1930s, or the cost of the *Border Star* posted or delivered daily for a year (B.St. 1930).

The recollections of Peter Winter highlight the camping tradition and the opportunities for resident involvement in tourism this provided.

> It [Coolangatta] was always a family holiday resort. But they camped anywhere. They used to camp on the beach front, in backyards. Those people brought not only their money to spend with the local tradesmen

... but [also with] the fishermen, the Boyd fishermen ... [who] made quite a sideline by cutting the poles for the tents and they used to have all this ready for them [to] hire their poles.

(interview with Peter Winter, 1995)

Attributes of the involvement stage prevailed. Coolangatta was largely tourist-oriented, but enterprise originated largely within the community.

Strains on local resources were experienced with such essentials as backyard water tanks, toilet facilities and garbage disposal being particularly affected. Conditions were primitive. Typically, ti-tree frames were: 'covered with canvas, hessian and even cornsacks to form the seaside "home" '. A 'couple of chaff bags' strung on ti-tree poles became a bed, scavenged kerosene cases became a food safe and 'scraps of galvanized iron salvaged from the council rubbish dump' became a galley (B.St. 5.1.40). Early campers manifested some of the improvising, making-do attributes of explorers.

Had Coolangatta's hospitality industry been based on lavish accommodation with the attendant mortgage risks and recurring costs, it might not have been sufficiently robust to withstand this period of economic crisis which dominated the 1930s. Atypically, those burdens did not exist. During the Depression, the railway workers had relatively secure government jobs. If their earnings were reduced, so were prices, and the cost of a Coolangatta holiday was not very high. Being based on low-cost holidaying, Coolangatta's growth was sustainable.

The following examples of events in the 1930s are indicative of the continuing strength of Coolangatta as a tourist destination, despite the economic conditions of the time:

- In the 1931/1932 Christmas period, for example, 22,573 vehicles had crossed the Loganholme bridge (B.St. 14.1.32).
- The Hotel Grande was built at a cost of £17,000 and land adjacent to camping areas was levelled to create more tent sites.
- During the 1932–33 summer holiday season 455 tents were pitched on council grounds and in excess of 500 in backyards.
- New Year's Eve 1933 brought 2,000 dancers to the Twin Towns – 700 to Jazzland in Coolangatta and 1,300 to the Empire Dance Pavilion across the border where the Sydney ABC Band was featured (B.St. 5.1.33).

Early visions and the perpetuation of the Cinderella syndrome

Early councils responsible for the Coolangatta area were hobbled by funding deficiencies. Rate-payers were few and defaults during the 1930s were common, while the burden of providing services for burgeoning numbers of visitors and of maintaining an emerging interstate highway was crippling. With a caution born of poverty, the council was conservative in expenditure. Consequently, Coolangatta was slow in extending electricity, sewerage and reticulated water to its residents and even slower in providing services for holidaymakers.

Council internal politics frustrated attempts to diversify Coolangatta's appeal. In 1935 Campbell Cleland, a Sydney businessman, in conjunction with an unnamed local businessman, proposed a huge undertaking to be financed by wide public subscription. The 200-acre amusement park would have contained an aquarium, a swimming pool, a motor racing and trotting track, a miniature rifle range, a ballroom, tennis courts and a host of carnival features. It would have been Australia's benchmark attraction but was rejected by the council in spite of strong Chamber of Commerce recommendation (B.St. 21.3.35, 28.3.35). Much later, an elaborate plan for an Aus. $4.3 million health club on Snapper Rocks was also rejected, but on this occasion local ambivalence played a more significant role. There was still a tension between a desire to move on a new tack and a desire to retain or recover the old tranquillity and simplicity.

Drawing attention to Canada's £40 million annual profit from tourism and America's 'something in the vicinity of [US $] 200,000,000', Jos Francis (B.St. 10.4.30) and the members of the House of Representatives (MHR) for Moreton put the case for a South Coast promotional body. In the same year (1930), Dr Earle Page, prominent MHR from across the border, in an address to a public meeting in Coolangatta (B. St. 28.8.30), projected tourism as a means of combating the 20–30 per cent unemployment. He referred to tourism as a 'trade' and, like Francis, cited Canada's enormous 'foreign capital' gains. To emulate Canada, he contended, Coolangatta would need 'a higher standard of hotel accommodation'. He attributed much of 'Canada's success to . . . wonderful hotels'. Thus he presented tourism as an internationally marketable product carrying great potential benefits for the ailing national economy. He was foreshadowing a capital-intensive industry aimed at a luxury market, an industry more akin to modern Surfers Paradise than the Coolangatta tradition. His accusation that 'The people of the Twin Towns seem to be suffering from a surfeit of beauty which apparently they did not want to share with others' (B.St. 28.8.50) may have been a misjudgement. What he read as selfishness might more properly have been interpreted as an unwillingness to go down a path so radically removed from their accustomed mode of sharing.

Within days (B.St. 4.9.50), the Coolangatta Chamber of Commerce, more attuned to financial dimensions than the general community, was studying a collection of American publicity brochures and planning their own 'comprehensive folder'. Len Peak, owner of two guest houses, later Gold Coast Mayor, and a realist, suggested they 'provide the attractions first and advertise them later'. His thoughts were later echoed in editorial comment (B.St. 14.4.32): 'For many years Coolangatta's natural charms attracted large numbers of visitors, but the attractions of five years ago are not sufficient to satisfy the wants of the holiday maker of today.'

The golden era: the second wave (1940s–1960s)

By 1940, wartime restrictions not withstanding, 'motor traffic was very heavy on Boxing Day' (B.St. 27.12.40). Leisure activity was not yet curbed and railway traffic was up 25 per cent with special trains from Brisbane and Ipswich transporting 6,000 passengers, 700 of them in a single extended train. An enormous 1,500 tents packed Coolangatta with a maximum of 20,000 people, campers and picnickers, at one time.

By 1940, camping sites were laid out in precise rows, their allocation was systematic, water supply was reticulated and sanitary arrangements, although crude and overtaxed, did not lag much behind expectations. Ridge-pole tents had been replaced by 15 × 15 marquees. With campers now frequently arriving by car and trailer, more elaborate camping equipment (mattresses, the wireless, the ice-chest and perhaps the family canary) became the order of the day. Women proudly displayed the colourful orderliness of their canvas homes (B.St. 5.1.40). Camping in this new form was a practical mix of pioneering independence and the transplanted environmental 'bubble' typical of organised mass tourism.

The nearing of the end of a long drawn out war brought a spirit of celebration unprecedented in Coolangatta's history. So great was the 1945 bigger-than-ever New Year influx, that several cafes were unable to cope, closing because of shortages of foodstuff and labour (TD 1.1.45).

As Coolangatta began to make adjustments after the disruption of the Second World War, a reconstituted Chamber of Commerce dedicated itself to 'improve its [Coolangatta's] attractiveness as a tourist resort' and their focus was again on the Canadian model (B.St. 17.10.45). Tourism was spoken of as an 'export industry'. By 1950 the *Border Star* carried such headlines as, 'Government Bid For Tourist Dollars' (B.St. 10.3.50) and 'Federal Eye on Dollar Tourist Trade' (B.St. 13.3.50).

An explosion of visitor arrivals followed the Second World War as tensions were released, fuel rationing relaxed, cars became available and some, especially ex-servicemen, attempted to catch up with the lost leisure of the war years. That was the period of Coolangatta's peak popularity. Rather than the luxury trade with a foreign clientele attracted by clever advertising, however, the Coolangatta formula was based on an elaboration of the simple, basic, inexpensive tradition established in the 1930's. It was based on the national reputation Coolangatta had established as a fun place for families and young singles.

The climb in arrivals peaked on Easter Sunday 1950 with 50,000 converging on Coolangatta for the Australian Surf Lifesaving Championships. It was proclaimed as 'the greatest day in Coolangatta's history' (*Tweed Heads and South Coast Daily* [TH&SCD] 10.4.50) and by Butler's criterion it was. Around 5,000 cars jostled for the limited parking space. It was a co-operative triumph for the surf clubs, the council, the Chamber of Commerce and the householders who billeted many of the 700 visiting lifesavers. In 1951, 'hundreds of visitors were being turned away from Coolangatta through lack of accommodation' (TH&SCD 8.9.51). By October that year all camp sites had been booked out for the summer holidays (TH&SCD 25.10.51).

This popularity owed much to the presence of a cluster of colourful innovators with a flair for providing extra attractions in the form of inexpensive, communal entertainment. The vogue of mass organised entertainment, as popularized in its most extreme form by Britain's Billy Butlin, may have been a carry-over of a war-time regimentation possible only in pre-TV times. In the era of post-war jubilation, the young were not too sophisticated for simple things like dancing the Hokey Pokey at Greenmount Beach under the direction of Doug Roughton. Hundreds travelled the length of the coast to dance to Billo Smith's orchestra at Jazzland, the picture shows were booked out and people were turned away from even a little theatre production at the Empire Theatre (B.St. 21.6.51). It was a time of going out *en masse* to activities as diverse as surf carnivals, beach beauty contests, radio picnics, beach concerts, beach boxing matches, beach treasure hunts and sandcastle competitions. Few resources beyond ingenuity, organisational skill, energy, and the beach were needed. At that time Coolangatta had all of those. Patrons of each guest house formed a club with its own action song to be performed at the inter-house games organised daily at the beach. Simple frolics like boys versus girls handicap cricket matches were part of the entertainment diet.

Over 25,000 came to Coolangatta on 1 January 1955, 15,000 to watch a heat of the Sun Girl Quest at Kirra (TH&SCD 3.1.55). Also immensely popular were the beach picnic entertainments co-ordinated by metropolitan radio stations and a variety of structured beach activities mainly at Greenmount. After the closure of the Empire Dance Pavilion at Tweed Heads, Jazzland was the south coast's undisputed headquarters of ballroom dancing.

Over this period, events were instrumental in boosting tourism, although it is not clear to what extent this was being employed as a deliberate strategy in the way events tourism has been employed more recently (Faulkner, 1993). In any case, the Coolangatta experience demonstrates that events tourism is not a new phenomenon.

Coolangatta was established as the premier destination, but structural evidence marking the achievement was absent. By 1955 the Kirra Hotel was under construction at a cost of £100,000 and was to include such innovations as stainless steel benches and a tiled floor in the bar. Coolangatta Hotel had been sold for a record £80,000 and was destined for remodelling and refurbishment (TH&SCD 1.1.55). But this was small-scale compared with the £500,000 Broadbeach Hotel then taking shape to the north near Surfers Paradise.

The geographical extent of Coolangatta's tourist market also expanded as improved transport made the journey possible for more and more interstate visitors. Increasing numbers of southern visitors were arriving, many of them to dodge the southern winter, thus reducing the seasonality of tourism. After small beginnings and a tragic accident in 1949 that claimed 21 lives, Coolangatta airport (formerly Bilinga aerodrome) was expanded in the early 1950s to be a fully functional facility for inter-state operations (interview with Warren Keats, 1995).

Ironically, it was the northern end of the Gold Coast rather than Coolangatta which was to eventually reap most benefits from the establishment of airline services. Ansett and Butler Airways were soon engaged in a price war bringing southerners on group packages to activities such as a Hairdressers' Convention or a bowls carnival in Coolangatta. Air travel was becoming the favoured mode of travel for southerners and Coolangatta was by-passed as the Surfers—Broadbeach area became the preferred destination (TH&SCD 14.5.55).

It was at this stage that the geography of the Coolangatta region began to play a role in limiting further development. To the south, the border with New South Wales had provided a barrier to expansion since the early days when gates were manned to control inter-state movements of livestock and agricultural produce, while hills and swamps discouraged development to the west. Meanwhile, Coolangatta airport claimed approximately 500 hectares to the north and the abundance of land suitable for development further north towards Surfers Paradise, provided an inviting alternative for investors (interview with Bill Stafford, 1995).

The vision fades (the lost slipper)

In 1960, the local member of State Parliament, E. J. Gaven, noted the popular belief that Coolangatta was 'over the hill' (*Daily News* 14.6.60). Campers and caravaners were opting out or being squeezed out by residential development, and were moving across the border to South Tweed Heads or farther south to the more pristine coastal areas of northern New South Wales. Trendy, youthful thrill-seekers were gravitating northward to the more risqué Surfers Paradise. Mr Jade Hurley (interview, 1995) found it 'astonishing that people get off the plane to go 25 kilometres north instead of 5 kilometres to the south. The beaches down this end are better than Surfers.' But the new tourists wanted more than beaches. The concrete, lights and publicity were giving clear signals about where it was. After the 1960s Coolangatta would be called a 'Town that's Lost Its Way' (*Gold Coast Bulletin* [GCB] 6.5.83). A. J. Maher, newly elected chair of the Chamber of Commerce, stirred local indignation by announcing that, 'Coolangatta needed a bomb under it in the next twelve months' (*Daily News* 6.9.65).

In 1995 with the benefit of hindsight, the *Courier Mail* could comment (22.4.95) that

> The story of Coolangatta and tourism is like a bizarre Cinderella tale told backwards. Here was the undisputed princess of Queensland tourism . . . until suddenly the glass slipper shattered and Coolangatta slipped back into a Cinderella role amid the ashes of the past.

A Sydney journalist could observe (*Sydney Morning Herald* [SMH] 8.1.94) that 'there is still an odd sense of limbo . . . as though the heart of the place stopped beating some time ago'. Like Bondi, Coolangatta had been part of tourism's post-war adolescence but its decline was a collapse as spectacular as its rise until it became an ageing settlement that Queensland retirees passed through on their way to the poker-machine excitement of the Tweed Heads Clubs. It was a 'down-at-heel township disappearing beneath sand like the wreck of the ship after which it was named' (SMH 8.1.94).

Creeping stagnation (1970s and beyond)

While Coolangatta has stagnated in accord with Butler's scenario, in other respects it is atypical. In particular, the influx of foreign investment normally associated with the consolidation phase did not take place and it is, indeed, this which is responsible for the stagnation. Nor, as a consequence, is there any community resentment against entrepreneurial invaders. Ironically, if there is any resentment, this has been turned inwards to the local landholders and potential investors who stood in the way of further tourism development.

There was an increasing but minor resentment against tourists who diminished the tranquillity of local areas, congested the streets and took the parking places. But, recognising that visitors were the community life blood as well as its guests, locals rescheduled their usage of shops, streets and beaches to avoid the peak demand times. This kind of irritation was not new. In 1925 (TD 6.1.25) a house-holder railed on the thanklessness of campers who had access to her water tank. In 1927, before graffiti was heard of, Archbishop Duhig, in opening the Coolangatta convent (*Coolangatta Chronicle* 3.1.27) deplored vandalism. Noisy New Year revellers always had their critics. More recently, Fred Lang (interview, 1995) reflects on the rising incidence of crime, drugs, drunkenness, violence – even broken glass on beaches and cyclists who monopolise walk-ways. Local charities which conducted the holiday carnival resented intrusion by professional showmen (B.St. 9.11.33). Shopkeepers resented hawkers (B.St. 5.5.32) and everyone resented the person who booked 100 tent sites and sublet them at eight times the council rate (B.St. 12.1.40). Bathers costumed inappropriately raised at least some local eyebrows.

Views on the directions of Coolangatta's development have always involved a tension between 'tradition' and 'modernisation.' Some believe that staying with 'tradition' and the failure to 'keep up' have been the main reasons for decline, while others have seen the effects of 'modernisation' as the reason for fading fortunes. Many who witnessed the heydays of the 1940s to the 1960s yearn to recapture something of the distinctive flavour Coolangatta offered, but recognise that society has changed. They themselves are part of an ageing population which is supplemented by the retiree influx from the south. While it is acknowledged that some degree of modernisation is necessary if the area is to retain a viable tourism industry, responses to the City of Gold Coast Planning Scheme (1994) reveal a resistance to changes that might transform Coolangatta into a destination akin to

Surfers Paradise. They object to the commercialisation of their resources and despoilation of the environment by the new style of development.

Meanwhile, to the north around Surfers Paradise, a new generation of flamboyant tourism entrepreneurs and innovators began to transform this area into the epicentre of tourism development on the Gold Coast. New hotels, night clubs, theme parks and other attractions were developed on a scale that was unprecedented in Australian tourism history. A tired and tawdry Coolangatta was rapidly losing the more youthful to the vital new attractions in the North and the families to the tranquillity farther south, where caravans could still be sited near the beach. The northern Gold Coast has been distinguished by its night clubs, its links with lavish theme parks, its Casino, its luxury accommodation and the foreign investors and clientele as well as its lack of wowsers. The diversity of attractions has compensated for having a beach inferior to those farther south. Its coming-of-age coincided with a national affirmation that tourism is indeed an export industry, and with the availability of technology and expertise in marketing previously unknown.

Coolangatta has failed to match the accommodation market needs. It has not moved in sympathy either with international or national trends in tourism. Mass international tourism is a reality that has brought a new dimension which Coolangatta had largely chosen to disregard. While the rest of Australia was experiencing an increase of international visitors of 183 per cent between the years of 1978 and 1987 (Faulkner, 1990: 30), Coolangatta was suffering stagnation and decline. Coolangatta lost ground even in relation to the domestic market, by failing to meet the rising expectations of those seeking low-cost accommodation.

The Butler model revisited

In response to the infrastructure deficiencies described above, there has been a recent increase in investment interest in the area among major national and international corporations. Part of Coolangatta's uniqueness in terms of the destination life cycle model lies in the fact that this has occurred at a stage when degeneration was quite advanced. This may be seen as either the beginnings of a rejuvenation phase, which may initiate a new cycle, or as the belated arrival of a consolidation phase. It is not surprising, considering its nature, its origins and its faltering if not failure, that it perturbed many residents. If this indeed was the beginning of a new cycle, local resentment so intense against renewal is strangely out of place with Butler's concept. The development was perceived as anything but a rescue.

In summary form, other important deviations from the Butler scenario in the Coolangatta case include:

• The discovery and involvement phases were truncated by the early arrival of mass tourism, which coincided with the extension of the railway access to Coolangatta.

- The development and consolidation stages did not initiate an influx of investment capital from outside has been instrumental in the failure of Coolangatta to take advantage of more recent growth in tourism demand at both the domestic and international level. Limits on the role of local government in infrastructure enhancement also affected the pace of development. The emphasis on camping as a mode of tourist accommodation meant that facets of the involvement stage became entrenched.

- The onset of stagnation and decline not only reflected the traditional mechanisms that have been identified by Butler and Plog (i.e. carrying capacities and obsolescence) but also these effects were accentuated by Coolangatta's proximity to the rapidly expanding and more modern facilities of Surfers Paradise.

Getting Cinderella out of the kitchen: the tensions of rejuvenation

Over the last decade, Coolangatta and the adjacent Tweed area appear to have been by-passed by the surge of development experienced elsewhere on the Gold Coast. The grounds for this perception go beyond the area's stagnation as a tourist destination, as social indicators suggest a more general malaise. Within the Gold Coast region, the Coolangatta/Tweed areas have the highest incidence of unemployment, the largest population of aged retirees and the highest proportion of welfare recipients (Australian Bureau of Statistics, 1995).

These observations, along with earlier observations regarding the history of the area as a tourist destination, highlight the magnitude of the challenge confronting community leaders in their deliberations on the future of the area. The Gold Coast City Council (GCCC) has developed a strategic plan for central Coolangatta which represents one view on how to get 'Cinderella out of the kitchen'. However, the viability of this or any alternative rejuvenation strategy will depend upon the resolution of tensions between the various stakeholders. Equally, there is a tension between tradition and modernisation that must be negotiated. The historical perspective developed in this study sheds some light on the nature of these tensions and the options that might be considered.

Regardless of its history, many of Coolangatta's assets as a tourist destination remain. Its excellent surfing beaches continue to attract both the sedentary 'sunlust' tourist and surfing enthusiasts in large numbers. The headlands and Tweed hinterland provide scenic and relatively unspoiled landscape features, while local clubs provide cheap food and rich entertainment menus. Meanwhile, the relatively cheap, but frequently obsolescent, tourist accommodation, and the proximity to the more modern attractions of the Gold Coast, means that Coolangatta continues to be popular with budget-conscious domestic tourists. Indeed, to many domestic tourists who are repelled by the high cost and increasingly international tourism emphasis of the Gold Coast's northern parts, the slower pace of Coolangatta's development is regarded as a virtue. The challenge confronting Coolangatta is to enhance its tourism market position by improving its infrastructure, without jeopardising its distinctiveness *vis-à-vis* the rest of the Gold Coast.

The City of Gold Coast Planning Scheme's (1994) vision of the development of Central Coolangatta as 'an integrated tourist node' (25–3) may be out of step with local opinion to the extent that it appears to advocate the Surfers Paradise model. Specifically, local business and community leaders' responses have encompassed the following range of views:

- To many, high-rise buildings are discredited by the excesses of Surfers Paradise, are associated with a human density that over-taxes the environment, and are further discredited by the failure of several earlier grand designs in Coolangatta;
- Coolangatta values the integration of tourism with the community rather than the exclusiveness of enclaves;
- Coolangatta has had a distinctive 'flavour' and this tradition is valued;
- Coolangatta values local ownership, control and profit retention;
- Coolangatta favours family clientele and recurring visits. In the words of one community leader 'We are first and foremost a family holiday resort' (GCB 2.10.95).

Coolangatta's tourism future may well depend on a few critical choices. The options available can be couched in terms of a series of alternatives that are as much a reflection of the area's history as they are of contemporary reality:

- *Domestic vs. international markets* The Australian family on holiday served as the backbone of tourism prosperity in Coolangatta's past. Its choice now is to stay with what it has done best or to 'catch up' with the northern end of the Gold Coast and provide the type of services, facilities and infrastructure which will help to claim its share of the growing international market.
- *Budget (domestic) vs. higher cost* Linked closely with the first alternative is the choice to continue the tradition of providing budget accommodation, as opposed to the 5-star variety. It was, indeed, the 1930s version of the battlers who carried Coolangatta through the Great Depression. Inherent in these alternatives is the aim of repeat visitation as opposed to becoming the destination for the 'once in a lifetime' holiday which can only be afforded by some. Craik (1991) alludes to the social equity implications of such a choice when she notes that if places of beauty are common property, why are the battlers deprived of access to it by the high cost of the sophisticated facilities that dominate the best places?
- *Nostalgia (e.g. targeting second honeymooners of the 1950s and 1960s) vs. other markets* As expressed by one interviewee, 'there are approximately 5 million homes in Australia and out of that 5 million homes, I would estimate that there

would be 1 million homes which would either have in a drawer, in a cupboard, in an old suitcase or a box under a bed, somewhere out in the shed a heap of photos of either mum and dad, grandad and grandma on their honeymoon or a regular yearly holiday at Coolangatta' (interview with Mr Jade Hurley, 1995). Coolangatta is rich in a history which has been shared by a large proportion of the baby boomers. In marketing terms these past good times, happy memories and holiday romances represent a major asset. If left much longer, however, the memories and the people who enjoy them will be gone.

Conclusion

If Coolangatta is to have a tourism future which will represent an improvement over its recent past, it can no longer allow the direction of change to be governed by accidents of history. A systematic approach to the identification and evaluation of options needs to be adopted, and plans must be developed for making the preferred option a reality.

Specifically, detailed market research must be carried out in order to ascertain and compare the potential of the various market segments implied by the options referred to above. The feasibility of catering for and attracting these markets should be investigated, and the infrastructure and investment requirements identified. While the task of addressing these requirements is substantial, it pales into insignificance when compared with the challenge of achieving the level of investment and/or consensus and commitment necessary to initiate rejuvenation. In this sense the inertia of Coolangatta might be construed as a major handicap. On the other hand, to paraphrase Plog, this history may contain the seeds of rejuvenation.

References

Special sources

Australian Bureau of Statistics (ABS) (1995) 'Estimated Resident Population, 30 June 1994', Qld Cat.No.: 131403.
'City of Gold Coast Planning Scheme', Council of City of Gold Coast, Feb. 1994.
Longhurst, R. (1994) *Nerang Shire: A History to 1949*, Albert Shire Council, Qld.
Sullivan, R. (1982) *The Changing Valley*, Tweed Heads: Twin Towns Printery.
Vader, J. and Lang, F. (1980) *The Gold Coast Book*, Milton, Old: The Jacaranda Press.

Newspapers

The Border Star (B.St) 1930, 1932, 1933, 1935, 1940, 1950.
The Tweed Daily (TD) 1915, 1945, 1950, 1925.
The Tweed Heads and South Coast Daily (TH&SCD) 1950, 1951, 1955.
The Daily News (Daily News) 1960, 1965.

The Gold Coast Bulletin (GCB) 1983, 1995.
The Courier Mail (*Courier Mail*) 1995.
The Sydney Morning Herald (SMH) 1994.
The Coolangatta Chronicle (*Coolangatta Chronicle*) 1927.

Oral Histories

Mr Peter Winter
Mr Bill Stafford
Mr Jade Hurley
Mr Fred Lang (Jr)

General references

Butler, R. (1980) 'The Concept of a Tourist Area Cycle of Evolution: Implications for Management of Resources', *Canadian Geographer* 24, 1: 5–12.

Choy, D. (1992) 'Life Cycle Models for Pacific Island Destinations', *Journal of Travel Research* 30, 3: 26–31.

Cohen, E. (1972) 'Toward a Sociology of International Tourism', *Sociology Research* 39, 1: 164–82.

Cohen, E. (1982) 'Marginal Paradises: Bungalow Tourism on the Islands of Southern Thailand', *Annals of Tourism Research* 9, 2: 189–228.

Cooper, C. (1994) 'The Destination Life Cycle: An Update', in A.V. Seaton, C.L. Jenkins, R.C. Wood, P. Dieke, M.M. Bennett, L.R. MacLellan and R. Smith (eds) *Tourism the State of the Art*, Brisbane: John Wiley & Sons.

Cooper, C. and Jackson, S. (1989) 'Destination Life Cycle: The Isle of Man Case Study', *Annals of Tourism Research* 46: 377–98.

Craik, J. (1991) *Resorting to Tourism*, Sydney: Allen & Unwin.

Douglas, N. (1994) 'They Came for Savages: A Comparative History of Tourism Development in Papua New Guinea, Solomon Islands and Vanuatu 1884–1984', unpublished PhD thesis, University of Queensland.

Faulkner, B. (1990) 'Swings and Roundabouts in Australian Tourism', *Tourism Management* 11, 1: 29–38.

Faulkner, B. (1993) 'Evaluating the Tourism Impacts of Hallmark Events', *Occasional Paper* No. 16, Bureau of Tourism Research, Canberra: ACT Australia.

Hovinen, G.R. (1981) 'A Tourist Cycle in Lancaster County, Pennsylvania', *Canadian Geographer* 25, 3: 283–6.

Meyer-Arendt, K.J. (1985) 'The Grand Isle, Louisiana Resort Cycle', *Annals of Tourism Research* 12: 449–65.

Pearce, D. (1989) *Tourist Development*, London: Longman.

Ross, G.F. (1994) *The Psychology of Tourism*, Melbourne: Hospitality Press.

Stansfield, C. (1978) 'Atlantic City and the Resort Cycle', *Annals of Tourism Research* April/June: 238–51.

Strapp, J.D. (1988) 'The Resort Cycle and Second Homes', *Annals of Tourism Research* 15: 504–16.

Tosh, J. (1991) *The Pursuit of History*, London: Longman Group.

Weaver, D. (1990) 'Grand Cayman Island and the Resort Cycle Concept', *Journal of Travel Research* 29, 2: 9–15.

Williams, M.T. (1993) 'An Expansion of the Tourist Site Cycle Model: The Case of Minorca (Spain)', *The Journal of Tourism Studies* 4, 2: 24–32.

7

TOURISM MARKETING AND THE SMALL ISLAND ENVIRONMENT

Cases from the periphery

Tom Baum

When we think about islands, in the tourism context, invariably the images that come to mind are those of sun-drenched, white-sanded, palm-fringed paradises – the stuff that cocktail and chocolate bar advertisements are made of. Such pictures are widely used by image-makers to represent all that is desirable in life and as a stark contrast to the physical and psychological realities of everyday existence. There is no doubt that, within the totality of island tourism, sun destinations such as the Bahamas, Bali, Hawaii and Majorca represent the dominant form and account for the vast majority of tourist arrivals. There are, however, alternative islands to those offering sun, sea and sand; destinations which are much more on the periphery of mainstream international tourism but for which tourism has been, is or is planned to be an important component within their profiles of economic activity. Such cold-water islands do not have the intrinsic climatic advantages of those we have already alluded to but, nonetheless, attract visitors for very different reasons – the natural environment, outward-bound activity, culture and heritage to name but a few. Destinations such as Iceland, the Shetlands, the Hebrides and the Falklands are examples of emerging locations which seek to attract visitors without the advantages which sun islands have at their disposal. Other such islands, generally rather closer to main centres of population, developed in popularity relatively early in the growth of modern tourism and, in many cases, have suffered relative decline in the face of competition from warmer alternatives. The Isle of Man; the Channel Islands; Bornholm, Gotland and the Aland Islands in the Baltic; and Prince Edward Island in Canada are examples of cold-water islands which experienced their tourism heydays between ten and fifty years ago and have been struggling to adjust and re-focus their tourism offering in the light of changing market demands since that time.

The purpose of this chapter is to examine tourism and tourism marketing to these cold-water islands and to explore the market situation of a number of destinations within this category. Island case studies will be presented in order to assist with this exploration, largely taken from on-going comparative research undertaken by the author.[1] As a theoretical framework, Butler's (1980) tourism area cycle of evolution will be used to explore the differing positions and stages of tourist development in the selected islands.

Island tourism

Interest in the study of tourism in the context of islands and small states is a developing area within the wider tourism field. This interest is represented by a rapidly growing literature. Three recent book publications address the area – Conlin and Baum (1995), Briguglio and colleagues' two volumes (1996a, 1996b), and Lockhart and Drakakis-Smith (see Butler 1993) – while a number of major conferences and many journal papers have also focused on this area.[2] The theme is certainly of considerable interest to academics and practitioners as well as the tourist. In addition, island communities themselves are constantly reappraising their relationship to tourists and tourism – of the various economic development options open to them tourism may or may not be the most suitable vehicle for sustainable growth and development. As a result, debate about the role of tourism within island communities is increasing (see, for example, Bird, 1988; Baum, 1996b).

What is the attraction of island tourism as a distinct field of study within the wider parameters of international tourism? First and foremost, islands are of significance, in tourism terms, because of the inherent attraction that they have for visitors which is of a scale beyond that of either the economic or geographical importance of most islands. There is an attraction and interest in islands, a fascination which draws visitors to islands and which acts as a stimulus beyond that which other tourism destinations, without the benefits of insularity, do not share. What is this fascination? This is discussed in some length in Baum's contribution to the Lockhart and Drakakis-Smith volume. In particular, a quote from Butler (1993: 71) is helpful.

> Their appeal may relate to the very feeling of separateness and difference, caused in part by their being physically separate, and perhaps therefore different from adjoining main-lands. Where such physical separateness is accompanied by political separateness, the appeal can be expected to increase, and given people's desires for the different while in pursuit of leisure, different climates, physical environments and culture can all be expected to further the attractiveness of island tourism destinations.

Butler certainly captures many of the attractions of islands here. Islands undoubtedly do provide a sense of adventure to travellers – the physical remove from the mainland, necessitating a conscious decision to cross the water, is one important dimension. In part, it is this element that distinguishes Douglas, Isle of

117

Man from Morecambe, Lancashire and the Bahamas from Florida. Losing the sense of a true water crossing, as Skye, Penang and Prince Edward Island have done, may also diminish their interest to visitors for whom 'getting there' becomes much more mundane and routine. Of course, the water is also a barrier to travel, for reasons of cost and convenience and this means that those tourists who do arrive in an island setting do so by design rather than by chance – they have made a positive choice to be island tourists!

Islands are also perceived by visitors to offer a significantly different environment to the pace and pressures of 'normal', particularly urban living. A particularly effective promotional video, produced by the States of Guernsey, evolves from the theme of escape from pressure, noise and turmoil to the tranquillity of the island setting. Islands are seen as slower paced, perhaps 'backward' in their culture, emphasising traditional, old fashioned values – a real chance to 'get away from it all'. Of course, such images do not match all island situations. Tourism development has, effectively, urbanised much of Oahu in Hawaii, especially in the Waikiki Beach area. Ibiza, in the Balearics, markets a vibrant all-action image to young visitors. But, by and large, the picture of difference, peace and 'another time' represents a key attraction for visitors to islands and is certainly evident in many locations, be they sun-drenched South Pacific destinations such as the Cook Islands and Vanuatu or cold-water locations such as the Scottish Hebrides or Baltic island destinations. Cape Clear Island, off the coast of west Cork in Ireland, represents an excellent example of how separation from the mainland does create or preserve distinctiveness. The town of Schull, on the mainland, is a thriving tourist town, increasingly influenced by global factors in terms of its tourism product, with a substantial number of businesses owned by outsiders from elsewhere in Ireland and Europe. English is the dominant language. By contrast, Cape Clear Island, just 7 sea miles away, remains a bastion of Irish language and culture and visitors are attracted there to experience the two as living, working dimensions of the community as well as an insight into Irish living as it may have been in the past. The tourism industry on Cape Clear is also predominantly locally owned and operated. This emphasis on old-time images is characteristic of much island tourism marketing – in addition to the example of Guernsey already mentioned, media collateral from both the Isle of Man and Prince Edward Island focuses on the image of tranquillity and a return to bygone eras.

Islands, to tourists, also represent a finite geographical environment, one with defined and frequently, relatively small delimiters which are easy to cope with physically in terms of internal travel and in psychological terms as well. By contrast, regions or districts which are part of larger land masses – Provence or the State of Arizona, for example, have few natural boundaries and political parameters mean little to those visiting. As a result, understanding the whole is much more of a challenge. There is, therefore, something particularly appealing about islands and island living which cannot be replicated on the mainland. Tourists, however, also face difficulties as a result of responding to their fascination for islands. Access, depending on the location of the island, can impose significant additional costs to an island location compared to those involved in mainland destinations. The necessity to use sea or air transportation can also impinge upon the flexibility which mainland travel permits. It can take the spontaneity out of travel, to the detriment of islands seeking to encourage visitors. Access may also be constrained by bad weather or industrial action, both representing examples of the vulnerability of islands to outside influences. Access, in fact, is an issue from both the point of view of visitors and islanders themselves. Transport companies, especially those without specific island commitment or ownership, frequently have little dedicated loyalty to individual islands and make pricing and scheduling decisions based on what they may see as the 'big picture'. Caribbean islands have been faced with these situations because of their dependence upon external carriers, notably US owned ones. The strength of both Aland Island (Baum, 1996c) and Iceland tourism lies in local control and ownership of access – the ferries and national airline respectively. Such ownership, however, may be unrealistic for many island destinations and they (and their visitors) remain under the control of major international air and sea carriers.

From the perspective of many island communities, tourism represents a major component within what is frequently a limited range of economic options. Small islands are frequently challenged by a limited diversity within their economic structures and tourism may be seen as one way of diversifying away from excessive dependence on a cash crop or single mineral resource. This has certainly been the case in a number of Caribbean islands but also in northern locations where, for example, the decline of the ground fishery has prompted both Iceland and Newfoundland to view tourism as an alternative worthy of further development.

A high level of dependence on tourism is, therefore, a characteristic feature of many small island communities. Whereas tourism in, for example, Germany represents less than 1 per cent of GDP, in the UK, 1.5 per cent and even in Spain just over 5 per cent, in Bermuda it accounts for approximately 50 per cent of the island's economic activity. The impact of tourism is felt in many ways. Obviously, the most direct is in terms of the generation of foreign exchange and investment. However, tourism to islands has a significant range of additional economic effects relating to employment and other forms of multipliers. The vagaries of the tourism industry hit harder in island destinations which do not have the capacity to generate alternative economic activities and where social support systems are relatively weak. At the same time, tourism does provide a number of side benefits to small islands which should not be overlooked; notably it can bring services and facilities which the island population could not otherwise sustain. Plans for an airport on St Helena, for example, have been greatly influenced by the benefits that it would bring to the tourism economy. The local population, however, may also see greatly enhanced lifestyle qualities as a result.

Tourism to islands (as to many other destinations) is frequently highly seasonal and seasonality impacts upon the quality of product and service delivery, the sustainability of employment and the viability of many business operations. Where tourism is but one component in an integrated economy (alongside agriculture, forestry and manufacturing, for example), there may be a sense of relief and rejuvenation within the tourism sector during off-season periods when those working in the industry move over to other industries. For other destinations, however, off season brings unemployment and, frequently, hardship. Few island destinations (with possibly some Caribbean exceptions) have really overcome the problems of seasonality challenges although most seek to do so.

Tourism is also more pervasive in its impact on the small island community than it is on larger mainland resort destinations. The influx of large numbers of tourists to an island destination is likely to have a profound effect on the community in cultural, social and environmental terms because of size considerations. Host–visitor contact is likely to be high and interactive, with both positive and negative outcomes. There is the opportunity for positive learning but also the dangers of mimicry as the local community, especially the young, adopt the less desirable attitudes and behaviours of their visitors. There are clear employment benefits but, in many tropical island environments, such work is low-skilled with highly paid positions frequently taken by expatriate labour. Tourism development investment is also frequently foreign although this is by no means the case in the cold-water context – locations such as Iceland, Aland (Baum, 1996c), Gotland and Bornholm (Twining-Ward and Twining-Ward, 1996) are examples of islands where virtually all of the tourism industry is locally owned.

Island tourism, therefore, has features and characteristics which set it aside from tourism development in other, mainland locations. These differences impact upon the marketing initiatives which island destinations must employ in order to attract visitors. There is a definite market position and advantage, as a result of location, which islands can adopt in creating a specific identity for themselves in the eyes of customers. Recognition of this has been relatively slow in impacting upon island tourism marketing initiatives and some islands have suffered as a result, seeking to compete with mainland destinations which have price and access advantage without exploiting what they have that is different and saleable.

Butler's tourism area cycle of evolution and tourism marketing

This chapter is about cold-water islands and the marketing issues relating to the tourism product and industry which is characteristic of such locations. Clearly, the development of tourism in small cold-water islands shows considerable differences to the development model which can be identified for destinations which are predominantly sun, sea and sand in orientation. Cold-water island destinations include some of the earliest locations in the development of modern, mass tourism – the English Channel Islands, the Isle of Man, the East Fresian Islands in Germany and Gotland, the Swedish Baltic islands, are all examples of islands which can trace their tourism history to the mid- or late nineteenth century. There are also cold-water islands which are only just emerging as tourism destinations – Greenland, the Faroes, and the Falklands represent good examples.

Butler (1980) adapted cycle theory and, in particular, the marketing model, the product life cycle, as a theoretical method to explain tourism development. Butler's tourist area life cycle (TALC) recognises changes in tourism demand, especially relating to evolving customer expectations within the tourism marketplace and combines this with the application of product life cycle theory to the supply side of tourism. TALC explains that visitor numbers at a destination evolve over time through a number of stages according to the provision of facilities, access, marketing and the carrying capacity of the destination. In some respects, the approach is a precursor of Steinecke's (1993) model representing the development of tourism, elaborated and extended by Baum (1995), which considers macro tourism trends as opposed to specific developments within individual resorts. Figure 7.1 shows TALC as an S-shaped curve which indicates the key stages of tourism development within a destination.

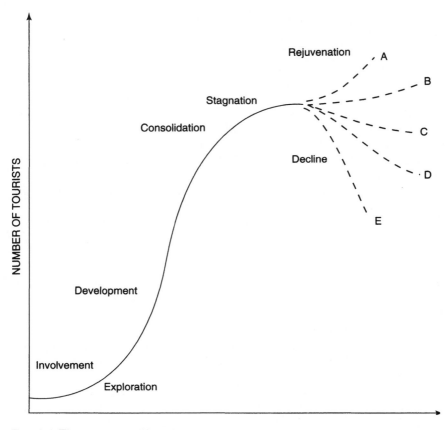

Figure 7.1 The tourist area life cycle

The *exploration* stage represents the initial discovery of the destination. The numbers of visitors involved are small and they represent particular and specialist tourist interests. At this time, there are no organised facilities for tourists who make their own travel arrangements and integrate closely with the local community representing allocentrics or explorers in Plog's (1973) and Cohen's (1972) typologies respectively. The impact of such visitors in social or economic terms is likely to be limited.

The next stage is described by Butler as *involvement* and is characterised by increased provision of tourist facilities in response to increasing visitor interest. This is the point, within the cycle, when local population participation in tourism is to be expected but is not always delivered as outside interests may truncate this stage in pursuit of their own objectives. However, the reverse may pertain – in the Isle of Man the continued importance of small businesses and operators within the sector suggests that this stage lingered beyond what we might expect from the model – well into the decline stage, in fact. During the involvement stage, contact between visitors and the community remains high. Initial marketing initiatives are launched along with the first components of formalised tourism organisation. The first indications of a defined tourist season begin to emerge.

The next stage in the Butler model is that of *development*. This reflects what Butler describes as 'a well-defined tourist market area, shaped in part by heavy advertising in tourist-generating areas'. Local involvement declines and the role of national and international organisations becomes more important. As a result, some local facilities disappear and are replaced by larger alternatives. Clear indications of change within the physical and cultural environment will be evident. This is the period of fastest growth within the cycle and brings with it regional and national planning. Some level of community resentment and opposition may also emerge.

Consolidation is the next stage identified within TALC and sees a slowing in the growth of visitor numbers. Large operators are likely to dominate the industry and there will be a close link and economic dependency between the area and tourism. However, relatively few new players will enter the arena. Visitors may outnumber the host community who may also show increasing resentment about tourism.

The *stagnation* stage is reached when arrivals numbers have peaked and aspects of the location begin to lose their attraction, either because of tired and dated physical plant or as a result of changes in consumer demands and expectations. Environmental, social and physical carrying capacities may have been exceeded and the destination will no longer be fashionable.

From stagnation, two routes or scenarios are offered by Butler through the TALC. *Decline* is a natural successor to stagnation if no action is taken to arrest the situation. Generally, the resort has been superseded by newer destinations elsewhere, facilities will close down and some may revert to local ownership. Some facilities may be converted to related uses or, possibly, out of the tourism arena altogether. Alternatively, *rejuvenation* may be undertaken which, generally, will involve a major refocusing and renovation of the attractions within a destination.

New products may be required, frequently man-made or, alternatively, previously untapped local resources may be exploited and developed.

TALC provides a very useful vehicle for analysing tourism development in retrospect but, along with all cyclic theories, has been criticised for simplicity and determinism in its approach so that it can be argued that it has limited application as an aid to marketing, planning, policy formulation or forecasting. Clearly, the application of the curve cannot be undertaken uncritically in relation to each individual destination case and the stages do not automatically follow on, one from another. Furthermore, as a 'here and now' tool, TALC may also be seen to have some problems – how does a location actually know where it is on the curve at any particular moment in time without the benefit of hindsight? A number of authors have, critically, applied the TALC to tourism situations, with varying success (see, for example, Haywood, 1986; Cooper and Jackson, 1989; Cooper, 1994).

TALC however, does have real value as providing a conceptual comparative framework which allows destinations to evaluate their own situation relative to that of other comparable locations, possibly competing in similar markets or constrained by comparable natural or market factors. Cooper (1994) sees three uses for the TALC: as an explanatory model; in support of marketing and planning; and as a vehicle to assist with forecasting. Twining-Ward and Twining-Ward (1996) make excellent use of the TALC in order to analyse the historical development of two comparable Baltic island destinations, Bornholm and Gotland, and from this are able to postulate specific development options for both locations. Within this chapter, TALC is used to position seven cold-water islands, participants in a comparative tourism study, in order that the benefits of hindsight from one island can be applied horizontally to other destinations within the project.

Lessons from the Edge: The North Atlantic Islands Programme

The application of the TALC will be considered in the context of the seven islands participating in the Lessons from the Edge programme. 'Lessons' is a project of the Institute of Island Studies at the University of Prince Edward Island, Canada, in association with NordREFO, the Nordic Institute of Regional Policy Research, Stockholm, Sweden. The programme represents a major initiative in the scholarship of islands and very small jurisdictions. It examines the political economy of seven northern island societies – the Aland Islands in the Baltic; the Faroe Islands; Greenland; Iceland; the Isle of Man; Newfoundland; and Prince Edward Island (the location of these islands is shown in Figure 7.2). What these islands have in common is their somewhat ambiguous political and economic relationships with 'mainland' land masses. With the exception of wholly independent Iceland, all of the islands have some degree of autonomy within wider political structures, and even Iceland's position is influenced by the

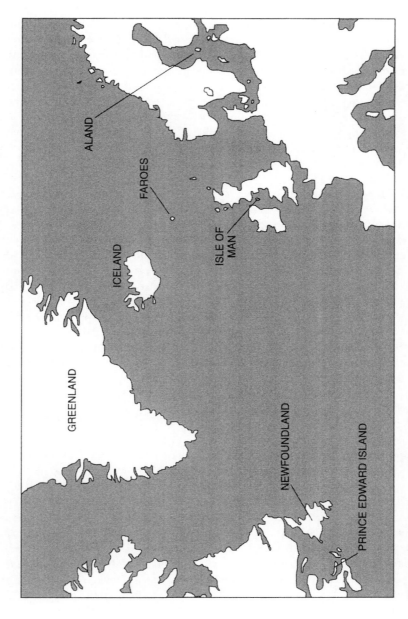

Figure 7.2 Islands in 'Lessons from the Edge'

country's position outside the European Union. Jurisdictional considerations were the prime rationale for the involvement of the seven island communities. In a wider sense, there has been relatively little comparative research relating to the islands of the far north, particularly in a transatlantic framework.

Within the 'Lessons' programme, a number of key economic sectors have been identified as the focus for detailed consideration and research. One of these is tourism. Methodologically, the tourism research programme considers the development of tourism in each of the seven island communities and its role, impact and potential within the economy and society of each. This consideration draws upon existing secondary sources as well as primary qualitative and quantitative data collection. Island-specific findings are then extrapolated to the wider, seven islands context and 'lessons' identified which may have common application within the total group or which may have specific implications for smaller clusters of islands within the total population. Finally, wider generic island tourism implications are extracted and presented.

The seven islands – heterogeneity and homogeneity

To a greater or lesser degree, as tourism destinations, all seven islands are faced with the benefits and challenges arising from their insularity which were considered earlier in this chapter. Seasonality is a major issue in all locations, caused mainly by climatic factors but also as a result of structural considerations – for example, school vacation dates in the summer greatly influence the tourism season in Nordic countries.

Likewise, access is of considerable importance as a tourism concern in all the islands even though the scale of access varies greatly from close proximity to the mainland guests to relatively remote locations away from key markets.

Although these common strands do exist between the islands in relation to tourism, there is also considerable diversity between them, both in direct tourism terms as well as with respect to their geography, history, economy and culture. At a simple level, they are diverse in terms of size, ranging from the very small (Aland, the Faroes and the Isle of Man) to the fairly big (Newfoundland and Iceland) and the world's largest island (Greenland). Populations also vary greatly from just 25,000 in Aland to 584,000 in Newfoundland. The islands range from the very remote (Iceland and Greenland) to locations close to the 'mainland' in the case of the Isle of Man, Aland and, in particular, Prince Edward Island, soon to be connected by bridge link to New Brunswick with some loss of 'islandness'.

Culturally and linguistically, the seven island communities are also diverse. Four have broadly Nordic connections (Aland, Faroes, Greenland and Iceland) although Greenland also has its own, distinctive, pre-colonial culture with more in common with native America than with Denmark, its colonial 'master'. The Isle of Man has Celtic as well as strongly British connections. The two Canadian islands, one the smallest province and the other the island-based half of a two-part province (Newfoundland and Labrador), have strong cultural and ethnic links to

Europe but have also developed their own character and characteristics which are markedly different from those of central and western Canada. Interestingly, from a cultural perspective that is important in the tourism context, six out of the seven islands have historical Viking connections, an association that has the potential to provide an important link between them and to act as a catalyst to the development of common tourism themes.

In terms of their tourism development, the islands also exhibit diversity and are at different stages within the TALC, from involvement and investigation, through development and consolidation to stagnation and decline with some evidence of rejuvenation on the agenda as well. Therefore, as a collection of destinations, the 'Lessons' project provides an excellent context within which to apply and interpret TALC.

Geographical location has had a major influence on the development of tourism in each of the seven islands and the character of the tourism industry, both in supply and demand terms. Geographical proximity to mainland markets, in two cases with major urban population concentrations, has undoubtedly been the main influence on the development of tourism in Aland, the Isle of Man and Prince Edward Island. As demand for tourism developed in adjacent originating areas, the islands were able to respond through exploitation of the 'island advantages' outlined earlier. In particular, the islands were able to respond to advantages based on common language and heritage, familiar culture and relative ease of access. Historically, the islands did not develop during the same period – the Isle of Man, as a tourism destination, is a Victorian and early twentieth-century phenomenon;[3] Prince Edward Island developed in this respect during the early years of this century (Adler; 1982); while significant tourism in the Aland Islands is much more recent, dating from the early 1960s (Baum, 1996c) However, there is much in common between tourism to these island communities in that in all cases it represents mass tourism – family vacations, highly seasonal in character, using local small business accommodation units (hotels, guesthouses, cabins, camping facilities) and dependent on a high level of repeat business. As we shall see, they are also destinations which have experienced mixed fortunes in recent years as consumer tastes and expectations change. These are also the islands which, even today and despite decline or slow-down in visitor arrivals, have the highest dependence on tourism in economic terms – Aland, by some considerable way with 26 per cent of GDP within a highly integrated economy derived from tourism; and Prince Edward Island and the Isle of Man at 6 per cent.

Tourism to the Isle of Man continues to be dominated by arrivals from other parts of the British Isles, both the UK and the Irish Republic (accounting, together, for well over 90 per cent of arrivals), and is, undoubtedly, the most mature of the seven islands in terms of its tourism development. The industry developed as a mass domestic destination from mid-Victorian times onwards, in competition with mainland resorts such as Blackpool, Ayr, Llandudno and Portrush. The heyday of tourism was, undoubtedly, between the two wars and

Table 7.1 Passenger Arrivals in the Isle of Man 1887–1996

Year	Arrials	Year	Arrivals
1887	347,968	1971	465,297
....		1972	499,658
1890	260,312	1973	529,224
....		1974	498,231
1900	351,238	1975	564,611
....		1976	533,011
1910	490,445	1977	485,248
....		1978	530,896
1920	561,124	1979	634,616
....		1980	568,676
1930	487,404	1981	456,643
....		1982	414,223
1940	25,841	1983	412,585
....		1984	377,586
1950	535,558	1985	351,240
....		1986	339,284
1960	473,704	1987	334,317
1961	520,783	1988	328,057
1962	446,578	1989	322,970
1963	435,248	1990	325,574
1964	460,643	1991	299,204
1965	462,124	1992	287,278
1966	408,694	1993	293,160
1967	488,642	1994	310,217
1968	494,699	1995	282,020
1969	532,809	1996	299,148
1970	494,863		

Source: Isle of Man Tourist Board

up to the late 1950s when, generally over 500,000 visitor arrivals per annum were recorded. Table 7.1 details tourist arrivals to the Isle of Man since 1887. Underlying decline, in arrivals terms, can be traced back to the early 1960s although the island's Millennium Year of 1979 brought record annual arrivals. Indeed, the 1992 figure of 282,278 visitors was the lowest (excluding the two world wars) since 1894 and less than 45 per cent of the 1979 peak.

The Isle of Man has, traditionally, focused on family holidays, based on its seaside towns and mountainous inland terrain. The demand cycle has always been highly seasonal and linked closely to school vacation periods in the main markets of northern England, Northern Ireland and the Republic of Ireland. In addition, the island has a specialist motor sports attraction, highlighted by the TT fortnight in early summer but supported by a series of other events throughout the year. This contributes significantly to the overall arrivals numbers but creates problems on the supply-side because of the concentration of demand over very short periods of time. The development of the Isle of Man as an off-shore financial centre has been accompanied by a significant growth in business tourism but the island's proximity to the mainland combined with the prominent role of technology in the financial sector means that much of this travel is short-term and brings only limited advantage to the island. Access is by ferry from both sides of the Irish Sea and this is the traditional vacation route. Air access is geared, in terms of scheduling and pricing, to the needs of the business traveller and is not particularly responsive to the needs of leisure visitors.

Tourism decline, in the Isle of Man, can be explained in a number of ways. Certainly, the situation there is one mirrored in seaside tourist resorts throughout northern Europe and reflects major changes in patterns of vacationing during the 1960s and 1970s as a result of low-cost access to sun destinations, particularly in the Mediterranean, as well as changing lifestyles and expectations from travel among a growing proportion of the traditional markets from which the Isle of Man drew its custom. Coincidental with changing market tastes were the rising affluence of the main market populations as well as the development of cheap transportation technology. The Isle of Man, partly as a consequence of these factors but also as a result of the ownership structure of the tourism industry within which the vast majority of enterprises are small, family-run businesses, has not been able to restructure in response to changing market conditions and has suffered chronic under-investment in, particularly, its accommodation stock. This has the effect of further undermining its attractiveness to tourists of today. With the exception of its niche market in the motor-sport field, the island is widely perceived to be old-fashioned and offering little by way of modern attractions. In TALC terms, therefore, the Isle of Man falls very clearly within the *decline* stage, although there have been a number of attempts and initiatives at *rejuvenation* over a considerable period of time. These have focused on activity-based tourism (in particular golf), conferences and related business sectors.

Prince Edward Island, like the Isle of Man, is highly dependent on regional and domestic markets for its core visitor base. These are represented by Atlantic Canadians, visitors from Ontario as well as New England or, as they are termed locally, tourists from the Boston States. Broadly, 70 per cent of visitors are Canadian; 27 per cent are from the United States; and just 3 per cent from overseas. Of these a sizeable proportion are from Japan, attracted by the island's setting as the home of the fictional Anne of Green Gables character, created by Lucy Maud Montgomery in the early years of this century. Figure 7.3 gives details of annual arrivals in Prince Edward Island from 1965 to 1995.

The island has traditionally been linked to Nova Scotia and New Brunswick by car ferries, both subsidised from federal and provincial sources. In 1997, a new fixed-link bridge opened, replacing the New Brunswick ferry and with charges fixed at equivalent to the ship crossing for the medium term. Air links with Prince Edward Island are low capacity and designed for resident and business use – they do not impact significantly upon the leisure traveller market.

Prince Edward Island developed, primarily, as a family beach resort, exploiting the excellent natural facilities of the island's North Shore. This market focus

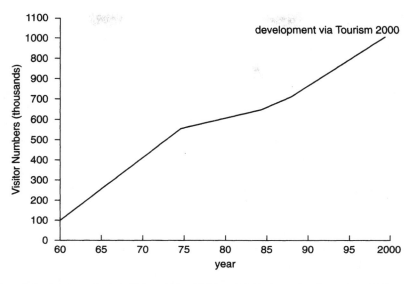

Figure 7.3 Tourist arrivals in Prince Edward Island 1965–95 and estimated to 2000
Source: Department of Economic Development and Tourism, Prince Edward Island.

resulted in accommodation stock which is largely built around self-catering cabins, low-cost hotels, bed and breakfast units, and camping. Attractions have traditionally also been family-oriented. Charlottetown, as the birthplace of Canadian confederation, has some success as a historical and cultural centre while *Anne of Green Gables* also features strongly as an attraction theme, aimed at family and overseas visitors. Figure 7.4 clearly shows that the number of visitors to Prince Edward Island has remained relatively static over the past twenty years, although there are some indications of a small level of growth in recent years. There are, however, signs of significant changes to the visitor profile. Whereas up to the mid-1980s, 65 per cent of guests reflected the traditional pattern of family-oriented tourists, with young children, more recent data suggests that there has been a shift to unaccompanied adults, with a bias towards the seniors market. Up to 65 per cent of visitors now fall into this category. Furthermore, the season, formerly focused on the June to August school vacation period has extended, significantly into October.[4] Both these changes have had major influences on the supply side in Prince Edward Island. There is under-utilisation of traditional, family-oriented facilities; other resources have had to review their periods of operation to cater for demand outside of the normal season; and new facilities have been developed to attract or in response to new visitor types. This is especially true of golfing facilities, which are now a major attraction on the island and the outcome of both public and private sector investment. This, in turn, has stimulated other supply-side sectors, notably higher grade accommodation, restaurants and retail outlets.

Prince Edward Island, therefore, represents a destination which is moving through a period of *consolidation* through to *stagnation* but is attempting to avoid *decline* through a conscious *rejuvenation* process. This is, by no means, a painless experience. While some businesses are very successful within the new tourism environment, others, with stronger roots in the traditional vacation product, are struggling. In all probability, many of these will cease to operate or will have to reorient themselves significantly in order to survive. Rejuvenation, therefore, does leave some casualties and, in this, has features in common with the decline phase. Despite its maturity, Prince Edward Island shows typical island characteristics in that it continues to be dominated by small players on the supply side, an interesting 'small island' variant from Butler's model.

By contrast to the other two mature island tourism destinations addressed in this chapter, the Aland Islands have a tourism history which is relatively recent. Table 7.2 shows that the 1960s and 1970s were a period of dramatic tourism growth for the islands, after which arrivals remained relatively constant until the late 1980s when they rose again significantly, only to fall back again after five or six years. Arrivals figures, however, must be interpreted with some caution because they include day-trippers who may spend as little as two hours in the destination as well as those on cruises who do not disembark at all. It is estimated that perhaps 50 per cent of all arrivals fall into this category.

The Aland Islands are an autonomous province of Finland, Swedish-speaking and located about half way between Stockholm and Turku on the Finnish mainland. The islands are members of the European Union but have considerable economic and political independence. Not surprisingly, Aland tourism is almost exclusively dependent on two markets. Close to 95 per cent of visitors are from these two countries 75 per cent from Sweden and 20 per cent from Finland.

Aland tourism is highly seasonal, with a major concentration in June, July and early August. It is also dependent, to a large degree, on low-spending, self-catering visitors and on family groups. A major attraction, for the Swedish market, is the opportunity to make duty-free purchases as part of the travel experience to Aland. The ferry companies (used by over 95 per cent of visitors) exploit this opportunity and earn the majority of their revenue from on-board sales. The tourism industry, including transportation, is 90 per cent Aland-owned, making for an unusually integrated industry structure (Baum, 1996c).

Tourism in Aland has reached a point of probable *stagnation* and the issue to be faced is whether this now translates into *decline* or *rejuvenation*. There is some official recognition of the potential problems facing the industry but there are few significant tourism development options open to the islands. One of the main problems relates to the power which the ferry companies hold over tourism development and marketing. Their interests are best served by large numbers using their services but not necessarily landing on the islands. Whether this scenario will change as a result of impending alterations to duty free status within the European Union remains to be seen.

Table 7.2 Tourist arrivals in Aland 1958–96

Year	Arrivals
1958	39,500
1959	64,000
1960	101,000
1961	140,000
1962	164,000
1963	180,000
1964	302,000
1965	500,000
1966	380,000
1967	334,000
1968	431,000
1969	460,000
1970	480,000
1971	640,000
1972	1,033,640
1973	1,135,338
1974	1,032,669
1975	1,098,286
1976	1,022,800
1977	1,049,463
1978	1,208,930
1979	1,252,201
1980	1,129,586
1981	1,077,166
1982	1,056,986
1983	1,081,265
1984	1,112,477
1985	1,095,151
1986	1,187,880
1987	1,209,398
1988	1,281,976
1989	1,453,274
1990	1,569,589
1991	1,642,527
1992	1,601,096
1993	1,428,430
1994	1,127,611
1995	1,120,759
1996	1,206,500

Source: Staistisk Arsbok for Aland 1997.

By contrast with the three more mature tourism destinations which we have considered thus far, the remaining four within the 'Lessons' project are relatively immature and underdeveloped as tourism locations. Iceland is, perhaps, the most mature of the four and has developed a significant niche market business, based on its natural resources of mountains, glaciers, fishing and thermal springs.

Table 7.3 Tourist arrivals in Iceland 1960–96

Year	Arrivals
1960	12,806
1961	13,516
1962	17,249
1963	17,575
1964	22,969
1965	28,879
1966	34,733
1967	37,728
1968	40,447
1969	44,099
1970	52,908
1971	60,719
1972	68,026
1973	74,019
1974	68,476
1975	71,676
1976	70,180
1977	72,690
1978	75,700
1979	76,912
1980	65,921
1981	71,898
1982	72,600
1983	77,582
1984	85,190
1985	97,443
1986	113,528
1987	129,315
1988	128,830
1989	130,503
1990	141,718
1991	143,413
1992	142,561
1993	157,326
1994	179,241
1995	189,796
1996 (est)	202,000

Source: National Economic Institute, Iceland and the Icelandic Tourist Board

Conference tourism has also been developed. especially in Reykjavik, which is an important element in what, otherwise, is a very seasonal industry. Iceland also has a long and rich cultural history, dating back to Viking times and certainly an attraction (if under-exploited) within the tourist marketplace. One of the problems is that there is little by way of physical manifestation for the tourist to experience about Viking Iceland – all the buildings were wooden and no longer survive – so that much is left to the skills of the interpretation that is delivered by guides and to

Table 7.4 Tourist arrivals in Newfoundland 1973–96

Year	Arrivals		
	Auto	Air	Total
1973	94,882	133,575	228,457
1974	99,073	146,130	245,203
1975	107,494	145,677	253,171
1976	99,270	142,746	242,016
1977	89,876	137,267	227,143
1978	100,500	139,800	240,300
1979	104,200	160,600	264,800
1980	102,080	166,778	268,856
1981	102,959	160,142	263,101
1982	89,557	132,378	221,935
1983	93,989	122,192	216,181
1984	88,947	153,186	242,133
1985	90,563	158,354	248,917
1986	96,838	171,626	268,484
1987	101,388	180,923	282,311
1988	108,042	200,941	308,983
1989	117,100	187,624	304,724
1990	114,334	175,910	290,244
1991	112,267	154,741	266,008
1992	110,778	153,432	264,210
1993	114,682	192,752	307,434
1994	114,629	214,800	329,429
1995	118,133	204,364	322,497
1996 (est)	110,500	196,000	306,500

Source: Department of Tourism, Culture and Recreation, Government of Newfoundland.

the imagination of the visitor. Tourism growth, from relatively small numbers, has been significant in recent years, reaching double-digit levels in many years. Figure 7.6 shows arrivals exclusive of cruise ship visitors for the period 1960 to 1995.

One of the main contributing factors behind Iceland's growth as a tourism destination has been the role played by Icelandair, especially in promoting stopovers between Europe and North America. For many years prior to airline deregulation, Icelandair offered one of the cheapest options for those travelling across the Atlantic and, despite changes in the airline industry, has maintained a strong and expanding route network and position in this market. Because of the strength of the national carrier (which is unusual in small island states), access costs do not present the same barriers to visitors that can exist with other, relatively remote island destinations.

In terms of the TALC, Iceland probably represents the *development* stage in that the tourism industry is clearly not in its immediate infancy but retains the potential to grow considerably before achieving maturity or consolidation.

Similar in aspirations and, in many ways product as well, is tourism in Newfoundland. This large island has only seriously turned to tourism and recognised its economic and development potential in very recent years, as a result of the decimation of the traditional ground fishing industry which threw about 30,000 people out of work, many of them living in remote locations away from the main population centres. The island offers diverse outdoor and adventure-style tourist opportunities, both land- and sea-based, as well as a distinctive historic and cultural environment. Tourism infrastructure is still relatively underdeveloped for a predominantly North American market, except around the capital, St Johns. Much of the existing tourist market is VFR-based, mainly from elsewhere in Atlantic Canada as well as the central Canadian provinces. Figure 7.7 shows arrival numbers in Newfoundland for the period 1973–95.

Tourism development in Newfoundland, looked at from a medium- to long-term perspective, reached a key point of development, in terms of arrivals some twenty years ago but has largely stagnated since then. Thus, it could be argued that in some respects tourism to the province reached its optimum some years ago and has not developed significantly since. However, this development was so 'stunted' and, taking into account a reasonable assessment of potential, so little reflects where tourism to the island could go in terms of additional arrivals, that it would be more satisfactory to see Newfoundland as representing early to middle stages within the *development* phase of TALC and that is certainly the assumption upon which recent economic development plans for the island have been based. It is argued that the former economic focus on natural resources, primarily ground fish, led to an inevitable neglect of development and marketing in the tourism area and that changes in the macro-economic environment have necessitated a reappraisal of tourism as a contributor to recovery.

However, such a reorientation cannot take place without recognition of some major impediments to the development of tourism. Newfoundland is very remote from its main tourism originating markets – it is located equidistant between Toronto in Ontario and Ireland in western Europe! As a result, access by land and sea or air is both long and costly and, with regard to air routings, is not a priority with major domestic carriers in Canada. The main ferry access route to the island still necessitates a 900 kilometre drive to the major population centre on the east coast.

Newfoundland's tourism resources, moreover, are widely dispersed and frequently located far from significant population and infrastructure centres. One of the main attractions for high-spend visitors to Newfoundland is the abundance of game fishing and hunting opportunities. Outrigger businesses, however, are constrained by local licensing regulations which severely limit the extent to which visitors can participate in these activities.

Greenland and the Faroe Islands are both tourist destinations in their infancy. The former probably represents the *involvement* stage within the TALC in that, traditionally, tourism has been placed low on the national agenda despite outstanding natural attractions. Annual arrivals were little over 5,000 in the early

1990s but reached approximately 16,000 in 1996 as a result of a growing product development and marketing focus. The island's strategic plan calls for an increase to 30,000–35,000 visitors per annum by 2005 and the market and facilities are being developed accordingly. However, tourism remains small by the standards of other islands within the study and is also high-cost in access terms, in internal transportation and in accommodation. The potential lies very much within specialist niche markets but there are some structural imbalances which need to be addressed in order to enhance the industry's situation, particularly in relation to transport costs.

The Faroes, likewise, represent tourism between the *exploration* and the *development* stage. Although the islands are by no means as remote as some of the other island destinations in the project, tourism has not really been developed to any great extent. In part, this reflects the level of financial subsidy from the Danish exchequer which, until recently, has been available to support the Faroese economy and has hindered diversification from the traditional fishing economy. There is increasing awareness and interest in the distinct cultural and linguistic environment which the Faroes have to offer but this remains very much potential rather than actuality. Access routes are relatively difficult and costs, considering the distances involved from potential main markets, are very high.

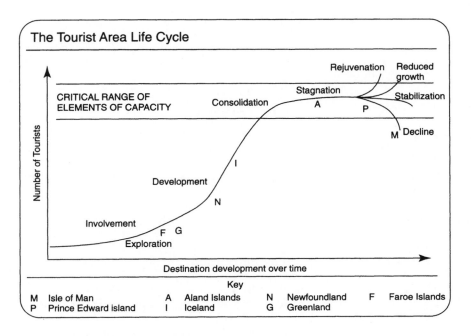

Figure 7.4 The tourist area life cycle and 'Lessons from the Edge' project
Source: After Butler (1980).

Conclusion

This chapter has addressed the application of the tourism area life cycle model to the context of cold-water island tourism, focusing, specifically on the participating islands in the Lessons from the Edge project. Tentatively, the seven islands have been shown to represent the various stages within TALC. This is shown diagramatically in Figure 7.4.

Clearly, the application of the TALC is not without its problems and, as with any conceptual model, it is important that these are fully recognised. However, from a marketing perspective, the TALC provides very useful assistance in understanding the market position and market potential of very different destinations. It also helps provide a framework for comparative analysis and the transfer of experiences, among the key objectives within the 'Lessons' project.

Notes

1 The author is Principal Tourism Researcher to Lessons from the Edge: The North Atlantic Islands Programme. This multi-disciplinary project is a joint initiative of the Institute of Island Studies University of Prince Edward Island, Canada and NordREFO, the research arm of the Nordic Council of Ministers. The project focuses on comparisons of the political economies of the Aland Islands, the Faroe Islands, Iceland, Greenland, the Isle of Man, Newfoundland and Prince Edward Island.

2 An island tourism bibliography, edited by the author of this chapter, can be accessed on the Small Islands Information Network (SIIN) on http://www.uei.ca/~siin.

3 Full arrivals data is available from 1887.

4 Information obtained in personal conversation with senior officer from the Department of Economic Development and Tourism, Government of Prince Edward Island, August 1996.

References

Adler, J. (1982) 'Tourism and Pastoral: A Decade of Debate', in V. Smitheram, D. Milne and S. Dasgupta (eds) *The Garden Transformed. Prince Edward Island, 1945–1980*, Charlottetown: Ragweed Press.

Baum, T.G. (1995) *Managing Human Resources in the European Tourism and Hospitality Industry: A Strategic Approach*, London: Chapman and Hall.

Baum, T.G. (1996a) 'The Fascination of Islands: The Tourist Perspective', in D. Lockhart and D. Drakakis-Smith (eds) *Island Tourism: Problems and Perspectives*, London: Mansell.

Baum, T.G. (1996b) 'Tourism and the Host Community – A Salutary Reminder', *Tourism Management* 17(2): 21–35.

Baum, T.G. (1996c) 'Tourism and the Aland Islands', *Progress in Tourism and Hospitality Research* 2: 111–18.

Bird, B. (1988) *Langkawi: From Mansuri to Mahatir*, Kuala Lumpur: INSAN.

Briguglio, L., B. Archer, J. Jafari and G. Wall (eds) (1996a), *Sustainable Tourism in Islands and Small States: Issues and Policies*, London: Pinter.

Briguglio, L., R.W. Butler, D. Harrison and W.L. Filho (eds) (1996b) *Sustainable Tourism in Islands and Small States: Case Studies*, London: Pinter.

Butler, R.W. (1980) 'The Concept of a Tourist Area Cycle of Evolution: Implications for Management of Resources', *Canadian Geographer* 24(1): 5–12.

Butler, R.W. (1993) 'Tourism Development in Small Islands', in D. Lockhart and D. Drakakis-Smith (eds) *The Development Process in Small Islands*, London: Routledge.

Cohen, E. (1972) 'Towards a Sociology of Tourism', *Social Research* 39: 164–82.

Conlin, M.V. and T.G. Baum (eds) (1995) *Island Tourism: Management Principles and Practice*, Chichester: John Wiley and Sons.

Cooper, C. (1994) 'The Destination Lifecycle: An Update' in A.V. Seaton *et al.*, *Tourism: The State of the Art*, Chichester: John Wiley and Sons.

Cooper, C. and S. Jackson (1989) 'Destination Lifecycle: The Isle of Man Case Study', *Annals of Tourism Research* 6(1): 377–98.

Haywood, K.M. (1986) 'Can the Tourist Area Life Cycle be made Operational?', *Tourism Management* 7: 154–67.

Plog, S.C. (1973) 'Why Destination Areas Rise and Fall in Popularity', *Cornell Hotel and Restaurant Administration Quarterly* 14(4): 55–8.

Steinecke, A. (1993) 'The Historical Development of Tourism in Europe: Structures and Developments', in W. Pompei and P. Lavery *Tourism in Europe*, Wallingford, Oxford: CABI.

Twining-Ward, L. and T. Twining-Ward (1996) *Tourism Destination Development: The Case of Bornholm and Gotland*, Nexo: Bornholms Forskiningscenter (Bornholm Research Centre).

8

PLANNING FOR STABILITY AND MANAGING CHAOS

The case of Alpine ski resorts

Richard D. Lewis and Suzan Green

This chapter explores two contrasting perspectives frequently found in tourism management, namely those of resort planner and resort entrepreneur. Following a general discussion pinpointing the paradigmatic foundations from which these two perspectives are derived, an overview of their conceptual underpinning for tourism applications is considered. A distinction is drawn between the operationalisation of the two different sets of concepts in practice and this opens out a wider window on to the ongoing tensions between planners and entrepreneurs.

A case study follows, which provides an analysis of the evolution of service provision in ski resorts in the French Alps. This study demonstrates how the planning and management of alpine resorts so far – and for the future – can most usefully be recognised as an *ad hoc* rather than as a rational process. An essential paradox is identified: any rational approach to planning which attempts to provide a structured framework within which human beings can operate and interact will also simultaneously create opportunities for entrepreneurial activity and the accompanying 'discontinuities' and 'intermittencies' which are difficult to account for through a rational approach (planning, prediction and control).

The chapter concludes that complexity theory offers an opportunity to produce a richer understanding of the reality of the management of tourism; particularly in view of increasing evidence that conventional approaches are inadequate to the achievement of objectives 'rationally' identified and pursued.

Two paradigms: two '*modi operandi*'

In drawing a distinction between the *modi operandi* of resort planners and resort entrepreneurs, two wholly different sets of assumptions are brought to bear. At one level, these translate theory into practice (the operationalisation of concepts) and at another, the living out of a world view (the embodiment of the *Weltanschauung*). The agents of these contrasting approaches – the planners and entrepreneurs – base their activities and knowledge on diametrically opposed views about the nature of:

- the practical dimension, as expressed in the process of management: control versus opportunism;
- the conceptual dimension, reflecting contrasting positions on the nature of rationality: determinism versus voluntarism; and
- the metaphysical dimension on the order of the universe: steady state versus chaos, complexity.

Since time immemorial, philosophers have considered the latter, struggling to make sense of the world around them and the people within it. However, it was only in the seventeenth century, that the ideas of metaphysicists and physicists came together in the work of Newton who synthesised the ideas of key thinkers such as Copernicus, Galileo and Descartes (Capra, 1982). This Cartesian/Newtonian view provided the foundation on which classical science (both natural and social) is based. This view is essentially deterministic, mechanistic and reductionist. It is predicated on the notion that an 'accurate' appraisal of the initial conditions of an event allow the precise and detailed prediction of subsequent events. In its purest form, it assumes a direct, cause and effect, linear, mathematical relationship between events.

This paradigm for interpreting the world remained largely unchallenged until the advent of Einstein's Theory of Relativity in the early part of the twentieth century, along with (later) quantum mechanics and Heisenberg's Uncertainty Principle. These developments called into question notions of linearity and causality and instead posited turbulence and potentiality as defining characteristics. They were the forerunners of work by influential theoretical physicists and mathematicians such as Lorenz (1963), Feigenbaum (1978, 1979) and Hawking (1988), who, working from very different starting points, shared an appreciation of the chaotic and lawless nature of the phenomena they observed around them. Each has contributed to the modern understanding of complexity theory in articulating notions of chaos in science where once order was assumed.

This new direction for the natural sciences has been uncovered in other areas of intellectual endeavour such as psychiatry (Mandell, 1985; Laing, 1969) psychology (Jung, 1972), sociology (Marcuse, 1964) and global socio-economic commentary (Toffler, 1970; Schumacher, 1973; Naisbitt, 1994; and Handy, 1994) in the search for an explanation about how to act, survive and prosper in the modern world.

Narrowing down even further, management and organisation theory has produced its own share of 'complexity' exponents. The contingency theorists (Fiedler, 1967; Likert, 1967; Lawrence and Lorsch, 1967). for example, were among the first explicitly to recognise the basic value of a complexity approach in arguing that any outcome will 'all depend' on the interaction of all the variables involved.

They also went so far as to acknowledge that identifying these variables is both time-consuming and difficult (if possible, which is doubtful). At this point, however, they were unable to get much beyond positivistic exhortation which proved helpful at the micro-level of an individual organisation, but lacked wider application as an heuristic device. The fundamental problem was that they were trying to provide chaos's answers with positivism's concepts. It was an endeavour doomed to failure, yet one repeated across the disciplines over the next three decades.

Table 8.1 summarises the key differences between the linear perspective which underpins the rational planning approach to tourism and the global perspective upon which the entrepreneurial approach is based.

It is clear that the conceptual foundations of a theory of chaos are truly multidisciplinary in origin and formulation. Like much of the work which informs our understanding of modern tourism phenomena, it is a theory borrowed and applied to this specialist field. Here, its relevance and usefulness are best understood when articulated through a comparison of the beliefs/actions of resort planners and resort entrepreneurs.

Table 8.1 A comparison of linear and global perspectives informing resort management

Linear perspective	Global perspective
Prevailing world view: Method defines outcome, you see what you expect to see and outcomes are circumscribed	*Prevailing world view:* Nothing is as certain as uncertainty, you intuit a spectrum from the anticipated to the unimaginable and outcomes are unbounded
Explanatory variables: rationality, equilibrium, stability	*Explanatory variables:* surprise, co-accidence, chaos
Focus: linear, convergent	*Orientation:* fractal, divergent
Process: goal-oriented	*Process:* nebulous
Means: purposeful and directed	*Means:* anticipatory and potentiating
Outcomes: bounded, forecast	*Outcomes:* ambiguous, unpredicted/ unpredictable
Projected results	Outcomes retrospectively interpreted
Characteristics: reductionist, recidivist	*Characteristics:* complex, holistic

These perspectives are operationalised through contrasting approaches applied to the management process. In the case of resort entrepreneurs, a global perspective is adopted leading to an entrepreneurial approach; for resort planners, a rational planning approach is informed by a linear perspective.

Planning for continuity: rational management

The arguments for rational planning, whether they be related generally to the task of managing (Taylor, 1911; Fayol, 1916; Urwick, 1933; Drucker, 1974), in a more narrowly defined field such as urban and regional planning (McLoughlin, 1969; Hall, 1992), or yet more tightly still in tourism planning (Pearce, 1989; Inskeep, 1991; Gunn, 1994) have been widely rehearsed over many decades. Here the focus is on alpine resort planning, for which rational models have been explicitly adopted. Thus for example, from 1951 the Alpine Research Institute, Innsbruck has been conducting a study on the resort of Obergurgl, and in 1974 the 'Obergurgl model' was used by UNESCO as the basis for a mathematical projection of possible developments of alpine resorts.

The essence of a rational approach is that the resort planner strives to control, integrate and allocate resources in order to achieve pre-determined objectives. Actions assume continuity; a working towards a long-term forecast of the future, based upon some form of prior rational predictive modelling.

Long-term development plans are built on policy prescribed by external bodies and the state, combined with the representation of local interests. Implicit in the planning approach is the expectation that resort progression will be iterative, with the assumption that the experience of the past few seasons will be repeated in those to come. When such long-term plans are being formulated, prospecting entrepreneurs who may have significant impacts on communities in the future, are, by definition, absent.[1]

In managing a resort season after season, decision-making is usually based on a consensus derived from a select group of individuals in the local community working to some form of plan. This may be a highly regulated and structured management plan, or simply a more informal arrangement between divergent interest groups within the resort (Jamal and Getz, 1995). Irrespective of constitution, resorts attempt to work within fixed budgets: available resources are known in advance and decision-making is concerned primarily with resource allocation between a narrow range of pre-determined alternatives. So, for example, decisions in a ski resort might focus on whether the surplus from last year should be used to install more snow cannons or refurbish a lift system.

Resort planners and managers recognise that there is likely to be change in the future and some of their resource allocation decisions are targeted towards future change through resort development. Such decisions, however, rely on a preconceived idea of what the future will be like. Although attempts may be made through contingency funds to respond to unforeseen future needs, resource-based limits and boundaries still exist.

Entrepreneurial discontinuity: phase-shift management

For readers unfamiliar with contemporary literature relating to entrepreneurial management, it may be useful to review a few salient points. Throughout this chapter the term 'entrepreneur' is used in its literal sense: it refers to individuals or groups of individuals who actively engage in the entrepreneurial process as defined by Bygrave and Hofer.

> The entrepreneurial process involves all functions, activities and actions associated with perceiving opportunities and the creation of organisations to pursue them.
>
> (1991: 8)

According to these authors, the development of a conceptual model should take into account the fact that the entrepreneurial process:

- is initiated by an act of human volition
- occurs at the level of the individual firm
- involves a change of state
- involves a discontinuity
- is a holistic process
- is a dynamic process
- is unique
- involves numerous antecedent variables
- generates outcomes that are extremely sensitive to the initial conditions of these variables

Although these features do not constitute a paradigm *per se*, they point the way towards a new management perspective and show clearly that there are areas problematic for contemporary tourism planning approaches. So, for example, when the entrepreneurially minded Tussauds group failed to secure planning permission from the urban and regional planning authorities for a new tourism development in the UK, they were forced to make their investment, secure their profits and pay their taxes elsewhere in Europe. This example highlights the disabling tension between the adaptive capacity of entrepreneurs and risk takers, and the bureaucratic straitjackets of planners and policy makers. Such problems are the symptoms of an underlying conflict between steady state and dynamic systems (Stevenson and Harmeling, 1990).

The entrepreneurial approach to management is quite different from that of the resort planner or resort manager charged with the implementation of management plans. The entrepreneur is constantly in search of opportunities, which are frequently found through innovation, intuition and a willingness to embrace unforeseen dynamic changes – a phase-shift management style. Another interpretation of phase shifts specifically in the tourism context is offered by Faulkner and Russell in their thought-provoking paper (1997) as part of a wider debate on chaos theory and as such is complementary to the arguments presented here.

Essential behavioural features of entrepreneurs in business are their inventiveness and willing engagement in problem-solving activities. In this process they invent new products turning them into business successes – they innovate. It is innovation which lies at the heart of an entrepreneurial approach.

Successful innovations rarely involve only minor changes in the way resources are managed; it is more often the case that they totally devastate the previous status quo, and indeed, at a later date, may themselves be consumed by the change which they initiated. An entrepreneurial approach is concerned with sudden change rather than progressive evolution, the so-called 'gales of destruction' described by Schumpeter, (Grossack, 1989) or the milder 'discontinuity' referred to here.

Entrepreneurs regularly challenge prior assumptions and old ways of doing things, and thus may appear outspoken and even eccentric. From Thomas Cook in the past, to Richard Branson today, they attract public interest through a media fascinated by their successes, their failures, and who they are.

Many of these are characteristics associated with Howard Head, innovator of the all metal ski which totally revolutionised alpine skiing. This product innovation was launched in the USA in 1952 and within ten years virtually all manufacturers of traditional wooden skis had gone out of business. Skiers using Head skis won nearly every competition, and high sales volumes guaranteed market domination in the USA and Europe. A few years later, fibreglass skis were introduced, causing a second product revolution. Howard Head went bankrupt, metal skis disappeared and new types of skiing evolved (Josty, 1990).

New alpine tourism experiences derive from both product and service innovations, the one providing opportunity for the other. They are resourced by surges of interest in the market and often are driven from sources outside the mainstream industry. Thus, with the advent of the snowboard in alpine tourism, strong influences have been exerted by young skateboard/surf enthusiasts and entrepreneurs in terms of the developments of culture (fashion, music, media, IT). This involvement pervades not only the design, manufacture and distribution of products, but also of services such as snowboard instruction and holiday provision (Green and Lewis, 1996).

When prospecting entrepreneurs bring fast-growth, market-led, innovations like the snowboard to holiday resorts, the impact can be dramatic. Resorts are faced with the choice of either resisting or welcoming change. In the first case, resorts will attempt to uphold the status quo established by current commercial operations; and, where required, maintain government support for existing resort operations and planned developments. In the second case, they will buy into risk based on a surge in the market, and, in so doing, experience an influx of boom-led resources. These resources are used to fund rapid developments in resort infrastructure combined with new commercial business activities. When this occurs, resorts experience a discontinuity between past operational performance and future seasons prospects.

This discontinuity in a resort brings a change in business culture and power structure. Decision-making – which was consensual in a steady state mode – now becomes individualistic, dominated by the entrepreneur. Carefully laid plans previously accepted by the local community become redundant and are discarded in the headlong pursuit of opportunity. This may not last for very long – only until a new controlling group becomes established – but it is characteristic that powerful entrepreneurs and their actions are remembered by local communities. European resorts are littered with ski runs, buildings and statues which bear their names.

Innovation and entrepreneurial activity frequently go way beyond tightly planned resource boundaries, exploiting opportunities which lie outside of the domain of forecast scenarios. Entrepreneurs always work in an environment of resource starvation and have no way of knowing with any degree of certainty how many resources will be required or how quickly their venture will grow. Such is the nature of risk, and the more innovative their idea, the less certain the outcome. When innovations are launched, the entrepreneurs' attention remains firmly fixed upon future resource accumulation rather than past resource allocation.

Table 8.2 compares essential differences between rational management achieved through a planning approach, and phase-shift management based on an entrepreneurial approach. It prefaces the case study, in which practical examples of each approach are illustrated.

Table 8.2 Comparison of planning and entrepreneurial approaches

	Planning approach: Rational management	*Entrepreneurial approach:* Phase-shift management
Management vision	Based on an 'objective' view	Based on a subjective view
Time horizon	Long-term goals and strategic vision	Short-term gains with long-term personal vision
Locus of control in the development process	Monitoring and regulation of development resources	Monitoring and influence of development process
	Stakeholding of interest groups	Personal stakeholding
Decision-making	Consensual	Individualistic
Resource management	Known resource allocation	Unknown resource accumulation
Business/resort progression	Iterative	Discontinuous

The case of alpine ski resorts

In tourism research, as elsewhere, robust conceptual understanding supported by equally robust research methodologies are still being sought within a 'linear' tradition. Much of the published work in tourism could still be categorised as of the 'field study' type (Dann *et al.*, 1988) and the establishment of a 'global' perspective remains unarticulated.

In the field of entrepreneurship investigation, Macmillan and Katz (1992) advance the view that a holistic approach – utilising a variety of complementary research methods – can be a particularly effective and fruitful research strategy. In this spirit, the twin strands of tourism and entrepreneurship have been pursued to provide a case study compiled from material derived from multidimensional methodologies in which both quantitative and qualitative techniques have been employed. This forms part of a continuing programme of reseach into aspects of Álpine tourism by the authors.

By force of history, then, the development of this case study has thus had to rely largely on 'linear' research methodologies for the most part. However, it has been compiled in a way which points to the the existence of discontinuity and instability, not only in the object of investigation, but also in the very process of that investigation itself.

Original research was conducted in 1994/5 examining the relationship between UK ski tour operators and French ski resorts (Lewis and Wild, 1995) and the published findings include details of the research methods employed. The research for this case study draws upon and extends this earlier work and includes further in-depth qualitative interviews, an analysis of the latest primary data obtained from resorts and a recent review of secondary data contained in industry reports.

The present study is more broadly based on the development of skiing in the European Alps. Evidence of the rational planning and entrepreneurial approaches in practice in ski resorts is initially outlined and analysed. The inevitability of complexity for resort planners and managers is then demonstrated showing how resort management proceeds in a rather more chaotic manner than is often acknowledged. Finally, consideration is given to the issues which emerge when efforts at structured, rational, decision-making collide with unstructured, opportunistic management.

The rational planning approach

Modelling resorts

There are many aspects of winter alpine tourism in general, and skiing in particular, which are amenable to a rational approach to the planning, development and subsequent management of resorts. Mass-market alpine skiing holidays essentially provide an experience with a relatively narrow focus. The vast majority of tourists go to alpine resorts to ski or to engage in other snow sports each and every day of their holiday. To achieve this, certain basics are necessary: a steep hillside, some snow (real or man-made), ski runs, ski lifts, ski security, a snow management system and ski/equipment hire/sales. To support these, more general tourism provision of accommodation, food, transport and entertainment is also required.

Many of these basic features can easily be quantified (e.g. lift capacity per hour, snow cannon volume per minute per litre, etc.) Their relationships can be mathematically modelled to provide 'idealised' resort provision defined in terms of optimum lift capacity to accommodation development, ratio of bars and restaurants to retail space and so forth. This is a methodology entirely suited to the allocation of known resources for management and development.

The many descriptive models of ski resort development as discussed by Pearce (1989), all reflect an increasing tendency towards a rational management model, with the post-war, purpose-built, integrated, French resorts of Courchevel 1850 and La Plagne providing examples of the ultimate realisation of this approach. Both academic and industry literature provide numerous examples of articles promulgating a logical, rational approach to both analysis and modelling. The underpinning concepts of linear regression, exponentialism, cause and effect, and solvable equations abound.

Modelling systems

Ski holiday provision is both intensively product dependent and at the same time highly service oriented. As noted in the previous section, ample basis for rational linear modelling is provided by tangible product forecasting (e.g. the relationship between number of child lift passes purchased and total available hire capacity for childrens' ski equipment in the resort). However, the question remains as to whether services, such as ski instruction, and factors at the boundaries of product/service provision, such as food on the mountain, are amenable to similar analysis.

A brief excursion into the development of ski instruction provides an answer.

The first acknowledged taught skiing technique was pioneered in the late 1890s in Austria by Mathais Zdarsky, and the development of ski instruction can be traced through the twentieth century from this time. In the early years, groups of instructors formed ski schools based on their own singular methods of instruction and technique. One of the best known was the Arlberg Ski School in Austria, whose Director, Hannes Schneider, bequeathed the legacy of the 'Stem Christie' turn, a teaching technique still widely used today throughout the alpine nations.

Forty years ago, in 1958, the publication of works like the 'New Official Austrian Ski System' (Müller) laid down the basis for subsequent development in ski instruction. It provided for a nationally recognised teaching qualification and created an organisation with powers of regulation and control over would-be or existing members. This was a system founded on multiple stakeholding by a variety of interest groups with management and decision-making by consensus.

By the late 1970s, each individual western European, alpine nation had agreed national training and qualification procedures. Despite national level agreement, the consensus reached was different in each country. Two basic systems emerged. One, subscribed to by France and Italy, favoured the attainment of a single qualification (an absolute standard) as a basis for entry into the cadre of professional ski instructors. The other identified different levels of attainment conforming to a three-tier qualification with relative standards set and complied with at each level. This was the system favoured by Austria and Germany.

Both systems undoubtedly produced professional ski instructors able to teach to a high standard. All the systems of instruction were based on rational principles, such as bounded progression (skill levels development), repetition (demonstrable competence) and an explicit goal-orientation. All achieved the same outcome for resorts creating an 'objective' means of managing one aspect of service provision (ski instruction) for tourists, whilst for local people they provided a means for regulating and controlling employment prospects.

Modelling markets

It is clear from the evidence that resort planning, management and development have become increasingly grounded in a linear perspective over time. Accordingly, there should be objective data to explain why resort planners and managers were convinced that this viewpoint was sustainable.

Two clear planning objectives were established for many post-war (Second World War) alpine developments. The first – the creation of a viable alternative economy to a failing alpine agriculture – was to be achieved through the mechanism of the second: the provision of easy and convenient mass access to winter sports.

A retrospective review of the post-war growth in European ski markets unequivocally demonstrates that these objectives were realised. Indeed, in many regions, ski markets have shown continuous long-term expansion with spectacular increases in size – particularly during the mid 1980s.

Market growth rates in France were reported to be as high as 14 per cent in some years (Services d'Etudes et d'Amenagement Touristique de la Montagne [SEATM] 1987), and Figure 8.1 reflects similar trends in Austria.

The growth observed during the 1970s and 1980s gave resort planners and managers the ability to predict future operational performance with confidence. It helped to reinforce an iterative approach to resort progression – 'the same as last season, but bigger and better' – based on an underlying assumption of linearity which remained largely unchallenged. Nevertheless, all the signs were that an explicitly entrepreneurial approach to resort management was being practised, even if not recognised or acknowledged.

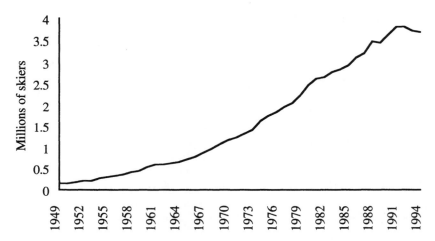

Figure 8.1 Tourist arrivals in the Austrian Tyrol, winter seasons 1949–94
Source: Amt der Tiroler landesregierung (1995).

The entrepreneurial approach

In the search for evidence of an entrepreneurial approach to management, it is important to recognise that entrepreneurial events are by their very nature ephemeral and quixotic. They frequently take place over a very short time; can often only be understood within the precise context in which they occurred, and are typically poorly recorded, making adequate recall difficult. This makes their 'substance' hard to grasp, since discontinuous, intermittent, episodes are just that. They are non-linear and non-linked in a clear and obvious way. Despite this, it is possible to point to certain phenomena, which, in terms of ski resort management, lend themselves to an interpretation in direct contrast to the rational approach described in the previous section.

A phase-shift view of resorts: the process of creation

Myriad examples exist to demonstrate that an entrepreneurial approach to ski resort management is now commonplace. This approach is characterised by discontinuity and 'phase-shifts' in resort planning, managing and development. As described elswhere, phase shifts refer to the changes a system undergoes once it is dislodged from 'a state of tenuous equilibrium whereby small changes involving individual agents may be enough to precipitate evolutionary change in the system through mutual adaptation of its constituents' (Faulkner and Russell, 1997).

In the previous section, the case of Courchevel 1850, in the Savoie region of the French Alps, was cited as the first widely acknowledged example of an inte-grated resort, purpose-designed and built to the 'Plan Neige' which was first conceptualised in 1964.

148

Yet in the first instance, the resort of Courchevel 1850, was an experiment, and the 'Plan Neige' never existed as a statutory policy document. Instead, it was an innovative set of ideas driven by three entrepreneurial politicians who shared a single, common vision: the successful development of tourism in Savoie. Whilst their political fortunes rested on the realisation of this objective, they nevertheless had the means with which to accumulate and direct resources to make this happen. Subsequent development of other resorts based on this untried model followed rapidly, each attempting to make a grander architectural statement than its predecessors. (Clary, 1993). The distinctive 'Galleon' at Aime 2000, La Plagne, France for example, provides unequivocal visible testament to this endeavour.

Posited as the product of a rational management decision-making process, these developments seem to be more plausibly understood as phase-shift outcomes, when the very basis on which they grew – the 'Plan Neige'- evaporated with the dissolution of the alliance of its architects.

A phase-shift view of resorts: the process of managing

Phase-shifts can also be driven by the temporal discontinuity inherent in the seasonal operation of resorts. The discontinuities thus created cannot successfully be accounted for in rational management models, since, although the discontinuities are regular, the outcomes are always irregular and unregulated. This can be illustrated by considering movements in the market for labour in alpine regions.

The massive influx of tourists to French alpine resorts each winter creates patterns of transhumance which are clearly evident every season. In some regions, where less than 10 per cent of the workforce is permanent, tourism can create more than 120,000 extra full time positions (French figures, winter 1993, SEATM). Such a huge change in the number and composition of the total workforce from year to year provides constantly renewed opportunities for entrepreneurs, who often develop business ideas away from the alpine environment, between seasons.

This can present problems for resort managers and planners, since they do not always know who and how many entrepreneurs and their associated task forces will return next season, nor indeed, what new business activity is likely to transpire (interview extract: Chambre de Commerce de Chambéry, 1994).

The reality of coping with transhumance season after season is that resort planning decisions and the empowerment to manage resorts soon become centred on a small, powerful group of individuals who live as part of the permanent, annual, resort community. This sets the dynamic for conflicts of power between protectionist, parochial, locals and opportunistic, entrepreneurial, outsiders. Uneasy truces and co-operation driven through by business imperatives override small-town politics on a day-to-day basis, but the underlying tension harbours the constant threat of disruptive 'incidents' emerging.

Even the most hard-line exponents of a rational approach would be likely to concede that 'human factors' are the most recalcitrant hostages to rational controls even in the most 'linear' of employment circumstances, such as a production line in a manufacturing organisation in a traditional, urban setting. When all the irregularities of remote location, transhumance, weather, exchange rates, etc. are factored in to the resort management equation, it is no surprise that the explanations offered up by steady-state models are less than adequate.

A phase-shift view of systems

Although the development of ski schools can be described as evolutionary and the delivery of contemporary ski instruction based on an 'objective' system of standardised practice; both have been born of phase-shifts driven by the actions of individualistic entrepreneurs in combination with critical incidents and key product innovations of a quite *ad hoc* and idiosyncratic nature.

In the early 1930s in Switzerland, a revolution occurred in skiing, largely brought about by the British, who abandoned the more sedate cross-country skiing on long narrow skis in favour of racing downhill, making fast slalom turns. This provided the impetus for a period of major product innovation which produced the first fixed-heel ski bindings.

In the 1936 Winter Olympics at Garmisch-Partenkirschen, Germany, Anton Seelos was banned from the slalom event because his method of unweighting his skis was deemed to give him an unfair advantage over other competitors. This was a critical incident in the development of skiing because less than five years later his style had been adopted by Olympic Teams throughout the Alps. This supplied a rationale for the formation and development of many new ski schools with directors such as Emile Allais in Chamonix, France, promoting the innovative technique.

Following the end of the Second World War, a new generation of ex-army 'alpine corps' instructors and former Olympic champions helped build the foundations of modern ski schools across the Alps. This expansion was accelerated by innovations borrowed from wartime developments in aircraft technology. Ski lifts were installed in resorts, and in the 15-year period from around 1950 to 1965, skis evolved from cumbersome, heavy, designs in wood, to light-weight, steel-edged, composites incorporating plastics and fibre glass.

In the 1970s, the '*ski evolutif*' method was developed in Les Arcs, Tarentaise, France. It was based on the assumption that novices would learn faster and better using short skis; and, as they mastered technique and speed, would gradually progress to using longer and longer skis until they reached an appropriate length for their height, weight and skill level. But this step-change was not new. Bass (1966: 18), reported that Sir Arnold Lunn, an early British ski entrepreneur and pioneer, had offered the same advice in 1933, stating that: 'beginners should hire skis of a length, which when stood upright, come up to their armpits'.

This recycling of ideas runs counter to the idea of linear progression and the rational sequencing of developments along predetermined lines. Instead, it points rather to the hop, skip, and jump nature of innovation in sport, underpinned by an entrepreneurial approach.

In sum, developments in skiing were often driven by enthusiasts and entrepreneurs in the true sense of the word. Second, their actions created leaps forward in the development of both the tourism product and the alpine resorts.

Third, periods of intensive change were often interspersed by periods of relative stability when innovations became accepted and adopted by expanding winter sports markets.

A phase-shift view of markets

A phase-shift view of ski markets is not typically recognised, as ordinarily only aggregated market data of a national scale is considered. Although data so aggregated can show a steady growth in market volume, if, instead, resorts are the unit of focus, then market progression is neither as steady nor as constant as a fitted linear growth curve might suggest. Figure 8.2 demonstrates this quite clearly, for the case of St Moritz, Switzerland.

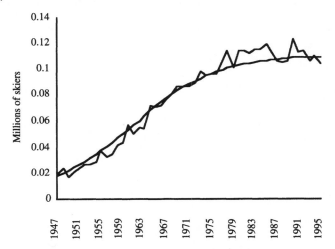

Figure 8.2 Growth of skiing in St Moritz, Switzerland, 1946–95
Source: ONST (Office National Suisse du Tourisme).

Further, when data is disaggregated to reveal the internal nature of the market – and for example, only domestic, as opposed to overall trends including figures for inbound tourists are considered – phase-shifts taking place in different parts of the market, in different ways, at different times, quickly become apparent.

So, for example, total numbers of winter holidays taken in the Alps showed an overall steady rise in the twenty years from 1970 to 1990. In France, however, this was not the case. After steady expansion in the total market in the first half of this period, growth slowed and then actually declined from 1983 onwards. This trend was even more pronounced for the domestic segment of the market (the French taking ski holidays in France), which experienced a sharp decline over the same period, as illustrated in Figure 8.3.

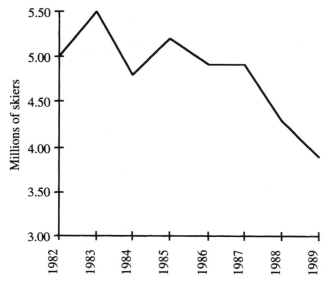

Figure 8.3 Ski market fluctuations 1982–89, France, domestic market
Source: SEATM.

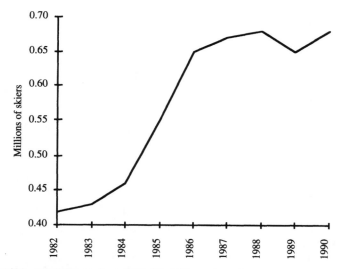

Figure 8.4 Ski market fluctuations 1982–90, UK, total market
Source: Lewis and Wild (1995).

152

In France, inbound tourist increases only accounted for a partial recovery of the projected domestic market figures and were not able to fulfil the promise held out by forecasts for the total market based on earlier years' linear growth. By contrast, as Figure 8.4 indicates, other markets were expanding. In particular, the UK ski market showed rapid growth over the period 1982–90.

Taken together, Figures 8.3 and 8.4 show variations in trends in different European ski submarkets. They suggest not only how undifferentiated market data can disguise the reality of internal shifts in the total European ski market, but also how particular perspectives can yield particular views of a contrasting and sometimes contradictory nature. So, for example, over the time period examined, a European-wide market view would show stagnation, a French market view would show decline and a UK market view would show expansion. Notwithstanding general arguments about the power of any set of statistics to portray a particular view of reality, differences in these micro markets are highly significant because of their implications for the change process.

In practical terms, the discontinuity uncovered through such a market analysis opens windows of opportunity for entrepreneurs when rising markets (as in the case of the UK) entered 'gaps' created by falling markets (French domestic market). Innovative responses to deal with new and changing balances between domestic and inbound tourists in French ski resorts have had far-reaching effects all over mainland Europe (e.g. air, ferry and rail transport networks) whilst in the UK proactive and opportunistic ski tour companies have not been slow to exploit the gaps and to establish new operations in/to France (Lewis and Wild, 1995).

Phase-shift developments in the market can thus be seen to generate unanticipated outcomes consistent with the precepts outlined in Table 8.1. Fundamental practical implications follow from this fact for tourism operations; for not only will pressure be brought to bear on the set of products and services on offer (e.g. ski instruction in a variety of languages) but the management of cultural diversity inherent in the nature of an international clientele will continue to remain complex with shifting balances in the composition of that group (e.g. culturally determined preferences concerning availability/nature of food products, child-care facilities etc., see Lewis and Green, 1995).

A rational world built on chaos

During the late 1980s, European alpine nations experienced sudden unexpected catastrophic falls in ski resort visitor numbers. This plunged some resorts, particularly the smaller ones at lower altitudes, into debt and even bankruptcy. In France the most difficult years were the winter seasons 1987/8, 1988/9 and 1989/90 (Barbier, 1993). This was a period during which resort planners and developers lost confidence in 'white gold' as a safe investment.

The emerging crisis during this period was blamed by many on a single overriding factor: three consecutive winters with poor snow falls – a simple cause-and-effect explanation. Were this reasoning adequate (which of course it was not) then it would suggest that winter sports provision was dependent on a highly chaotic system, and one which has defied all attempts at linear rationalisation: the weather (Lorenz, 1963).

If, on the other hand, the crisis for ski resorts had been attributed to the interaction between a number of critical factors, then this would have pointed the way to a more complex set of relationships which were at least contingent and, more likely, random in nature. On the whole, those rationalists who considered multivariate causes for the problems of resorts came up with a further two explanations: world recession and instability in foreign exchange rates: chaotic systems both, founded on speculation and 'confidence', and not really rational at all.

In an attempt to respond to this perceived market crisis, ski resort managers and planners embarked on a repositioning of their resorts to offer an increasing variety of winter tourism activities whilst at the same time trying to solve 'the weather problem' by increasing investment in snow cannons and other techno-logical systems. Both plans of action, whilst appearing to be rational, introduced increasing levels of disorder in the system. In the first case, encouraging innovation and diversification enhanced the opportunites for entrepreneurs to direct future developments in the industry, with the accompanying chance of 'gales of destruction', as referred to earlier.

Second, by investing in more snow-making machinery and offering 'guaranteed' snow to the market, resorts and ski tour operators colluded in grounding their future financial success more firmly than ever before in the one factor neither could control – the weather (Lewis and Wild, 1995). The capacity of snow cannons to compensate for unfavourable weather is limited because this technology is itself only effective under a narrow range of climatic conditions, thereby rendering such 'guarantees' offered to the market uncertain, particularly in lower-lying resorts.

These examples serve to illustrate the reality of business in the alpine environ-ment. Changes initiated by the ski industry and their subsequent impacts on resorts were essentially grounded in a non-linear world and tourism businesses and managers now operate within an increasingly 'chaotic' environment.

How, then, can rational systems manage this unpredictable turbulence? The example of off-piste skiing illustrates the problems and forms a starting point for the next section.

A chaotic world rationalised

In attempting to re-stimulate the downhill ski market after the universally acknowledged 'poor seasons' for resorts, the ski industry worldwide encouraged innovations and engaged in new marketing initiatives to revive the image of skiing. In particular, 'adventure skiing' was aggressively promoted through the ski press and through the production of sensational ski videos encouraging tourists to venture off-piste. Ski manufacturers launched product innovations, like wide skis, specifically to enable increasing numbers of less experienced and less skilled skiers to explore unpisted areas.

As a result of the increased attractiveness of skiing 'out of bounds', alpine resort management now includes the whole mountain domain rather than simply isolated resort villages and connecting pistes. Resorts have been compelled to provide designated off-piste sectors and off-piste 'runs'. To date, in Europe, these remain unpatrolled in any systematic way by 'la Securité des Pistes', and notices warn skiers that they go into such areas at their own risk.

In the past, resorts have managed pisted runs which are prone to avalanche danger either by closing them in periods of high risk, or by making them safe through controlled detonations. Guaranteeing protection from avalanche danger is virtually impossible for ski resort authorities in off-piste areas, and any rescue from an avalanched area off-piste is particularly difficult (*Enquête à l'Alpe*, 1996).

In the absence of any legislation to regulate such risk, one method of managing the safety of increasing numbers of skiers in ever larger areas of mountain terrain is to provide an off-piste guiding service – an attempt to systematise the unpredictable. In order to do this, all ski instructors in the French Alps, for example, now have to undertake a specialist, one-week, off-piste course as part of their formal training. This allows them to guide and instruct clients within the bounds of terrain serviced by ski lifts, but not on glaciers or in more remote areas.

Higher risk off-piste skiing of the enticing nature featured in the extreme ski media, is only sanctioned (for insurance purposes at least and then, not always) when undertaken with a qualified Guide de Haute Montagne. To acquire this revered status, the guides follow an intensive and very comprehensive training programme to obtain the UIAGM (Union Internationale des Associations de Guides des Montagnes) qualification which is recognised throughout Europe.

The current supply of UIAGM guides in European resorts is inadequate even to meet the demands of domestic markets. When inbound tourist numbers are taken into account, the situation is nothing short of acute, as the following illustration for UK skiers venturing abroad shows.

The majority of UK ski tour operators selling holidays in the 1996/7 season mentioned the possibility of guided off-piste skiing in resorts, in their brochures. Yet few resorts have more than a handful of qualified UIAGM guides available to meet the increasing demand every season for their services.

For a UK market volume of approximately three quarters of a million skiers per year, there were 100 UIAGM British Guides with a further 15 aspirants (qualified, apart from an obligatory period of 'teaching practice' – usually one year) and 13 trainees. Of those qualified, around 15 are regularly involved in off-piste ski-guiding and/or ski-mountaineering (interviews with UIAGM representatives, 1994, 1996). Generally contacts are established either directly with clients in the UK, or via one of the specialist UK off-piste operators which have emerged in the last three to five years (e.g. Fresh Tracks, Ski Powder Byrne).

The under-supply of qualified guides in general, and English-speaking guides (for the English) in particular, is a 'wicked' (Rickards, 1988) structural problem in an industry desperate to capitalise on demand-generated products. On average, it takes four to five years to achieve the UIAGM qualification, based on an individual with prior mountain experience or other 'relevant' qualifications in outdoor pursuits, sports or leisure. Even a concerted effort to recruit and train adequate numbers of suitably qualified guides, would require several cohorts over several years – for the skills shortage lies not only in those yet to be recruited, but also in those qualified to train them.

Due to this under-supply and the expense of off-piste guiding, many skiers are venturing off-piste without a guide (and hence insurance or legal protection from liability) encouraged by images of freedom and thrill-seeking promoted in ski videos. Despite efforts to regulate them (such as lift access being restricted only to groups with a guide e.g. la Vallée Blanche, Chamonix, France) bounded management within such an unbounded environment is all that is possible.

Only a small amount of skier behaviour can feasibly be controlled through systematic planning – for example closing lifts to prevent overcrowding on the slopes – the rest (e.g. missing skiers) can only be managed on an *ad hoc* basis. (Green and Lewis 1997b). The challenge is to find flexible ways of coping with the unexpected and anticipating the unimagined. This is only likely to come about through an elaborated conception of the entrepreneurial approach based on a global perspective.

Conclusion

In the introduction to this chapter, two contrasting perspectives were articulated and formed the basis for a conceptual framework (Table 8.1) for the *modi operandi* of rational planning and entrepreneurial approaches to management. Some of the contrasts, contradictions and potential areas of conflict arising from these applications were summarised in Table 8.2. These were based on a comparison of tourism planning theory and general entrepreneurship theory since tourism entrepreneurship as a field of study is, as yet, only embryonic.

This work has demonstrated that there are real differences between the rational planning and entrepreneurial approaches which must be explored more fully to gain a greater understanding of the ways in which resorts are managed. In turn, greater conceptual development is required to both interpret and inform managerial practice.

The last two sections of this case study provide evidence of 'linearity' in a chaotic system, but additionally show how complexity is also present. Moreover, they show that complexity is always an integral aspect of the development and management of alpine ski resorts. As Herbin (1995: 102) correctly identified:

> tourist resorts are pulled between the divergent individual interests and group strategies of landlords, developers, hoteliers, estate agents, traders, summer residents and others.

In so far as the 'others' category includes prospecting entrepreneurs bringing innovations to resorts, dynamic changes with unpredictable outcomes will continue to occur.

The authors do not suggest that a rational planning approach is redundant; for the entrepreneurial approach can too easily be misconstrued as being 'anti-management' (Kaplan, 1987). Rather, the fact is acknowledged that resorts progress through a combination of linear passages and chaotic episodes which no amount of rational forecasting can predict (Faulkner and Valerio, 1995), nor planning, control.

The recent introduction of many innovations in winter sports activities in alpine resorts (such as snowboarding, parapenting and the Skikart) puts down a marker for resort managers. It is clear that an increasing proportion of winter tourists in the future will simply not be satisfied with 'restricted' descents down a mountain, following pathways dictated by resort authorities. They may do this some of the time, but not always. Instead, more and more of them will increasingly demand access to an unbounded experience that is innovation-led, individualistic, and constantly changing. This heralds a new era for resort management.

In the emergent behaviour of their clientele, then, a metaphor for emergent management practice is contained. Straight lines (downhill descents, linearity – a rational planning approach) are a partial, necessary and desirable part of the picture. Global terrain access (off-piste adventuring, heliskiing and boarding, parapenting and microlites, domain possibility – an entrepreneurial approach) potentially provides a total, 3D experience and a necessary way forward for the successful management of alpine resorts.

Complexity/chaos theory encompasses all potential dimensions of this management challenge. Within this whole, linearity provides a partial and necessary but not sufficient explanatory feature. Rational interpretations deriving from a linear perspective can hence only normally provide incomplete solutions to complex problems and thus the vitality of the rational approach is limited.

Yet neither extreme of the rational–chaotic continuum in practice, will satisfy either resort managers or visitors. The compromise position – 'planned chaos' – however, generates a paradox. Rational planning for resort development directed at enabling an entrepreneurial approach, will simultaneously disable the chaotic conditions in which an entrepreneurial approach can thrive.

So, for example, the planned development of purpose-built snowboard parks in resorts is designed to both create an environment which caters for and encourages the innovation, but which at the same time controls and regulates its use. Building snowboard parks has thus had the effect of enclaving the activity, and constraining the freedom and the 'open mountain adventure' aspects of the sport which are central to its very spirit and rationale (Green and Lewis, 1997a).

The very innovation designed to attract change adopters to resorts and so rejuvenate ailing markets can be comprehensively undermined by change managers fears about unbounded risk and their actions to minimise it. In planning a structural framework for innovations, the benefits they were intended to secure are at risk of being forfeited.

In so far as the planner intervenes to manage the original innovation, its potential future direction and development, and indeed the eventual outcomes for the sport as a whole, become bounded by those actions.

Such a paradox is commonly found wherever managers are confronted with the management of change. It is vital to realise that: 'Paradox does not have to be resolved, only managed' (Handy, 1994: 22). To manage paradox, it first has to be identified and, second, its nature needs to be understood. Current research methods are barely yet capable of addressing these questions, let alone illuminating them.

The directions for future research which may help in the understanding of the *ad hoc* nature of tourism planning, development and management, lie, conceptually, in complexity theory and practically, in the management of uncertainty. Tourism theory must now fully embrace developments in general management and entrepreneurship theory and turn to new methodologies to guide practice. To expedite the utility – for research as well as practice – of an entrepreneurial approach in the tourism context, the words of Bygrave (1993: 279) should be heeded.

> If we want to understand entrepreneurship, our research methodology must be able to handle nonlinear, unstable discontinuities. Chaos may not be able to provide us with a mathematical theory for entrepreneurship, but it is probably more relevant than linear regression analysis. Entrepreneurship, after all, is the science of turbulence and change, not continuity.

Note

1 In tourism literature the term 'entrepreneur' is often used very loosely and frequently it is assumed to refer to any small business owner. This false assumption leads to much confusion. In a resort there may be many businesses, small and large, but there will only be a few entrepreneurs and typically many will come from outside the local community.

References

Special sources

Amt der Tiroler Landesregierung (1995) *Der tourismus im winterhalbjahr 1994/5*, Fachbereich Statistik, Veröffentlichung Nr. 70, September, Innsbruck.

Barbier, B. (1993) 'Problems of the French Winter Sport Resorts', *Journal of Tourism Recreation Research* 18: 5–11.

Bass, H. (1966) *Winter Sports*, London, Stanley Paul.

Clary, D. (1993) *Le Tourisme dans l'éspace français*, Paris, Masson.

'Enquête à l'Alpe d'Huez: Le hors-piste menace' (1996) *Le Skieur* 4: 24–26.

Green, S. and Lewis, R.D. (1997b) 'License to Thrill', *Skier and Snowboarder* Spring, Mountain Marketing.

Lewis, R.D. and Green, S. (1995) 'On the slopes', *Leisure Management* 15, (12).

Lewis, R.D. and Wild, M. (1995) *French Ski Resorts and UK Ski Tour Operators: An Industry Analysis*, Centre for Tourism Research, Sheffield Hallam University.

Rickards, T. (1988) *Creativity at Work*, Aldershot, Gower.

SEATM-CEMAGREF (1987) *Les Loisirs de montagne: Le marché des stations de sports d'hiver en 1987*, Industry Report.

General references

Bygrave, W.B. (1993) 'Theory Building in the Entrepreneurship Paradigm', *Journal of Business Venturing* 8: 255–80.

Bygrave, W. B. and Hofer, C.W. (1991) 'Theorising about Entrepreneurship', *Journal of Entrepreneurship Theory and Practice* Fall: 7–38.

Capra, F. (1982) *The Turning Point: Science, Society and the Rising Culture*, London, Flamingo.

Dann, G., Nash, D. and Pearce, P. (1988) 'Methodology in Tourism Research', *Annals of Tourism Research* 15: 1–28.

Drucker, P. (1974) 'New Template for Today's Organisations', *Harvard Business Review* Jan./Feb.: 22.

Faulkner, W. and Russell, R. (1997) 'Chaos and Complexity in Tourism: In Search of a new Perspective', *Pacific Tourism Review*, 1, (2): 93–102.

Faulkner, W. and Valerio, P. (1995) 'An Integrative Approach to Tourism Demand Forecasting', *Tourism Management* 16 (1): 29–37.

Fayol, H. (1916) *Administration industrielle et générale*, translated by C. Storrs, London, Pitman (1949).

Feigenbaum, M. (1978) 'Quantitative Universality for a class of Nonlinear Transformations', *Journal of Statistical Physics* 19: 25–52.

Feigenbaum, M. (1979) 'The Universal Metric Properties of Nonlinear Transformations', *Journal of Statistical Physics* 21: 669–706.

Fiedler, F.E. (1967) *A Theory of Leadership Effectiveness*, New York, McGraw-Hill.

Green, S. and Lewis, R.D. (1996) 'Culture Club', *Skier and Snowboarder* Nov., Mountain Marketing.

Green, S. and Lewis, R.D. (1997a) 'From Innovation to New Sport Development', *Journal of Tourism and Recreation Research* Jan. (Forthcoming).

Grossack, I.M. (1989) 'Joseph Alois Schumpeter', *Business Horizons* Sept.–Oct.: 70–76.

Gunn, C.A. (1994) *Tourism Planning: Basics, Concepts, Cases*, New York, Taylor and Francis.

Hawking, S.W. (1988) *A Brief History of Time*, New York, Bantam Press.

Hall, P. (1992) *Urban and Regional Planning*, London, Routledge.

Handy, C. (1994) *The Empty Raincoat: Making Sense of the Future*, London, Hutchinson.

Herbin, J. (1995) *Mass Tourism and Problems of Tourism Planning in French Mountains*, in G.J. Ashworth and A.G.J. Dietvorst (eds) *Tourism and Spatial Transformation*, London, CAB International.

Inskeep, E. (1991) *Tourism Planning: An Integrated and Sustainable Development Approach*, New York, Van Nostrand Reinhold.

Jamal, T.B. and Getz, D. (1995) 'Collaboration Theory and Community Tourism Planning', *Annals of Tourism Research* 22 (1): 186–204.

Josty, P.L. (1990) 'A Tentative Model of the Innovation Process', *Journal of Research and Development Management* 20 (1): 35–45.

Jung, C.G. (1972) *Synchronicity: An Acausal Connecting Principle*, London, Routledge and Kegan Paul.

Kaplan, R. (1987) 'Entrepreneurship Reconsidered: The Antimanagement Bias', *Harvard Business Review* May–June: 84–89.

Laing, R.D. (1969) *Self and Others*, London, Tavistock.

Lawrence, P.R. and Lorsch, J.W. (1967) *Organization and Environment*, Boston, Harvard Business School, Division of Research.

Likert, R. (1967) *New Patterns of Management*, New York, McGraw-Hill.

Lorenz, E. N. (1963) 'Deterministic Nonperiodic Flow', *Journal of the Atmospheric Sciences* 20: 448–64.

Macmillan, I.C. and Katz, J.A. (1992) 'Idiosyncratic Milieus of Entrepreneurial Research: The Need for Comprehensive Theories', *Journal of Business Venturing* 7: 1–8.

Mandell, A.J. (1985) 'From Molecular Biological Simplification to More Realistic Nervous System Dynamics: An Opinion', in J.O. Cavenar *et al. Psychiatry: Psychobiological Foundations of Clinical Psychiatry* 3: 2, New York.

Marcuse, H. (1964) *One-dimensional Man*, London, Routledge and Kegan Paul.

McLoughlin, B.J. (1969) *Urban and Regional Planning: A Systems Approach*, London, Faber.

Müller, O. (1958) *The New Official Austrian Ski System*, trans. R. Palmedo, London, Nicholas Kaye.

Naisbitt, J. (1994) *Global Paradox*, London, Nicholas Brealey.

Pearce, D. (1989) *Tourism Development*, 2nd edn, London, Longman.

Schumacher, E. F. (1973) *Small is Beautiful*, London, Blond and Briggs.

Stevenson, H. and Harmeling, S. (1990) 'Entrepreneurial Management's Need for a More "chaotic" Theory', *Journal of Business Venturing* 5: 1–14.

Taylor, F.W. (1911) *Principles of Scientific Management*, New York, Harper.

Toffler, A. (1970) *Future Shock*, London, Bodley Head.

Urwick, L. (1933) 'Organization as a Technical Problem', reprinted in L. Gulick and L. Urwick (eds) (1937) *Papers on a Science of Administration*, New York, Columbia University Press.

Part II

TOURISM AS AN AGENT OF CHANGE

INTRODUCTION
TO PART II

Bill Faulkner

While the case studies contained in Part I focused on tourism management responses to change in various settings, those in this section emphasise the actual or potential role of tourism as an instrument of change; that is, in particular, where tourism can play a role in the rejuvenation of economically depressed areas or in the modernisation process of developing countries. As we shall see, however, the benefits that can be derived from tourism in both these contexts depend upon the establishment of an effective management regime and, in some cases, tourism can actually be an antidote for resisting changes that threaten the local cultural or environmental heritage.

The on-going process of economic restructuring in developed countries has resulted in some areas of these countries becoming economically marginalised. Some regions are marginalised because the products they produce are no longer competitive in markets that have been opened up to competition by the relaxation of trade restrictions. Others have lost ground because they have failed to modernise their productive infrastructure or because the goods or services they produce have themselves been superseded. Elsewhere, regions have struggled to maintain their population base because their main industries have modernised and the mechanisation of the means of production have resulted in a reduction of employment opportunities. A combination of these conditions have prevailed in the Canadian prairie provinces referred to by Weaver in the first chapter of this section. In particular, Manitoba and Saskatchewan have suffered from the mutually reinforcing effects of a narrow industrial base and rural depopulation associated with farm mechanisation. To arrest the decline of these two areas, their economic base needs to be diversified, and it is argued that the development of ecotourism resources can contribute in this regard.

Two case studies, one by Lipscomb on village tourism in the Solomon Islands and the other by Faulkner on an agrotourism project in Central Java, look at the potential of tourism as a basis for economic development in these areas. Both studies highlight some of the problems and obstacles to tourism development in developing countries. The necessity of sound planning and management

163

practices to ensure that tourism actually enhances the welfare of local residents is emphasised and the relevance of sustainable tourism development principles to the achievement of this objective is recognised. This point is reinforced in the Solomons case study, in particular, where reference is made to the 'four legged chair' analogy to emphasise how the four key interrelated activities involved in tourism development (product development, marketing, infrastructure development and training) must proceed apace in order to avoid 'lopsided' development. The Sodong agrotourism project in Central Java illustrates how tourism can be instrumental in preserving the cultural heritage of an area. By attaching economic value to the retention of traditional practices that would otherwise become increasingly unviable, tourism can ensure that culturally significant activities can be quarantined from the impacts of modernisation. In this sense, tourism might be construed as an antidote to change.

In her West Indian islands case study, Frances also alludes to the commonly perceived role of tourism as a vehicle for economic development in developing countries. However, in doing so she highlights a parallel between tourism and the role of the plantation system in the genesis of the dependency status of these islands. Tourism's record as an instrument for economic development has been somewhat flawed in many developing countries and the West Indies are no exception in this respect. This observation brings into focus the necessity of local resident participation and empowerment in the tourism development process if sustainable outcomes are to be achieved. The application of Pretty's (1995) model of participation to the analysis of the West Indian island tourism situation highlights how readily the plantation heritage of these islands has been exploited by both local elites and metropolitan (multinational) tourism interests to perpetuate exploitive dependency relationships. However, this framework also shows how the relationship between the tourism sector and the islands' population might evolve towards the more meaningful levels of participation ('self-mobilisation') described in the model. Apart from providing a foundation for economic development, therefore, tourism might also have the potential to act as a catalyst for increasing the control of local people over their economic destiny. The role of tourism within the economic planning strategies of developing countries might therefore be construed as either a vehicle for increased dependency or economic emancipation, and movement towards one or the other pole will depend upon the policy and management regimes that are established.

The 'tourism as an antidote to change' theme is revisited in Stabler's chapter, which examines the interrelated roles of conservation and tourism in the regeneration of the inner city area of Temple Bar, Dublin. Here, as in many cities of the Western world, the under-utilisation of older and sometimes historically significant buildings has combined with the suburbanisation of people and economic activities to produce urban blight in areas that might otherwise have heritage and tourism value. In the Temple Bar study, this issue is analysed in terms of the economist's perspective, which highlights deficiencies in evaluations of publicly funded urban heritage restoration programmes. In particular, attention is drawn to the importance of taking into account such factors as the economic value component and externalities, while the application of cost benefit analysis is essential to enable the net economic benefits of conserving historic buildings to be isolated. The potential income from tourism activities features in this equation. The Temple Bar case also emphasises how important the participation of those living and running businesses in the area is in ultimately determining the success of the regeneration programme. Public sector involvement is often essential to provide impetus to the restoration process, but its longer-term viability depends on commercial and public good outcomes.

Crookston's description of the restoration of historic precincts in the two Middle Eastern cities of Dubai (United Arab Emirates) and Salt (Jordan) provides another perspective on the nexus between heritage conservation and tourism in the urban context. In this chapter, more attention is given to the architectural and physical planning aspects of this process. Apart from providing insights into the role tourism is playing in the regeneration of both local economies and the physical fabric of these cities, this chapter also highlights the importance of matching tourism markets and heritage product in a way which ensures that inherently fragile historical resources are preserved. As in the case of Temple Bar, the importance of public sector investment as a catalyst for attracting the commercial linkages is also emphasised. Again, however, the author points out that income from commercial activities is essential if the public sector is to avoid the financial burden of long-term maintenance costs.

Reference

Pretty, J. (1995) 'The many interpretations of participation', *In Focus* 16: 4–5.

9

ECOTOURISM AND ITS POTENTIAL CONTRIBUTION TO THE ECONOMIC REVITALISATION OF SASKATCHEWAN AND MANITOBA

David B. Weaver

This chapter examines the actual and potential role of ecotourism in contributing to the sustainable development of the Canadian prairie economy, as represented by the provinces of Saskatchewan and Manitoba (see Figure 9.1). While it is recognised that these two provinces have evolved along distinctive socio-economic and political trajectories, and will likely formulate ecotourism policies as separate competing units, there exist enough commonalities to merit their consideration as a single regional entity. The first section of the chapter is contextual, describing the weaknesses and strengths associated with their economy, and outlining the underlying factors and consequences related to those characteristics. The second section begins with a generic discussion of the ecotourism phenomenon, then considers the potential role of ecotourism in contributing to the economic enhancement of mainly rural destinations such as the two provinces. The actual extent of ecotourism in Saskatchewan and Manitoba is described in the third section, which then discusses the constraints and opportunities affecting the future expansion of the sector. In the final section I make recommendations, and present a framework within which the ecotourism industry of the prairie provinces, as well as other jurisdictions, can be planned and managed.

The economy of Saskatchewan and Manitoba

Characteristics of the economy

Saskatchewan and Manitoba both display the classic dependent hinterland economy, their historical development having been dictated by external circumstances and decisions, within national and international centres of power, over which they had little influence. (The third prairie province, Alberta, is not included in this discussion since its large oil reserves have resulted in its clear differentiation as an economically advantaged province.) Geographically, the southern portion of the two provinces is a region of extensive agriculture based on wheat and cattle, whereas the north is a resource frontier dominated by mining and forestry (Figure 9.1). A weak manufacturing sector is restricted to larger urban centres such as Winnipeg, Regina and Saskatoon by small domestic markets and relative isolation (Barr and Lehr, 1982). Government, from the municipal to federal level, is an extremely significant source of employment throughout the region.

It is the primary sector, therefore, represented by agriculture, forestry and mining, which has traditionally defined the region and influenced all other aspects of the economy and society. In Saskatchewan, which has historically relied upon the export of farm-based commodities, the C$4.6 billion agriculture sector directly accounts for 18 per cent of the labour force (compared with 5 per cent for North America as a whole), and close to 50 per cent if indirect influences are taken into consideration (SRTEE, 1992; Stabler and Olfert, 1992; Statistics Canada, 1993a). More recently, the province has emerged as an important producer of potash, uranium and oil. Manitoba, with relatively advanced secondary and tertiary sectors in Winnipeg (metropolitan population over 700,000), is less dependent upon agriculture, which contributed C$2.1 billion to the provincial economy in 1991, and directly accounted for 8 per cent of the labour force (Statistics Canada, 1993a).

Unfortunately, such primary sector commodities are highly vulnerable to changes in external market demand and, in the case of agriculture, to the drought and frost risks of a marginal continental climate. Problems with the agricultural economy, accordingly, have been apparent since the earliest days of European colonisation. However, the global overproduction of wheat during the 1980s and the concomitant failure of GATT to restore order in world grain markets resulted in several successive years where costs of production were at or below prices received. This severe cost–price squeeze was not adequately compensated for by existing federal and provincial income support programmes (Stabler et al., 1992). Such problems, of course, were not unique to the Canadian interior plains (Buttel and Gillespie, 1991), but the higher-than-average dependency of that region upon the farming sector did result in more severe consequences, as described in the next two sections.

Social consequences

The relative isolation and primary sector dependency of the region have contributed to a pattern of population stagnation which became evident almost immediately after the completion of the initial agricultural settlement process. As shown in Table 9.1, the population of Saskatchewan and Manitoba has increased

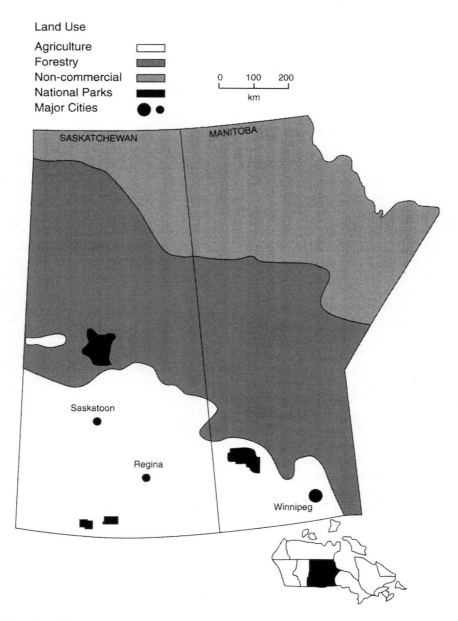

Land Use
Agriculture
Forestry
Non-commercial
National Parks
Major Cities

0   100   200
km

SASKATCHEWAN   MANITOBA

Saskatoon
●

Regina
●

Winnipeg
●

Figure 9.1 Land use, Saskatchewan and Manitoba

Table 9.1 Population data, Manitoba and Saskatchewan 1931 and 1991

Variable	Year	Manitoba	% of total	Saskatchewan	% of total
Population:	1931	700,000	6.7% of Canada	922,000	8.9% of Canada
	1991	1,091,000	4.0% of Canada	989,000	3.6% of Canada
% Increase		**55.8**		**7.2**	
Rural population:	1931	384,000	54.9	631,000	68.4
	1991	305,000	28.0	366,000	37.0
% increase		**−20.6**		**−42.0**	
Farm population:	1931	256,000	36.6	564,000	61.2
	1991	78,000	7.1	148,000	15.0
% increase		**−69.5**		**−73.8**	
No. of farms	1931	54,199	–	136,472	–
	1991	25,706	–	60,840	–
% increase		**−52.6**		**−55.4**	

Sources: Dominion Bureau of Statistics, 1936a, 1936b; Statistics Canada, 1992; Statistics Canada, 1993b; Colombo, 1996.

only modestly during the past sixty years compared with the rest of Canada, due to persistent employment-related out-migration and a failure to attract and retain a proportional share of immigrants. With regard to the more serious problem of inter-provincial migration, Manitoba and Saskatchewan have experienced net respective losses of 39,533 and 69,396 between 1987 and 1991, resulting in an erosion of their share of the national population.

The province-wide statistics, however, mask significant internal variations in population performance. In general, larger cities and their rural commuting zones have expanded at a rate comparable to the national average. In contrast, the farm population has plummeted, inducing an overall decline in the rural population (see Table 9.1). Especially hard hit are the towns and villages which depend upon the farm market (Brierley and Todd, 1990; MacLean and Rounds, 1991; Stabler *et al.*, 1992). As a consequence of youth out-migration and the in-migration of retired or semi-retired farmers, these agricultural service centres are acquiring an ageing population profile. For example, whereas 12.8 per cent of Saskatchewan's population consisted of older adults (i.e. those 65 years of age or older) in 1988, the comparable proportion for the combined total town and village population was about 20 per cent (Nilson and Weaver, 1992; Nilson *et al.*, 1996). The declining, widely dispersed and economically inactive (retired) population which is emerging in the rural south of Saskatchewan and Manitoba is not conducive to the rationalised delivery of expected services and, barring unexpected developments, the region can expect a gradual deterioration in quality of life indicators. Northern communities, while displaying if anything a younger than average population profile, are more subject to the 'boom and bust' cycles which emerge in response to rapidly changing market conditions and resource depletion.

While this rural decline has without doubt been accelerated by the current cost–price squeeze and the marginality of the climate, the underlying role of the overall modernisation process over the past sixty years must be emphasised. The substitution of capital for labour in agriculture, the urbanisation of rural tastes, the acquisition of vehicles and the upgraded road system, have all combined (as throughout the entire 'developed' world) to reduce the population required to produce agricultural goods and to rationalise the provision of services to that population (Stabler and Olfert, 1992). Typical of the attitude toward such changes is a pervasive sense of powerlessness over forces which appear to be anonymous and seemingly inexorable.

Environmental consequences

In the south, it has been necessary for farmers to increase production in order to survive the cost–price squeeze. This can be achieved by acquiring additional property (e.g. from a bankrupt neighbour), intensifying the application of pesticides and fertilisers on existing croplands, and by converting marginal wet-lands, grasslands and woodlands to crop production. The latter two scenarios, in particular, are forms of 'predatory cultivation' which have obvious and serious environmental implications, yet their practice has been implicitly encouraged by an array of federal and provincial acreage-based programmes and policies which have discouraged the retention of land in an 'unproductive' natural state. Between 1951 and 1981, the amount of cultivated land in the Canadian prairies increased by 25.7 per cent, while 'improved' pastures expanded by 129.1 per cent. Concurrently, the area occupied by 'unimproved' or wooded farmland declined by 28.9 per cent and 66.9 per cent respectively. As well, about 40 per cent of wetlands have been eliminated (Turner *et al.*, 1987). Ironically, even while this intensification has been occurring, there has been considerable abandonment of marginal farmlands in areas along the border with the northern resource frontier which have proven to be non-viable for agriculture.

Environmental problems in the north are more subtle in that the forestry sector depends upon the maintenance of a superficially natural landscape, and the mining industry occurs in isolated nodes. Like agriculture, these sectors are dependent upon unpredictable external markets and decision-makers who are more concerned with short-term profitability than environmental integrity. Associated problems include unsustainable rates of harvesting, the fostering of single-species forest plantations, under-representation of the old-growth succes-sion stage, clear-cutting practices which induce erosion, and the dispersal of radioactive mine tailings from areas of uranium extraction. While the far north still consists mainly of wilderness which has not experienced these problems, the forestry and mining resource frontier is encroaching, both along access roads to new mining areas, and through technologies which make feasible the exploitation of smaller, slower-growing trees.

The ecotourism phenomenon

Given the economic and related social and environmental consequences already described, it is not surprising that both provincial governments are actively attempting to identify and implement sustainable rural development strategies which will contribute to a more diverse and stable provincial economy. Though there is recognition that the primary sector will most likely continue to dominate that economy, there is also a realisation that the potential for expansion within that sector is limited, and that diversification into the more stable manufacturing and tertiary sectors is desirable. The 'problem' with manufacturing and most services (including the cutting-edge, 'high tech' sector [Markusen, 1985]), however, is their preference for an urban location where critical masses of labour, market and affiliated services are readily available. The challenge then, is to identify non-primary activities which are best carried out in a rural environment, and it is here that tourism, and ecotourism in particular, is of interest.

Definition

There is no consensus on the nature of ecotourism, as evidenced by the myriad definitions encountered within the literature. However, there does seem to be a high level of agreement that ecotourism:

- is primarily focused upon attractions related to the natural environment;
- is oriented toward the appreciation of these attractions for their own merit;
- is non-consumptive in the sense that wildlife is not deliberately removed from its habitat (Shaw and Mangun, 1984); and
- attempts to minimise the concomitant negative environmental, socio-cultural and economic impacts associated with its pursuit.

Within this framework of agreement, it is still possible to group the various definitions of ecotourism into two basic categories. *Active* ecotourism (Orams, 1995) maintains that the sector should have a positive, enhancing effect upon the resource, and that the experience should be transformational for all participants with respect to attitudes about the environment (Ziffer, 1989; CEAC, 1991; Butler quoted in Scace, 1993). As such, active ecotourism is very much tied into the environmentalist ethos associated with the advent of the so-called Green Paradigm (Knill, 1991). In contrast, *passive* ecotourism definitions omit this transformational imperative, and maintain merely that the activity should not harm the resource (Ceballos-Lascurain, 1988; Rymer, 1992). This chapter argues for a definition of ecotourism encompassing both perspectives, since restriction to the

171

active manifestation would result in an elitist pursuit involving only a minuscule portion of the market.

Similarly, whereas it is frequently argued, in a form of 'spatial elitism', that ecotourism is best accommodated within wilderness or strictly protected areas, this chapter argues that ecotourism can also occur in relatively modified environments, such as farmlands, reservoirs and planted forests, which are more accessible to the vast numbers of actual and potential passive ecotourists. As a final point of clarification, the one-way travel thresholds which are commonly employed to distinguish domestic 'tourist' from 'non-tourist' trips (McIntosh *et al.*, 1995) are not employed in this chapter; instead, ecotourism is defined more by the activities pursued than the origins of the participant, thus allowing for even relatively short trips from home to be counted as a component of ecotourism.

The attraction of ecotourism

From an environmental perspective, the popularity of ecotourism derives from its symbiotic relationship with nature. The argument goes that the economic return from ecotourism justifies the retention of land in a natural state which would otherwise be more profitably utilised (at least in the short term) for ecologically destructive agricultural or forestry purposes (Sherman and Dixon, 1991), as has happened in Saskatchewan and Manitoba. Socio-culturally, the attractiveness of ecotourism is based upon its widely acknowledged status as a form of *alternative tourism* which:

- ideally emphasises local control over the sector;
- engenders strong backward linkages within the local economy;
- fosters architecture which is compatible with the local vernacular;
- emphasises the local 'sense of place';
- attracts types of visitors more sympathetic to local sensibilities; and
- plans holistically on a sustainable, long-term basis (Weaver, 1993).

Such considerations are especially important in hinterlands such as Saskatchewan and Manitoba where external decision-making tends to be the norm.

It is the economic argument, however, which captures the interest of most governments and entrepreneurs. Evidence suggests, at least at the international level, that ecotourists tend to spend more than the 'average' tourist (Wilson, 1987; Eagles, 1992; SWWA, 1996b). Combined with this is anecdotal evidence that ecotourism is growing at a more rapid rate than the overall tourism sector. To be sure, the common assertion that ecotourism has been expanding at an annual rate of 20 per cent or more (e.g. Ecotourism Society, n.d.; SWWA, 1996b) is almost certainly an exaggeration based on the unwarranted extrapolation of exceptional performances by isolated segments of the industry. More plausible indications of lower but still robust growth, however, can be obtained from an extensive survey of American adults which monitored growth in participation across a broad array of recreational activities between 1982 and 1995 (Cordell *et al.*, 1995). The strongest growth was recorded in the ecotourism-related activity of birdwatching (+155.2 per cent; from 21.2 million to 54.1 million), followed by hiking (+93.0 per cent; 24.7 million to 47.7 million) and backpacking (+72.7 per cent; 8.8 million to 15.2 million). Similar results were obtained in a Canadian survey (Filion *et al.*, 1993), which found that the number of days occupied by domestic non-consumptive wildlife trips increased by 49 per cent during the 1981–91 period (from 56.7 million to 84.3 million days), during which the overall population increased by only 15 per cent.

Ecotourism in Saskatchewan and Manitoba

The above considerations have no doubt played a significant role in the decision of government in both provinces to pursue ecotourism strategies. This section examines the actual extent of ecotourism in Saskatchewan and Manitoba, within the context of the overall tourism industry. The constraints and opportunities faced by the provinces in developing the ecotourism sector are then discussed.

The overall tourism context

Tourism is a relatively minor but expanding component of the prairie economy. In Saskatchewan, tourism accounted for 4.1 per cent of the Gross Provincial Product in 1992, up from 3.6 per cent in 1988 (SED, 1991). It has been estimated to support 18,000 jobs directly and indirectly, and to generate between C$650–850 million per year, 77 per cent of which is derived from the provincial market (SRTEE, 1992; SED, 1993). In Manitoba, tourism generated C$800 million in 1990 (70 per cent generated from within the province), or 3–4 per cent of the GPP (Manitoba Environment, 1993). Several major discrete sectors of the prairie tourism industry can be identified, as follows, recognising that local 'non-tourists' would account for many of the respective participants:

- recreational urban attractions and events;
- social visits with friends and relatives;
- recreational visits to provincial and national parks;
- incidental visitors in transit, as along the Trans-Canada highway;
- patrons of northern lodges and southern vacation farms; and
- patrons of other rural tourism enterprises.

Of particular relevance to rural areas are the park systems, lodges, vacation farms and transient visitors. The Manitoba and Saskatchewan provincial park networks constitute by far the largest component within the rural tourism base. Other important sectors include hunting, which in Saskatchewan accounted for 129,000 hunter-days in 1990 and C$1.4 million in licenses, and fishing, which generated C$71 million in expenditures by anglers in 1985, 87 per cent of which was in the north (CSPAF, 1994). Of interest is the recent advent of Reserve-based gaming, which mirrors a North American-wide trend (Harris, 1994), but which is also highly controversial with respect to its long-term economic and social impacts (Gabe *et al.*, 1996).

Current extent of ecotourism

Because there is no consensus regarding definitions of ecotourism, and because neither provincial government has ever methodically gathered data on anything defined as 'ecotourism', estimates of the extent of the activity within Saskatchewan and Manitoba cannot be more than crude approximations. Clearly, ecotourism-like activities have been carried out for many years on an informal basis, particularly within wilderness and protected areas. However, contemporary visitation data for the latter provide little indication of the magnitude of ecotourism, since probably the majority of visits to such entities are multi-purpose recreational episodes only marginally related to ecotourism, in that nature is more the pleasant venue than the inherent attraction.

More tangible evidence may be obtained by examining visitor patterns to attractions which are more explicitly related to ecotourism. These would include such relatively recent phenomena as Wilderness Provincial Parks (the mandate of which is conducive to a high level of ecotourism participation), and high-profile community-based initiatives such as the Redberry Lake Pelican Project near Hafford in Saskatchewan (Hawkes, 1993). Regionally significant sites include the Fort Whyte Centre (95,000 visitors in 1990 and 180,000 in 1994), Oak Hammock Marsh (83,000 in 1989) and the Narcisse Snake Den (9,500 in 1990), all near Winnipeg (Weaver *et al.*, 1996), and the Last Mountain Game Sanctuary between Saskatoon and Regina. The attraction with perhaps the highest inter-national profile is Churchill, Manitoba, where an estimated 20,000 tourists arrive each year mainly during the 37-day polar bear viewing season from early October to early November, to be in the company of some 20 per cent of the global polar bear population (Johnston, 1995). So far, this level of viewing activity has not been linked to any undue stresses in the polar bear population.

Beyond such higher-profile areas and sites, there exists a much less publicised and less formal private sector ecotourism 'industry'. A recent survey of the thirty or so known small-scale private sector ecotourism site operators in Manitoba (Weaver *et al.*, 1996) indicated that operators had little formal training in ecotourism, relied on informal marketing mechanisms such as 'word of mouth', were engaged in ecotourism largely as an 'add-on' to other tourism or non-tourism activities, and were frustrated by a wide range of perceived constraints associated mainly with the policies of the provincial government. The latter could be grouped as 'inadequate government support', 'hindrances to development' and 'lack of internal government co-operation/co-ordination'. The picture, then, is of an 'unorganised and incipient' sector far removed from the glamorous image of ecotourism often conveyed in the literature. Similar characteristics apply to the

100-strong vacation farm sector within the two provinces. A recent study of Saskatchewan operators (Weaver and Fennell, 1997) revealed 'wildlife viewing' as the most commonly cited client recreational activity, although very few could be described as ecotourism-specialised. A major challenge facing these facilities is the potential conflicts arising from the fact that many vacation farms accommodate both wildlife viewing and recreational hunting in order to achieve the critical visitor base required to survive as a tourism operation. Operators accommodating such a mixture did not indicate conflicts, but further investigation is necessary to determine whether this is due to the implementation of appropriate strategies, or because the mixture does not result in conflicts.

Efforts are under way, both directly and indirectly, to foster the development of ecotourism in the Canadian prairie provinces. In Saskatchewan, the provincial conservation strategy, though not mentioning ecotourism directly, proffers several recommendations which would indirectly support the expansion of ecotourism-related activity:

- the province should strongly support the Canadian Heritage River System, a joint federal–provincial initiative which aims to preserve the heritage and ecological integrity of high profile waterways;
- the protected area network should be expanded so that about 12 per cent of each provincial biome is formally protected;
- co-operative development of multiple land use objectives should be established for flora and fauna; and
- strict tourism guidelines and enforcement programs should be developed for ecologically and culturally sensitive sites (SRTEE, 1992).

More directly, Saskatchewan is in the process of formulating a provincial eco-tourism strategy, having progressed to the stage of identifying a network of ninety ecotourism sites where significant wildlife-viewing opportunities are available (SWWA, 1996a). Implementation is now being encouraged through the following recommended stages:

- formation of a Sustainable Tourism Council to co-ordinate all aspects of the ecotourism product;
- introduction of an ecotourism accreditation program and development of a sustainable tourism policy for Saskatchewan;
- development of nine pilot ecotourism tour packages (i.e., product development initiative);
- a marketing and public relations initiative to market and promote the provincial ecotourism product;
- a community participation initiative to encourage and facilitate local involvement; and
- a human resource initiative to provide training programs for guides and interpreters as well as product providers (SWWA, 1996b).

Manitoba is also intending to implement an ecotourism strategy, and has to date completed an inventory of sites similar to what Saskatchewan has achieved (Matrix Management, 1991).

Opportunities and constraints

The emptiness of the prairie provinces, while decried as a hindrance to economic development, may ironically prove to be a lucrative asset for ecotourism in a world which is experiencing a net increase in human population of 90 million per year. Related assets associated with the natural environment include considerable diversity within five distinct biomes, a perception of minimal pollution and richness in wildlife (and especially bird) species, including several which are rare (such as the whooping crane and the piping plover). However, among the constraints which offset these opportunities are the extensive modification of the natural environment (as described earlier), the issue of Aboriginal land rights (elaborated upon later), and innate environmental liabilities such as the severe winters, restricted seasonality of most birds, and susceptibility to drought and fire. Furthermore, the utilisation of rare and endangered species and spaces in promotional campaigns is naive, misleading and dangerous, since such precarious resources cannot accommodate anything beyond a minimal visitation level, and thus cannot form the basis of a sustainable and economically significant ecotourism industry.

Existing networks of protected areas, vacation farms and lodges also offer opportunities as well as constraints. Approximately 5.5 per cent of Manitoba's surface area satisfies the criteria of the World Wildlife Fund's Endangered Spaces Campaign (Manitoba Environment, 1995), while Lawton (1995) has identified twenty-seven categories of protected land in Saskatchewan, encompassing 6.5 per cent of the province. Virtually all of these lands are suitable and available in theory for ecotourism, but most categories are not presently utilised in any significant way as venues for that activity. Potential constraints include the existing use of certain protected areas for grazing (e.g., Prairie Farm Rehabilitation Agency pastures) and hunting. Similarly, the vacation farm and lodge networks are already mobilised to accommodate hunting and fishing-oriented activities which may not be compatible with ecotourism. Beyond these networks, the diffusion of ecotourism into agricultural and forestry areas will be hindered by similar incompatibilities. For example, farmers may seek compensation from ecotourists, as they currently do with hunters, for damage allegedly caused to crops by wildlife. Extensive clear-cutting in the commercial forest zone will reduce the areas available for quality ecotourism activities.

The issue of Indian rights has enormous implications for ecotourism throughout the region. Levels of participation in ecotourism are currently low, though great potential exists because of the relatively natural condition of many Reserves and northern settlement hinterlands, the seeming ethical compatibility of traditional Aboriginal lifestyles with ecotourism, and the attraction of that culture to the non-Aboriginal tourist market, especially within Europe and Asia. That this potential is not restricted to the relatively small area currently occupied by Reserves and other native settlements is clear from recent developments concerning the broader application of Aboriginal entitlements. The Sparrow case in 1990, for example, resulted in a directive from the Supreme Court of Canada to the government that indigenous people must be included in the co-operative management of natural resources. This extension of Indian influence reinforces the increased tendency to assert treaty-based hunting and fishing rights on 'unoccupied Crown Lands' (Berg *et al.*, 1993). Ironically, this environment of enhanced Indian opportunity has spawned a concomitant environment of uncertainty among non-indigenous ecotourism operators, many of whom rely on leasing arrangements for their Crown land-based operations. In the Manitoba survey, considerable ambivalence was expressed by operators over the perceived threat to long-standing privileges (Weaver *et al.*, 1996). Clearly, the Canadian situation offers an intriguing parallel to the Mabo and Wik controversies in Australia.

A framework for ecotourism management: ecotourism context zones

The formal efforts which are currently in progress to foster ecotourism within Saskatchewan and Manitoba are both necessary and useful, as long as the limitations which effectively preclude the development of a large industry are taken into consideration. Especially important to recognise is that many of the opportunities and constraints are not equally distributed, and that the internal geographical variations within each province suggest the logic of distinctive ecotourism 'sub-strategies' which take into account this spatial context. In other words, to plan and manage ecotourism sites in isolation from this surrounding context is ludicrous and ultimately self-defeating. The following framework is an attempt to augment the efforts already underway by representing this diversity in the form of five generalised *ecotourism context zones* (Weaver, 1997) which are particularly relevant to the Canadian interior (see Figure 9.2). Table 9.2 provides a synopsis of ecotourism-related characteristics for all zones, which grade along a continuum from the most developed to the least developed landscapes. The following text provides further commentary about the zones, and attempts to avoid unnecessary duplication with Table 9.2.

Urban context zone

Urban areas, entailing built-up areas with at least 5,000 residents, are rarely cited as venues suitable for ecotourism. However, these spaces have a great potential based on both supply and demand considerations, the former in the provision of extensive networks of semi-natural and other open spaces (e.g. floodplains, parks), and the latter in their proximity to great numbers of casual ecotourists mainly of

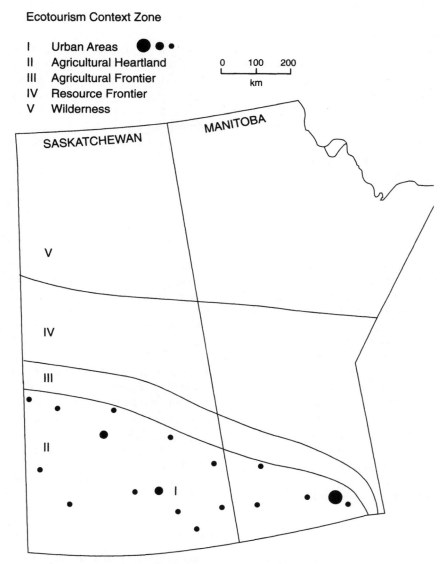

Figure 9.2 Generalised ecotourism context zones in Saskatchewan and Manitoba

Table 9.2 Ecotourism context zones for Saskatchewan and Manitoba, and associated characteristics

Variable	Urban	Agricultural heartland	Agricultural frontier	Resource frontier	Wilderness
Ecotourism space	-linear 'semi-natural' and 'modified' (floodplains, power corridors, trails)	-fragmented 'semi-natural' (marginal land) -ubiquitous 'modified' (crop/grazing land)	-ubiquitous 'semi-natural' (second growth) -fragmented 'natural' (parks)	-ubiquitous 'semi-natural' (harvested forests) -fragmented 'semi-natural' (parks)	-ubiquitous 'natural'
Market	-local	-regional, urban	-regional/provincial, urban	-provincial, urban	-provincial, international
Economic status	-negligible	-local 'add-on' effect	-potentially dominant	-significant 'add-on' effect	-potentially dominant
Opportunities	-affordable, accessible -recruitment for 'harder' ecotourism -educational function	-vacation farm network -extensive but obscure protected lands -low population density -sustainable agriculture more widespread	-other primary sector activities retreating or not established -extensive reversions to semi-natural state -grid roads allow access	-lodge network -exploited areas kept in somewhat 'natural' state	-no competition with other users at present -opportunity for wide dispersion of activity
Constraints	-population and land use pressures -competition from other recreationalists	-predatory cultivation -deterioration of resource due to island effect -competition with hunters	-competition with hunters	-primacy of forestry and mining -mining contamination, erosion due to forestry -unsettled land claims -competition with hunters	-expansion of resource frontier -unsettled land claims -low multiplier effect

the passive persuasion. While certain of the high profile urban ecotourism spaces (e.g. Wascana Centre, Regina) do have considerable potential to attract non-local ecotourists, the economic benefit from ecotourism in the urban context zone is minimal. Instead, such areas, in addition to fulfilling the recreational needs of 'local ecotourists', will serve as a 'recruiting ground' for developing a clientele of future ecotourists who will travel into the rural areas of the provinces, thereby contributing to the transfer of revenue from urban to non-urban areas. As well, urban ecotourism sites are especially attractive in offering formal educational opportunities to local school groups and others.

Agricultural heartland

Beyond the immediate urban vicinity, adjacent and intensively utilised agricultural lands constitute the most accessible venue for the largest proportion of the population. It would be delusory to suggest that ecotourism will ever constitute more than a fringe component of the heartland economy, although the sector could provide a critical source of supplementary income for farmers faced with the cost–price squeeze, while providing an incentive for the preservation and restoration of quasi-natural spaces. Potentially, ecotourism could contribute to the expansion of the vacation farm base to levels far exceeding the current 100 or so. In addition, there is further scope for the establishment of community-based initiatives such as the Redberry Pelican Project.

Agricultural frontier

The agricultural frontier can be either expanding or contracting, and in the case of Saskatchewan and Manitoba, the latter scenario prevails as agricultural activity retreats from extensive marginal environments. As far as economic scenarios are concerned, it is reasonable to argue that the frontier could potentially accommodate ecotourism as a dominant economic pursuit. This is not only because much of the area has proven to be unsuited for farming, but also because other primary sector pursuits such as forestry and mining are not established on anything other than a small scale, reversion to a semi-natural state is already proceeding on a large scale, and a relatively high quality network of access roads and services has been retained as a consequence of its previous settlement.

Resource frontier

As in the agricultural heartland, ecotourism is unlikely to achieve any broadly significant economic status in the medium term, given the pre-eminence of the forestry and mining sectors. However, ecotourism could become locally important as an add-on, and in some cases as a specialisation, in the lodge network and within indigenous communities.

Wilderness

One paradox of ecotourism is that the exploitation of the wilderness, even for the theoretically benign purposes of wildlife appreciation, threatens to undermine the very assets which give value to this setting in the first instance. This is partly because of the role ecotourism can inadvertently play as a pioneer activity which opens the area to more intensive, and possibly less benign, forms of tourism (Butler, 1980). If the wilderness quality is to be preserved, then ecotourism cannot become a large-scale revenue generator within this context zone without conferring upon it what could be termed a tourism-related resource frontier status. Some ecotourism, of course, can and should be accommodated, though on a dispersed basis to avoid this disruption.

Conclusions

This chapter suggests that ecotourism can play a useful secondary role in the economic rehabilitation of marginal hinterlands such as Saskatchewan and Manitoba. Unlike most of the literature, which concentrates on protected areas and wilderness as the primary venues affected by ecotourism, this contribution views ecotourism as a much more ubiquitous activity which can and should be pursued across the broad range of modified and less modified landscapes, thus disseminating the benefits more widely. Urban areas and the agricultural frontier, in particular, are identified as the crucial future areas of ecotourism within Saskatchewan and Manitoba, with the first context zone providing accessible opportunities to most of the population while functioning as a recruitment ground for more involved forms of ecotourism, and the second context zone constituting the core of an extensive ecotourism industry. It is in the latter, as well as in the wilderness zone, that ecotourism could very well emerge as a dominant future economic activity. Clearly, this notion of context zones also usefully augments the nodally based ecotourism planning norms which are currently being pursued, by recognising that the opportunities and constraints associated with ecotourism are not equally distributed, and that ecotourism sites must be affected by what is occurring in surrounding areas. The concept is applicable to any jurisdiction, though each would have its own combination of context zones, depending on its peculiar circumstances. Smaller, long-settled entities such as Prince Edward Island, for example, would likely only possess the urban and agricultural heartland forms, while coastal jurisdictions may find the addition of a marine context zone to be appropriate. In any case, it would be unrealistic to posit that ecotourism by itself can reverse the chronic economic and social problems which beset many rural regions. Context-specific planning and management, however, can ensure that the optimum utilisation is made of available landscapes.

References

Barr, B.M. and Lehr, J.C. (1982) 'The Western Interior: The Transformation of a Hinterland Region', in L.D. McCann (ed.) *A Geography of Canada: Heartland and Hinterland*, pp. 251–93. Prentice-Hall Canada, Scarborough, ON.

Berg, L., Fenge, T. and Dearden, P. (1993) 'The Role of Aboriginal Peoples in National Park Designation, Planning, and Management in Canada', in P. Dearden and R. Rollins (eds) *Parks and Protected Areas in Canada: Planning and Management*, pp. 225–55. Oxford University Press, Toronto, ON.

Brierley, J. and Todd, D. (1990) *Prairie Small-Town Survival: The Challenge of Agro-Manitoba*. The Edwin Mellen Press, Queenston, ON.

Butler, R.W. (1980) 'The Concept of a Tourist Area Cycle of Evolution: Implications for Management of Resources', *Canadian Geographer* 24: 5–12.

Buttel, R. and Gillespie, Jr, G. (1991) 'Rural Policy in Perspective: The Rise, Fall, and Uncertain Future of the American Welfare-State', in K.E. Pigg (ed.) *The Future of Rural America: Anticipating Policies for Constructive Change*, pp. 15–40. Westview Press, Boulder, CO.

CEAC (1991) *A Protected Areas Vision for Canada*. Canadian Environmental Advisory Council, Ottawa.

Ceballos-Lascurain, H. (1988) 'The Future of Ecotourism', *Mexico Journal* January: 13–14.

Colombo, J.R. (1996) *The 1997 Canadian Global Almanac*. Macmillan Canada, Toronto.

Cordell, H., Lewis, B. and McDonald, B. (1995) 'Long-term Outdoor Recreation Participation Trends', in J.L. Thompson, D.W. Lime, B. Gartner and W.M. Sames (eds) *Proceedings of the Fourth International Outdoor Recreation and Tourism Trends Symposium and the 1995 National Recreation Resource Planning Conference*, pp. 35–42. University of Minnesota, Minneapolis, MN.

CSPAF (1994) *Saskatchewan Long-Term Integrated Forest Resource Management Plan*, second draft. Canada–Saskatchewan Partnership Agreement on Forestry, The Canadian Forest Service and Saskatchewan Environment and Resource Management, Prince Albert, SK.

Dominion Bureau of Statistics (1936a) *Seventh Census of Canada, 1931. Vol. I: Population Summary*. J.A. Patenaude, Ottawa, ON.

Dominion Bureau of Statistics (1936b) *Seventh Census of Canada, 1931. Vol. VIII: Agriculture*. J.A. Patenaude, Ottawa, ON.

Eagles, P. (1992) 'The Travel Motivations of Canadian Ecotourists', *Journal of Travel Research* 21: 3–7.

Ecotourism Society (n.d.) *Ecotourism Statistical Fact Sheet*. The Ecotourism Society, Alexandria, VA.

Filion, F., DuWors, E., Boxall, P., Bouchard, P., Reid, R., Gray, P.A., Bath, A., Jacquemot, A. and Legare, G. (1993) *The Importance of Wildlife to Canadians: Highlights of the 1991 Survey*. Canadian Wildlife Survey, Environment Canada, Ottawa.

Gabe, T., Kinsey, J. and Loveridge, S. (1996) 'Local Economic Impacts of Tribal Casinos: The Minnesota Case', *Journal of Travel Research* 34, 3: 81–88.

Harris, T.R. (1994) 'Winner Take All: Expansion of the Gaming Industry', *Western Wire* Fall: 18–20.

Hawkes, S. (1993) 'Watchable Wildlife: The Redberry Pelican Project', in S. Hawkes and P. Williams (eds) *The Greening of Tourism: From Principles to Practice, A Casebook of Best*

Environmental Practice in Tourism, pp. 59–63. Centre for Tourism Policy and Research, Simon Fraser University, Vancouver, BC.

Johnston, M.E. (1995) 'Patterns and Issues in Arctic and Sub-Arctic Tourism', in C.M. Hall and M.E. Johnston (eds) *Polar Tourism: Tourism in the Arctic and Antarctic Regions*, pp. 27–42. John Wiley & Sons, Toronto, ON.

Knill, G. (1991) 'Towards the Green Paradigm', *South African Geographical Journal* 73: 52–59.

Lawton, L.J. (1995) *A Status Report of Protected Areas in Saskatchewan*, 2nd edn. Policy and Public Involvement Branch, Saskatchewan Environment and Resource Management, Regina, SK. Technical Report 95–1.

McIntosh, R.W., Goeldner, C. and Ritchie, J.R. (1995) *Tourism: Principles, Practices, Philosophies*, 7th edn. John Wiley & Sons, Toronto, ON.

MacLean, A. and Rounds, R. (1991) *An Analysis of the Population of Agro-Manitoba*. Rural Development Institute Report Series 1991–3, Brandon University, Brandon, MB.

Manitoba Environment (1993) *State of the Environment Report for Manitoba 1993*. Winnipeg, MB.

Manitoba Environment (1995) *State of the Environment Report for Manitoba 1995: Focus on Agriculture*. Winnipeg, MB.

Markusen, A. (1985) *Profit Cycles, Oligopoly, and Regional Development*. MIT Press, Cambridge, MA.

Matrix Management. (1991) *Manitoba Wildlife Viewing Tourism Study*. Matrix Management Consultants, Winnipeg, MB.

Nilson, R. and Weaver, D. (1992) 'Planning for Social Change: Major Demographic Trends in Saskatchewan', *Recreation Canada* 50, 2: 28–33.

Nilson, R., Weaver, D. and Yoshioka, C. (1996) 'Leisure and Aging in Southern Saskatchewan', *Great Plains Research* 6, 1: 85–104.

Orams, M. (1995) 'Towards a More Desirable Form of Ecotourism', *Tourism Management* 16: 3–8.

Rymer, T. (1992) 'Growth of U.S. Ecotourism: Its Future in the 1990s', *FIU Hospitality Review* 10: 1–10.

Scace, R.C. (1993) 'An Ecotourism Perspective', in J.G. Nelson, R. Butler and G. Wall (eds) *Tourism and Sustainable Development: Monitoring, Planning, Managing*, pp. 59–82. Heritage Resources Centre Joint Publication No.1, University of Waterloo, Waterloo, ON.

SED (1991) *A Tourism Strategy for Saskatchewan*. Saskatchewan Economic Development, Regina, SK.

SED (1993) *Tourism Saskatchewan Indicators Report 1993*. Saskatchewan Economic Development, Tourism Information Unit, Regina, SK.

Shaw, W. and Mangun, W. (1984) *Nonconsumptive Use of Wildlife in the United States*. US Dept. of the Interior, Resource Publ. 154, Washington, DC.

Sherman, P.B. and Dixon, J.A. (1991) 'The Economics of Nature Tourism: Determining if it Pays', in T. Whelan (ed.) *Nature Tourism: Managing for the Environment*, pp. 89–131. Island Press, Washington, DC.

SRTEE (1992) *Conservation Strategy for Sustainable Development in Saskatchewan*. Saskatchewan Round Table on Environment and Economy, Government of Saskatchewan, Regina, SK.

Stabler, J.C. and Olfert, M.R. (1992) *Restructuring Rural Saskatchewan: The Challenge of the 1990s*. Canadian Plains Research Centre, University of Regina, Regina, SK.

Stabler, J.C., Olfert, M.R. and Fulton, M. (1992) *The Changing Role of Rural Communities in an Urbanizing World*. Canadian Plains Research Centre, University of Regina, Regina, SK.

Statistics Canada (1992) *91 Census: Agricultural Profile of Canada. Parts 1 and 2*. Catalogue Nos. 93–350 and 93–351, Ottawa, ON.

Statistics Canada. (1993a) *91 Census: Industry and Class of Worker*. Catalogue No. 93–326, Ottawa, ON.

Statistics Canada (1993b) *91 Census: Profile of Urban and Rural Areas, Part A*. Catalogue No. 93–339, Ottawa, ON.

SWWA (1996a) *Ecotourism in Saskatchewan Report I: State of Resource*. Saskatchewan Watchable Wildlife Association, Saskatoon, SK.

SWWA (1996b) *Ecotourism in Saskatchewan Report II: A Working Strategy*. Saskatchewan Watchable Wildlife Association, Saskatoon, SK.

Turner, B., Hochbaum, G., Caswell, D. and Nieman, D. (1987) 'Agricultural Impacts on Wetland Habitats on the Canadian Prairies, 1981–85', *Transactions of the Fifty-second North American Wildlife and Natural Resources Conference* pp. 206–15.

Weaver, D.B. (1993) 'Ecotourism in the Small Island Caribbean', *GeoJournal* 31, 4: 457–65.

Weaver, D.B. (1997) 'A Regional Framework for Planning Ecotourism in Saskatchewan', *Canadian Geographer* 18: 357–65.

Weaver, D.B. and Fennell, D. (1997) 'The Vacation Farm Sector in Saskatchewan: A Profile of Operations', *Tourism Management* 18, 6: 357–65.

Weaver, D.B., Glenn, C. and Rounds, R. (1996) 'Private Ecotourism Operations in Manitoba, Canada', *Journal of Sustainable Tourism* 4, 3: 135–46.

Wilson, M. (1987) *Nature-Oriented Tourism in Ecuador: Assessment of Industry Structure and Development Needs*. Forestry Private Enterprise Initiative Working Paper No. 20, North Carolina State University, Raleigh, NC.

Ziffer, K. (1989) *Ecotourism: The Uneasy Alliance*. Conservation International, Los Angeles, CA.

10

VILLAGE-BASED TOURISM IN THE SOLOMON ISLANDS

Impediments and impacts

Adrian J. H. Lipscomb

Among early writers on the problems of Less Developed Countries (LDCs) was J.H. Boeke, who proposed a dualistic model of development. Boeke focused his attention on Java, but suggested that 'the economic problems of Indonesia are typical for a large and important part of the world' (1953: vi). He theorised that the less developed state of the world's poorer countries resulted from the conflict between their traditional village economic activities, which he believed to be fatalistic and unresponsive, and Western, materialistic economic thought imposed by the colonial powers which dominated the LDCs of his era. Effectively Boeke's use of the term dualism encompassed the relationship between what we now commonly term the formal and informal sectors of such countries' economies.

Boeke's theories became unfashionable in time, and his arguments were criticised as being somewhat teleological and simplistic in their use of cultural factors to explain economic disparities. With the more recent promotion of tourism as a viable economic activity in some LDCs, however, this formal/informal dichotomy is worth reassessing, particularly in light of the facts that, first, tourism is a product of the affluence and improved travel systems of the developed countries, and is not an activity familiar to inhabitants of most LDCs, and, second, that traditional villages in many LDCs are nevertheless coming more and more frequently in contact with tourism and tourists through the growth of adventure tourism, cultural tourism and ecotourism.

A relevant case study in this context is the Solomon Islands in the south-west Pacific. This chapter will attempt to explore the impediments and impacts associated with the development of a village-based tourism industry in the Solomon Islands – a form of tourism which seeks directly to impose a formal economic activity on to a culture in which informal activities have previously been dominant.

The history and economy of the Solomon Islands

History

The Solomon Islands is located about 1,500 miles north-east of Brisbane, Australia, and due east of Papua New Guinea. It is the third largest archipelago in the South Pacific, covering almost 1.35 million square kilometres of sea area, and comprising an extended, scattered, double chain of about 1,000 islands. It has a population of almost 380,000, approximately 95 per cent of whom are Melanesian, with the rest being Polynesian, Micronesian, European and Asian.

Colonialism features prominently in the history of the Solomon Islands, as it does with most small South Pacific nations. A British protectorate was established there in the late 1800s, and the Solomons played a pivotal role during the war in the Pacific in 1942–3 when Henderson Airfield, near Honiara on the island of Guadalcanal, became a strategic node controlling the sea lanes of communication between the North Pacific and the Coral Sea, culminating in the naval battle of Iron Bottom Sound. The fighting raged fiercely on land, sea and air throughout the archipelago, and the wrecks of ships and planes from that time now comprise many of the dive sites which are among the main tourist attractions of the area.

The Solomon Islands was granted independence from Britain in 1978, and is now a member of the British Commonwealth.

The economy

With a large *village-based* population, much of the Solomon Islands' economic activity takes place in the non-formal, or subsistence, sector. However, this is increasingly coming under threat for a variety of social and structural reasons, notably greater access to Western consumer goods, discontent – particularly among the youth – caused by increasing exposure to Western media (many villages now have VCRs and radios), and more frequent travel by villagers to the capital, Honiara. Nevertheless, the subsistence village lifestyle provides a 'safety net' for the wider population.

Statistics on economic performance in the Solomon Islands are scanty. The 1994 *Annual Report of the Central Bank of the Solomon Islands* commented:

> the measurement of GDP data in Solomon Islands has deteriorated in both coverage and timeliness over the last several years . . . Although official GDP data are not available for Solomon Islands in the last two years, several estimates made by international institutions and agencies notably the World Bank and the IMF indicated that GDP growth over the last several years has been around 3 per cent compared to the population growth of 3.5–4.0 per cent. According to these sources, per capita GDP has declined in the last ten years or remained stagnant at the level they were a decade ago.
>
> (CBSI, 1995: 9)

As a developing nation, the Solomon Islands is the recipient of large amounts of foreign aid and assistance, particularly from Australia, New Zealand and Great Britain. In 1994 Foreign Exchange Current Receipts totalled SI$715 million, of which nearly SI$69 million derived from 'Foreign Aid, International Organisations, Gifts, Donations, Pensions, Legacies and Churches and Charitable Institutions' and a further SI$26 million derived from 'Foreign Government Services' (CBSI, 1995: 94). Other less specific categories of international assistance may lift this percentage even higher.

The formal economy depends overwhelmingly on timber and fish exports. The value of timber exports in 1994 was SI$277 million and that of fish exports was SI$99 million; these two combined formed 76 per cent of total exports for that year (CBSI, 19: 99). Both these industries are overwhelmingly controlled by foreign interests: Japanese and Chinese in the case of fishing and Korean, Malaysian, and Australian in the case of logging. Both these industries have also been heavily criticised for their deleterious environmental impacts. The 1994 *Annual Report* of the CBSI (1195: 17–18) noted, 'the current unsatisfactory record of logging activities in Solomon Islands is a direct result of the lack of clear and enforceable standards of operations ... Actual production levels are estimated to be three times greater than the sustainable level.' In the case of commercial fishing, complaints of overfishing, undersized catches, and illegal fishing in areas claimed by customary rights are frequent. Furthermore, complaints that resource owners do not get a fair share of economic return from commercial logging and fishing are also commonplace.

Clearly, tourism must be considered among the limited range of alternative industries available to the Solomon Islands, should logging and fishing decline in coming years.

The tourism industry

The Solomons' tourism industry is in its infancy, and is a late developer in the South Pacific region. Visitor arrivals in the Solomon Islands in 1994 totalled a mere 11,900 (Statistics Division, Solomon Islands Government, cited in SITA, 1995). Neighbouring Vanuatu, New Caledonia, and particularly Fiji and Tahiti, are significantly more advanced in the development of tourism infrastructure and capitalisation.

The limited development of the Solomons' tourism assets continues to present a major deterrent to foreign investment in the industry. Nevertheless, as a free enterprise, market-oriented country, it has been largely the entrepreneurs who have controlled the direction of the industry. To date these entrepreneurs have, with minor exceptions, been expatriate developers, and they have concentrated on five main categories of tourism activities, as shown in Table 10.1. Indigenous enterprises have, in general, been rare, and have invariably been small scale, almost exclusively in category 2c – *small village-based or family-based resorts and rest houses*. These indigenous enterprises depend greatly on the nation's natural

Table 10.1 The five main categories of tourism industry activity presently being
undertaken in the Solomon Islands

---

1. The dive industry
2. Accommodation houses and resorts, comprising:

   a. the larger hotels and resorts, mainly based in Honiara, such as the Solomon
      Mendana, the Honiara Hotel and the King Solomon's Hotel
   b. the smaller 'up-market' resorts such as Uepi Island Resort, Vulalua and
      Tuvanupupu
   c. the small village-based or family-based resorts and rest houses such as Maqarea,
      Michi, and Zipolo Habu

3. Organised tour operators, especially Second World War remembrance tours
4. The cruise ships
5. The airlines (the largest carrier, Solomon Airlines, is, of course, government-owned)

---

tourism assets as attractants for tourists, and their establishment is only marginally
deterred by the general lack of asset development – indeed, it could be argued that
the lack of asset development works in favour of small local operators by limiting
the degree of competition presented by foreign entrepreneurs.

Tourism assets and liabilities

The tourism assets of the Solomon Islands comprise little of what would normally
be expected of a richer, industrialised nation aspiring to develop a tourism
industry. As with most LDCs, the Solomon Islands' assets consist of its inherent
attractions, rather than its conventional resources. Put simply, these are: the
islands, the water, the people and the history.

The magnificence of these assets cannot be overstated, and they provide a basis
for an impressive tourism industry. They are, however, heavily offset by tourism
liabilities and constraints. These include:

* limited available local capital to finance tourism enterprises;
* the absence of any domestic tourism market possessing significant disposable
  income (with the minor exception of the expatriate market – that is, short-
  term development workers, volunteers and contracted employees based
  temporarily in the Solomons, originating primarily from Australia, Britain
  and NZ);
* continuing environmental degradation resulting from over-logging and overfishing;
* inadequate internal transport infrastructure and private sector tourism
  development, which in turn hampers further significant development;
* remoteness from major international air routes, and the consequent high
  cost of air travel;
* the presence in most places of malaria, and the probable growing awareness
  in the travelling public of the greater significance of all communicable diseases;

- lack of training and expertise among indigenous operators;
- inadequate financial and human resources to market the Solomons as a destination;
- the initial economic advantage gained by Fiji and some other Pacific nations who developed their tourism industries earlier, thereby giving them a competitive edge;
- political and security problems on Bouganville Island (a part of neighbouring Papua New Guinea) which overflow to the Shortland Islands (the westernmost extent of the Solomon Islands) and deter tourism development there; and
- frequent customary land disputes which hinder the development of tourism enterprises in many areas.

These have been, and will continue to be, major impediments to all forms of tourism development in the Solomon Islands.

Tourism as a 'benign' alternative

Tourism in the Solomon Islands has been promoted in some quarters as a more environmentally benign alternative to logging and fishing (e.g. Berry, 1995). International organisations such as the World Wide Fund for Nature (WWF) have focused on village-based tourism with this end in mind, particularly in the Western Province. Michi village in the Marovo Lagoon is a prime example of a village seeking to develop its tourism potential specifically in order to reduce its reliance on logging. Under the guidance of WWF advisers, it has built three picturesque leaf houses on a little island near the village, for use by 'ecotourists'. It provides food, canoe transport, tour guides and cultural experiences.

Tourism in this context is proposed as a way for villages to earn much-desired cash without having recourse to selling timber or fishing rights to multinational corporations. Whether it can fulfil this role in the longer term will depend upon:

- what sort of tourism is pursued, and whether the cultural and environmental impact of the chosen tourism activities are less harmful than the logging or fishing they are meant to displace; and
- whether the political will exists in the longer term to pursue logging and fishing in sustainable ways, for without such political will the ability of tourism to ameliorate these industries' harmful environmental impacts can only be a stalling device.

Village-based tourism

The concept of *village-based tourism* in the Solomon Islands is relatively simple. The main tourism potential of the Solomons is associated with its lagoons – particularly the Marovo, Roviana and Vona Vona Lagoons around New

Georgia. Many villages edge these lagoons, and the waters are calm and relatively safe for tourists to engage in limited water-borne expeditions. In such a context, village-based tourism usually involves tourists travelling from village to village by motorised canoe. They undertake local bushwalking, snorkelling, historical and nature tours, cultural displays, arts and crafts, and so on at each place, organised by local villagers. Unlike conventional tourism where they return each night to a hotel or resort which reinstalls them in a more familiar culture, they sleep in one of the villages visited and thereby have an opportunity to gain deeper insights into the village lifestyle, make friends, and acquire a more meaningful cultural experience. They have the options of travelling in small organised groups, or individually, with transport they organise themselves. The accommodation is generally simple and in traditional Solomons-style, that is, leaf houses with split bamboo floors. Amenities are few, adequate cooking facilities are usually provided – although eating with the villagers is generally regarded as a good way to get to know them – and few villages presently provide toilet facilities more substantial than a 'deep drop' toilet. In some villages, visitors are required to squat on mangrove roots on the water's edge, like the locals, presenting some concerns for health and hygiene.

Village-based tourism is aimed at a particular type of tourist – one who seeks contact with the natural environment, exposure to a cultural and historical context, and active participation in outdoor and cultural activities. This is a target market which broadly encompasses adventure tourism and cultural tourism, and includes most backpackers. Significantly, however, it is not necessarily restricted to those travellers with low disposable incomes. A large part of the Solomon Islands' potential market comprises middle-to high-income earners who wish to experience 'something different', the same sorts of people who go trekking in Nepal or 'on safari' in Africa.

The basis of village-based tourism is that, generally speaking, it should be planned around existing social, cultural and environmental elements of the villages involved. It is, after all, these which form the basis of the tourism asset itself; if that authenticity is degraded or modified to suit the demands of tourism, the attractiveness of the village as a tourist destination is eroded. In some ways, this is a radical deviation in an industry which has often sought to change the status quo of a destination, at some stage in its life-cycle, to suit its own perceived needs.

Village-based tourism conforms very closely to Butler's *exploration stage* in the life-cycle of a tourist destination:

> The Exploration Stage is characterised by small numbers of tourists . . . allocentrics and . . . explorers making individual travel arrangements and following irregular visitation patterns . . . They can also be expected to be non-local visitors who have been attracted to the area by its unique or considerably different natural and cultural features. At this time there would be no specific facilities provided for visitors. The use of local

190

facilities and contact with local residents are therefore likely to be high, which may itself be a significant attraction to some visitors. The physical fabric and social milieu of the area would be unchanged by tourism, and the arrival and departure of tourists would be of relatively little significance to the economic and social life of the permanent residents.

(Butler, 1980: 6–7)

There is a strong argument to suggest, therefore, that village-based tourism should be regarded as a 'sunset' industry – once it gets much beyond the exploration stage it moves directly to the stagnation stage, and will either cease to exist or it will 'rejuvenate' into some other form of tourism. Needless to say, the potential of village-based tourism to persist at that exploration stage for a long period of time, producing a small income for the village, is in itself a compelling argument for the development of this sort of tourism.

The argument for village-based tourism

Fagence and Craig-Smith (1993) have discussed the unique problems confronting tourism planning in the South Pacific region, given the extant economic, social and geographic constraints. They advocate an unorthodox approach to tourism planning along the lines proposed by Gunn, in which, *inter alia*, selected areas are chosen for development in innovative ways. They conclude that, if the region is to 'maximise its tourism potential it needs to be developed idiosyncratically and not as a reflection of other tourism areas' (1993: 147). Clearly, village-based tourism is one such alternative.

On the other hand, Fiji is changing 'from relying on the old low-income, Australian and New Zealand visitors to a new high-income, high-profit tourism from East Asia and North America' (Grynberg, 1995). As a consequence, Fiji is moving more to the top end of the market with four- and five-star hotels and resorts, and they are planning for a projected and unprecedented tourist growth rate of 6–7 per cent per annum, the achievability of which is highly questionable. There are, however, major differences between the industries in Fiji and the Solomon Islands, not least of which relate to infrastructural development, accessibility and the incidence of malaria. It makes sense for the Solomons not to try to compete with Fiji in the same market-place but to try to establish its own niche market. This is not to say that conventional tourism enterprises such as large-scale, foreign-funded resorts and hotels should be deterred in the Solomon Islands. If undertaken sympathetically, and in such a way as to facilitate skill transfer and possibly part equity by Solomon Islanders, selected foreign-capitalised enterprises can be of considerable benefit to the local economy and the local people.

Nevertheless, it is clear that conventional tourism has already impacted upon South Pacific nations in deleterious ways, and the Solomon Islands is no exception. Sofield (1993a) has reviewed the literature on this subject and summarised the main impacts into six categories. He suggests that tourism in the South Pacific has generally:

- contributed little or nothing to the decentralisation of the population;
- generated less indigenous employment than often anticipated;
- led to 'de-agriculturisation';
- created problems of seasonality and dangers of sole source market dependency;
- had unwelcome distributional consequences; and
- exaggerated hopes for a country's balance of payments because leakage has often been high.

Nevertheless, Sofield argues that tourism *is* often more appropriate than other activities which have, perhaps ironically, been the recipient of the bulk of foreign aid. In recent years, however, significant aid money has found its way to the Solomon Islands' tourism industry; whether the areas within the tourism industry at which this aid has been directed are the most appropriate remains a vexed question, and will be discussed briefly in this chapter.

Impediments and impacts

Marketing village-based tourism

Tourism planning and marketing presents unique difficulties for LDCs. The main problem is that the prime target market, even for village-based tourism, is an international one, not a domestic one. Village-based enterprises clearly lack the capacity to market themselves internationally. If marketing is to be done, therefore, it must be done as a part of a nationally focused and government-funded campaign.

But resources are scarce for tourism marketing, and it is not an activity likely to attract funding from aid donors. Few donor countries would presently consider providing aid dollars for tourism marketing in the knowledge that it is to be spent primarily in the target markets of Australia or New Zealand. The Solomons must, therefore, source their main marketing funding in other ways. It is to their credit that they have gained some little extra revenue from a far-from-ubiquitous tourist bed tax. Even so, their marketing budget is negligible (it is also interesting to conjecture on the prospect for success in imposing a tourist bed tax on leaf house accommodation in villages).

It is an irony indeed that the Solomon Islands has one of the finest nature-based tourism products in the world, but almost no resources to market it in the country which is its greatest potential market: Australia. Conversely, and equally ironically, Australia has a will to help develop the Solomons' economy in environmentally sustainable ways via its aid organisation (AusAid), and it possesses extensive tourism marketing expertise in Federal and State Tourism

Commissions . . . but sadly the need and the solution have never been matched together.

Problems of planning and operation

In developed countries, tourism is a typical and mainstream economic activity. Almost all inhabitants of developed countries have been tourists themselves at some stage; tourism is inherently a 'Western' phenomenon. However, it is rare for a poor villager in an LDC to have been a 'tourist' – the culture of tourism is therefore less fully understood, and it is difficult to be an effective tourist operator if you have never been a tourist yourself. Unlike most other economic activities – such as agriculture and fishing – in which the indigenous technical knowledge (ITK) has some relevance to a more commercially oriented operation, there is little ITK in most villages of direct relevance to tourism operation. Villagers want to attract tourists, but they have difficulty understanding why tourists want to visit their villages and what it is they really want to do. A common misapprehension is that if they build a traditional leaf house for tourists, regardless of accessibility or location, then tourists will come to visit – and the villagers get angry when they do not. Culturally sympathetic training in tourism operation is therefore crucial, and some writers suggest that village women must inevitably play a dominant part in this (see, for example, in the context of Papua New Guinea, van den Berg et al. 1995).

Training is particularly important in the areas of food preparation, and health and hygiene. The typical Solomon Island meal comprises a large mound of boiled white rice with some slippery cabbage (similar to spinach) on top, and, if affordable, some cheap tinned tuna (tinned is considered superior to fresh!) sprinkled on top. The Western palate rapidly grows tired of such fare, and even hardy ecotourists will eventually demand a more familiar diet.

While the very nature of village-based tourism requires minimal asset development, some concessions to Western ways are advisable. These relate primarily to the design and siting of the leaf houses for tourists (to give them privacy and to minimise the mosquito problem), water supply (tourists invariably use more water than locals), and toilet facilities (the Western tourist is unaccustomed to using a mangrove root as a toilet seat). Much of the material for a leaf house can be obtained from the rainforest, but some capital expenditure is needed to purchase water tanks, minimal plumbing, fibreglass canoes, outboard motors and other equipment. Unfortunately, few villages have the financial resources for this. There are, of course, different economies of scale and 'break even' points for small village-based enterprises. Capitalisation is inevitably less, labour and operating costs are negligible, and consequently the 'break even' occupancy rates can sometimes be as low as 10 or 15 per cent rather than the 50–70 per cent for larger, more highly capitalised resorts. And, significantly, the present tourist numbers coming to the Solomon Islands are insufficient to maintain viability in the fast growing number of small village-based operations that are springing up, particularly in Western Province. Without increasing numbers of tourists to frequent the village-based enterprises, there is a danger that the villagers will lose interest in maintaining their tourist leaf houses, they will lose enthusiasm and they will 'burn out'. This points directly to the crucial requirement to begin marketing the village-based product to its target market at the same time as the product is being developed.

A further problem concerns communication. Tourists usually need the security of knowing their accommodation is booked. If tourists arrive at a village unexpectedly, the villagers will probably not turn them away, and accommodation will be found somewhere, unless there is to be a funeral or some other special event. But this uncertainty may be inconvenient for both villagers and tourists – and the villagers may be too polite to express their reservations. Telephones, of course, are not found in the villages, and mail is slow and unreliable. The alternative is radios or radio-telephones, which are generally beyond the means of most villages, and would add considerably to the capital costs of establishing such a business. So this too remains a continuing, unresolved problem.

All the above-mentioned difficulties are further exacerbated by the fact that, as in many former colonies, the Solomons' bureaucracy is bound up in regulation and 'red tape'. There are three tiers of government – National, Provincial and Area Council – from which approvals must often be sought, even for village-based projects. The public service at all three levels is not known for its speed or efficiency. Development applications, business proposals and building permits, all get 'bogged down' in the system, causing confusion, costly delays, and sometimes cancellation of projects. The Solomon Islands, to use Sofield's words:

> [has] bypassed the pioneer stage of development (i.e. that early stage of development which is usually epitomised by a lack of regulation and an ease of *getting things done*) and moved directly from a scattered village-based society with no central power structure to one where the colonial power imposed its authority and created a government apparatus that paid scant attention to traditional practices.
>
> (Sofield, 1993b: 730–1)

This 'central power structure' is, in turn, influenced by other interactive cultural impediments and attitudes.

Cultural impediments to village-based tourism

Perhaps the one phenomenon which is the greatest impediment to development of any sort in the Solomon Islands, and especially to village-based tourism, is the fact that most land in the Solomons is owned by customary title. The *Solomon Islands Tourist Development Plan 1991–2000* noted, in this respect:

'tourism facilities and infrastructure represent a form of land use. Disputes and problems may arise within land owning groups and between different groups, as well as between the latter and developers. These are complicated by the fact that people often have unrealistic expectations due to lack of a proper understanding of not only the potential benefits from tourism, but also its constraints and general working processes.

(Solomon Islands Ministry of Tourism and Aviation and the Tourism Council of the South Pacific, 1990: 43)

Ironically, the customary land tenure system is also the one main feature of Solomons' culture which has operated to ensure that land ownership has been retained by Solomon Islanders rather than foreigners. Nevertheless, it can present significant problems for business enterprises, be they locally or foreign-owned, particularly when further exacerbated by associated cultural impediments.

Jealousy is common in the Solomons, and it too can be a major impediment to development. Jealousy between individuals, between villages or between islands. Jealousy is, of course, not unique to the Solomons, but most Solomon Islanders seem to accept that as a national trait it is peculiarly strong within their islands. It is associated with that other, wider, Melanesian system of affiliation and obligation known as the *wantok* system. This literally means the *one talk* system (Pijin language) and implies that people will look after their kith and kin (or, more specifically, those who speak the same language – there are eighty-six separate languages in the Solomon Islands) first and foremost, sometimes at the expense of the wider national community. This has implications for the political system, the public service, and in any circumstance where someone is given power. It is a system akin to what Western commentators would call nepotism, and has a major impact on where, when and how development takes place.

Associated with this is the *Big Man* system, a traditional village system of power and hierarchy which now has wider applications in politics and business. Certain men are recognised as 'successful' in their communities – they may have no formal authority or power, but simply be men who lead because others wish to follow. With this recognition comes automatic respect, influence and authority, as well as certain obligations, and they may occasionally need to defend their implicit power from rivals. Because they have power, they may wish to obstruct developments they see as inimical to their own or their *wantok's* interest, albeit clearly for the wider good. This too, can have major implications for tourist developments. Undoubtedly, the *wantok* and the *Big Man* systems are among the main causes of the corruption which is endemic in the Solomon's political system at all levels, and which, in turn, is also a major handicap in business. Suggestions of political corruption are very common in Solomon Islands' media. Some are proved, some are unprovable and some are merely hinted at. Indeed, because of the cultural differences, what is seen as corruption in many Western societies is often seen as merely a part of the acceptable *wantok* or *Big Man* systems in the Solomon Islands. A typical result is a front page report in the *Solomon Star* of 10 November 1995 headlined: '$7m Scam Surfaced'. This story, based on 'a reliable confidential report', names three ministers, the Minister for Commerce, Employment and Trade, the Minister for Finance and the Minister for Home Affairs who, together with senior government officials, allegedly received SI $7 million in bribes between 1993 and 1995 from Integrated Forest Industry Ltd. a subsidiary of a Malaysian logging company, *Kumpulan Emas*. Such stories are plentiful in the Solomon Islands.

Another cultural impediment to enterprise development is the phenomenon whereby villagers start projects with high enthusiasm and earnest activity but rapidly burn out for want of continuing support. The stamina of the participants is rapidly dissipated once they encounter minor obstacles or problems. This could be ascribed to the desire for consensus rather than majority rule in village politics; objections, criticisms and misgivings are not openly thrashed out during the planning stage. Hence, when problems *are* encountered after the project has been started people's true commitment is tested and many are often found to have given only token assent. The role of the Big Man is therefore crucial, and the success or failure of any project depends on the Big Man's strength of character and personality to pull the community together towards the common objective. If he falters, then commonly so will the project, unless someone else is willing to step in. This is a cultural phenomenon which has manifested itself in such a way as to be a major impediment to business decision-making, particularly at a village level. Such national traits are, of course, not unique to the Solomon Islands. Roces and Ruces (1986), for example, have noted similar phenomena in the Philippines: rapidly waning community enthusiasm for a community project is known as *ningas kugon*, or a 'grass fire' – which burns itself out quickly; *dilihensiya* or *balato* roughly equates to the *wantok* system; and *palakasan* has a similar meaning to the power of the Big Men, or what Europeans would call 'working the Old School Tie Network'.

Christianity is a relatively new feature on the cultural landscape of the Solomon Islands. Villages generally subscribe to one or other of the Christian faiths *en masse* – usually Catholic, United Church or Seventh Day Adventist (SDA) – and rivalry is strong. Many of the villages wishing to develop tourist enterprises are SDA, and these are presented with a unique problem resulting from the nature of their religious observation. Tourists usually expect to be able to undertake activities every day, and sometimes have difficulty coping when the village 'closes down' for one day each week: Saturday the Sabbath. This leads potentially to all sorts of misunderstandings and disappointments if tourists are confined to their accommodation, with nothing to do, and no food provided.

There are two attitudes which could be adopted concerning this problem: first, that the sorts of tourists attracted by village-based tourism are culturally aware, and wish to experience the totality of village life. If the village is SDA, then the reality is that no work takes place on the Sabbath. Tourists should consequently conform to the local way of life and have no expectations of the village on that one day of the week. The second attitude is that most tourists are visiting for a relatively short period of time, and getting to and from the Solomons is expensive. Consequently their time is valuable and they do not wish to set aside a day of inactivity for religious reasons to which they themselves do not subscribe. There is some validity to both attitudes. Ideally, however, if an SDA village wishes to embark on a tourism enterprise then they should make arrangements with a nearby non-SDA village to take over their tourism responsibilities on the Sabbath. This may have a secondary benefit in that it also spreads the profits of tourism and the work experience more widely. Unfortunately, relations between villages of different faiths often will not allow this sort of arrangement. So it must remain a reality that tourism development in non-SDA villages is easier than in SDA villages.

The impact of tourism

The specific effects that tourists have on host communities generally have been well-documented (e.g. Mercer, 1995; Pearce, 1995). Such impacts in the Solomons are common to many LDCs, and are not restricted to village-based tourism. Physical problems include congestion and environmental degradation – anchor damage to coral, and the collection of rare and endangered species of molluscs by tourists are prime examples in the Solomons. Sociological problems caused by tourism include acculturation, ethnocentrism and cultural trivialisation. To minimise the possibility of conflict, particularly with visiting yachts, the Western Province Government published in 1995 a guidebook for visitors to Gizo, the provincial capital, incorporating a section entitled 'Hints for Visitors'. That section is reproduced at Figure 10.1, and it is instructive to note the range and degree of impacts discussed. Nevertheless, resentment caused by the 'demonstration effect' is particularly noticeable in the Solomon Islands, and there is a generally-held view that all *Tie Vaka* (foreigners) are rich, well-educated and highly mobile. Unfortunately, this attitude also leads to its corollary, which produces negative self-images and subtly downgrades all things to do with the Solomon Islands in the eyes of many Solomon Islanders.

Perhaps the most noticeable cultural impact of tourism on the Solomon Islands is the disappearance of many of the islands' cultural and historical artefacts. These include *bakiha*, the carved doughnut-shaped clamshells which were the equivalent of 'money' in the 'time before White Man', and ancient skulls dating to head-hunting days. *Bakiha* and skulls were often stored together in secluded *tambu* or 'sacred' sites, and are now becoming extremely rare in the Solomon Islands. Because of tourism and other introduced activities and beliefs (notably Christianity) these artefacts are sometimes undervalued by villagers who sell them to tourists or visiting 'yachtees'. Conversely, in some villages they are valued both for their cultural worth and for the ongoing income they can generate from tourists who are willing to pay a fee to visit the *tambu* sites where they are kept.

CULTURE AND HISTORY ARE IMPORTANT!

It is an offence to remove historical or "tambu" objects from the Solomon Islands. According to the Western Province Preservation of Culture Ordinance 1989, it is illegal to buy or sell traditional artifacts (ie. items from the past or tambu items). It is also an offence (under the Protection of Wrecks and War Relics Act 1980) to tamper with, damage, or remove any relics or wrecks dating to World War II.

Please help keep the material culture of the Solomon Islands
intact & where it belongs!

HINTS FOR VISITORS

HINT Number 1: Body Language
- Open displays of affection or emotion, such as kissing or hugging, are frowned upon in the Solomon Islands. Holding hands is, however, acceptable between members of the same sex. It is not uncommon to see Solomon Island men holding hands as they walk up the street.
- It is considered impolite in the Solomon Islands to pass between two people talking together. If it is not possible to go around, then the accepted posture as you pass between them is to "scrunch" up the body, and say "sorry".

HINT Number 2: Modesty
Please respect the local custom and do not wear revealing clothing or swimwear (such as bikinis for women or tight swimming costumes for men) in public places.

HINT Number 3: "Tambu"
The word "tambu" means forbidden or sacred or sometimes "no entry". Tabu sites where skulls and shell money were stored are sometimes prohibited to Europeans. The Golden Rule is: IF IN DOUBT – ASK!

HINT Number 4: When swimming or fishing . . .
- DO NOT spear-fish, collect shells or coral (including Black Coral)
- DO feel free to buy shells from villagers if offered (but not the donut-shaped traditional shell money or "bakiha").
- DO ask permission to fish using a fishing-line.
- DO NOT drop anchors on coral reefs. Use buoys if provided. Anchor on sand, not on coral gardens.

HINT Number 5: When visiting villages or islands by yacht . . .
- DO seek permission from the village chief or landowner before anchoring or visiting a village.
- If permission is not granted, DO NOT insist on staying, there are many other places you will be welcomed to visit, and there may be a good reason why you cannot stay at that particular time.
- DO enquire if a mooring fee is required, and do not tie up until you have negotiated a fee.

HINT Number 6: Tipping
Tipping is not expected in the Solomon Islands. However, almost invariably, any money offered will be accepted.

Figure 10.1 A page of hints for tourists and visiting 'yachtees' prepared by the Western Province Division of Culture, Tourism, Women's Affairs and Environment in the Solomon Islands

The Second World War was an important time in the history of the islands, and was a precursor to popular movements towards national independence. Its cultural significance to Solomon Islanders is significant, and many old men still march proudly on Remembrance Day. Wrecks of aeroplanes and ships dot the reefs and beaches around the islands, and relics are frequently unearthed on the old battlefields. These include ammunition, rifles, helmets and even old Coca Cola bottles which were discarded by the American troops. They are prized by many tourists as souvenirs, and much has disappeared over the half century since the war ended. In an attempt to control this trade in war souvenirs the National and some Provincial governments have enacted legislation to prevent their removal, and random searches are made of tourists' luggage on their departure from Henderson Airfield. Nevertheless, many Solomon Islanders consider these relics to be a lucrative source of cash, and sell them to tourists or swap them for T-shirts or other cheap goods.

These sorts of impacts are not, of course, unique to village-based tourism but are associated with tourism generally in the Solomon Islands. It is clear, however, that the greater the degree of contact between tourists and villagers, the greater the scope for impact, one upon the other. On the other hand, village-based tourism by its very nature, attracts tourists who are more socially- and environmentally-aware than conventional tourists, and their impact may consequently be more benign. This in itself is a major argument for the development of village-based tourism.

Conclusion

It is an unusual analogy, but one which I have found helpful in the Melanesian context, to compare the development of a tourism industry in an LDC with the task of lengthening a chair's legs. A chair has four legs. If some legs are lengthened without modifying all four legs equally, the chair will become lopsided and unstable. So it is with the tourism industry. The four legs can be equated to the four main interrelated activities or needs of a developing tourism industry: product development, marketing, infrastructure development, and training. If one or more of these activities is ignored or inadequately addressed the tourism industry as a whole cannot develop to its full potential – the chair will be lopsided, unstable and less functional. The analogy can be extended in other ways too. If four different craftsmen are asked to manufacture the four legs, and there is little or no communication between them, then the likely outcome will be a chair with four different sized legs – again it will be lopsided and unstable.

Co-ordination and communication between the four main players – the four 'craftsmen' of the chair analogy – require money and expertise. Indeed, in the Solomon Islands' case, there has historically been a certain tension between some of the players. As a result, the four legs of the tourism chair are seemingly being constructed in semi-isolation and they are growing at very different rates and with different funding and resource bases. There is an especially urgent need for greater co-ordination between the two 'front legs': product development and marketing. At the moment, for example, there are substantial amounts of money (New Zealand aid) being spent by the Ministry of Tourism to help develop village-based enterprises, but very little is being spent by the Solomon Islands Tourist Authority (SITA) to market this particular product to the appropriate overseas niche markets.

Sofield concluded his paper on indigenous tourism development with an admonition that:

> an appropriate path for village-based tourism in the Solomon Islands must be sought. If not identified, then Solomon Islands faces the prospect of foreign domination, local elite control, and/or government institutionalisation of the tourism sector, as has occurred in many other small developing nations.

(1993b: 747)

The truth of this statement is self-evident to anyone who is familiar with the Solomon Islands – but it is a truth which is valid for the Solomon Islands' economy generally, not just the tourism sector. Many parts of the Solomons, and particularly its areas with high tourism potential, such as the Marovo Lagoon, have been overwhelmed in the last decade by social impact studies, environmental impact studies, development plans, feasibility studies, implementation plans and other academic intrusions (most done by well-intentioned Western consultants), none of which appears to have produced any lasting benefit for the tourism industry or the society. This could, in part, be attributed to inexperience and lack of finance, but is, I suggest, primarily due to factors implicit in the above-mentioned 'chair analogy' – there has been too much emphasis on product development and not enough on marketing, training and infrastructure development. Whether an adequate balance can ever be achieved, given the competing interests and the abundant impediments, is debatable, and there is a strong argument to suggest that (for all the reasons outlined in this chapter) the growth of tourism generally, and village-based tourism specifically, will remain limited in the Solomon Islands.

It is an unfortunate truth that the economic prospects for the Solomon Islands generally are poor, and such conditions are always ripe for plunder. Significant changes are required at both the micro- and the macro-level to set a scene suitable for the development of appropriate forms of tourism; it is questionable whether such a small and commercially inexperienced nation is capable of such changes in the time-frame available. If it is not, then there is the danger of a continued duality between a limited formal sector encompassing conventional forms of tourism which provides little for the village inhabitants, and a magnificently varied and culturally rich, informal sector which dabbles in tourism to little economic benefit. Sad to say, as Sofield has suggested, such a scenario is more likely to engender economic imperialism and social and economic elites than development in the true sense of the word. The traditional lifestyle based largely on village gardening and small-scale fishing may provide a safety net for the economic welfare of the population should the formal economy decline further in coming years. But tourism of some sort will undoubtedly persist in the Solomon Islands, with village-based tourism as one component. This component will likely develop at a sedate pace, and will not impact greatly, for the good or the ill, upon the economy or the culture – and that may, perhaps, not be such a bad thing.

References

Berry, T. (1995) 'Solomon Islands: The Economy and the Potential for Growth in Village Level Tourism and Timber Milling', paper presented at the 24th conference of economists, September, University of Adelaide.

Boeke, J.H.(1953) *Economics and Economic Policy of Dual Societies as exemplified by Indonesia*, H.D. Tjeenk Willink, Haarlem.

Butler, R. (1980) 'The Concept of a Tourism Area Cycle of Evolution: Implications for Management of Resources', *Canadian Geographer* 24: 5–12.

CBSI (Central Bank of the Solomon Islands) (1995) *Annual Report, 1994*, CBSI, Honiara.

Fagence, M. and Craig-Smith, S.J. (1993) 'Challenge to Orthodoxy: Tourism Planning in the South Pacific', in *Building a Research Base in Tourism, Proceedings of the National Conference on Tourism Research*, at the University of Sydney, Bureau of Tourism Research, Canberra.

Grynberg, Roman (1995) 'But Where are the Investors?' *Pacific Islands Monthly*, May: 19–22.

Gunn, C. (1988) *Tourism Planning*, 2nd edn. Taylor and Francis, Philadelphia.

Mercer, D. (1995) 'Native Peoples and Tourism: Conflict and Compromise', in W.F. Theobald (ed.) *Global Tourism: The Next Decade*, Butterworth-Heinemann, Oxford.

Pearce, P.L. (1995) 'Tourist–Resident Impacts: Examples, Explanations and Emerging Solutions', in W.F. Theobald (ed.) *Global Tourism: The Next Decade*, Butterworth-Heinemann, Oxford.

Roces, A. and Ruces, G. (1986) *Culture Shock!* Times Books International, Singapore.

SITA (Solomon Islands Tourism Authority) (1995) Brochure – 'Solomon Islands Tourism', SITA, Honiara.

Sofield, T.H.B. (1993a) 'Ecotourism as an Appropriate Form of Development in the South Pacific', Conference Papers Vol. 2, *Fifth South Pacific Conference on Nature Conservation* and Protected Areas, 4–8 October.

Sofield, T.H.B. (1993b) 'Indigenous Tourism Development', *Annals of Tourism Research*, 20 (4): 729–50.

Solomon Islands Ministry of Tourism and Aviation and the Tourism Council of the South Pacific (1990) *Solomon Islands Tourism Development Plan 1991–2000*, Vol. 1, Summary.

van den Berg, P., Berman, M., Dinnas, A. and Lehman, B. (1995) *Handbook for the Women's Training Workshop for Independent Village Guest House Management*, The West New Britain Provincial Tourist Bureau, PNG.

11

TOURISM DEVELOPMENT OPTIONS IN INDONESIA AND THE CASE OF AGRO-TOURISM IN CENTRAL JAVA

Bill Faulkner

Like many countries in both the developing and developed world, Indonesia has recognised the potential role of tourism as a catalyst for economic development and, ultimately, improved living standards for communities in many regions. This is reflected in the country's five-year development plans, where tourism is designated a priority area for economic planning and development purposes. The enthusiasm for tourism may be counter-productive, however, if inappropriate tourism development options are chosen in order to pursue targets which are unrealistic and/or unsustainable in the longer term. The aim of this chapter is to explore some of the issues involved in planning tourism development in this country and to elaborate on these by discussing the specific case of the Sodong agro-tourism project in Central Java.

As indicated in Figure 11.1, Sodong is located 20 kilometres south of the provincial capital, Semarang. The northern part of Central Java around Semarang is generally regarded as a transit area for travellers, rather than as a tourist destination in its own right. Indeed, while statistics suggest that the average duration of stay among international visitors to Central Java as a whole varies between just under three days (for visitors from Hong Kong and France) and over ten days (Australia), the corresponding figure for the Semarang area has been estimated at just one day (Statistik Pengeluaran dan Pandangan Wisatawan, 1993). The Yogyakarta area to the south provides the main magnet for international and domestic visitors alike because of its better known attractions, which include the historic sites of Borrobodur and Paramban. Within this context, the challenge confronting those involved in tourism development in the Semarang region is how to increase its critical mass of tourist attractions in order to encourage visitors to either stay for a longer period during their transit or visit the area as a destination in its own right. The Sodong project is seen to be an important element of this strategy.

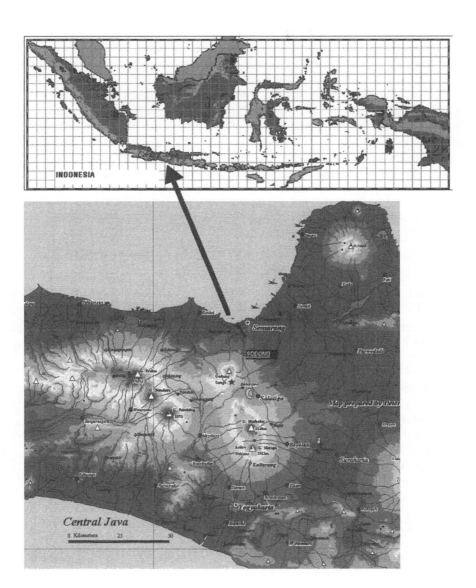

Figure 11.1 Sodong, Central Java, Indonesia

The chapter begins with a brief overview of tourism development in Indonesia, where the government's aspirations regarding the performance of the tourism industry are described, and the constraints and challenges affecting the realisation of its targets are identified. As much of the recent comment on managing the growth of tourism suggests, sustainability has become the 'new paradigm' for planning in this area (Bramwell and Lane, 1993), and it is arguable that this philosophy is particularly relevant as a framework for analysing tourism development problems in developing countries. Insights are therefore drawn from the diverse literature on the principles of sustainable tourism development to establish a foundation for identifying and evaluating tourism development options in Indonesia. The Sodong agro-tourism project is then examined as an example of an approach to tourism development which has considerable potential as a means for achieving sustainable tourism development outcomes in this context. However, the examination of this case study also reveals the obstacles and risks involved in such projects, and how sustainable development objectives devised by planners can be frustrated by the pressures of the commercial environment.

Tourism development in Indonesia: goals, assets and challenges

Indonesia is a country of 200 million people, which is emerging as a major economy in the Asian region. Sustained economic growth of around 6 per cent over the last twenty-five years has been accompanied by increasing urbanisation of the country's population and improved standards of living. While oil and gas exports have provided the main impetus for economic development to date, the need for a more diversified economic base has been recognised and tourism is seen as having the potential to play an important role in this respect. Indeed, tourism was identified as a key industry development area in the Indonesian government's Sixth Five-Year Plan (*Repelita VI*, 1994–98) and tourism has been targeted to become the next most important foreign exchange earner after oil and gas. This would require tourism to overtake the clothing and textiles sector, which is currently the second most important exporter after oil. In response to these broader economic goals, the Indonesian Department of Tourism, Posts and Telecommunications has set a target of between 6 million and 6.5 million international visitors in 1998. This compares with an estimated 4.3 million visitor arrivals in 1995 and thus requires an annual average increase of around 12 to 15 per cent over the 1996–98 period.

While this target is ambitious, there are many features of Indonesia which have a substantial potential for tourism development. Geographically, Indonesia is an archipelago comprising 17,000 islands. Beaches, islands and marine environments therefore stand out among the landscape features that are relevant to tourism, and these are complemented by mountains, volcanoes and rain-forests. However, it is the cultural landscape of Indonesia which is arguably the main tourism asset of many parts of the country.

With the succession of religions that have influenced the country's history over the last 2000 years (including Buddhism, Hinduism and Islam), a rich cultural heritage is reflected in the historic sites, art/craft and performing arts (dance and music) of many regions. Meanwhile, despite the industrialisation that has occurred over the last twenty years, 75 per cent of the population are still engaged in agriculture, and the continuation of traditional village lifestyles and land use practices in many areas provides opportunities for cultural tourists to pursue their interest in such communities. With the combined effects of the modernisation process referred to above and the possibility of adverse impacts associated with poorly managed tourism development, however, the tourism resource value of traditional villages could be very quickly eroded. This and related issues will be discussed in more detail subsequently – initially through the examination of the principles of sustainable tourism development and then in the context of Sodong agro-tourism case study.

Indonesia's geographical proximity to the major growth economies of Asia has contributed to the substantial growth in international visitors received in recent years and the prospects of strong growth in the inbound sector hinge on the continued growth of these markets. Asia-Pacific countries (including Singapore, Japan, Taiwan, Malaysia, South Korea, Hong Kong, Philippines and Thailand) comprised 70 per cent of arrivals in 1995 and four of these countries stand out as the major growth markets for Indonesia in the 1990s. These are the Philippines (with 35 per cent growth between 1990 and 1995), Taiwan (28 per cent), South Korea (26 per cent) and Thailand (20 per cent). However, despite the strong growth of visitors from these countries, two other Asian countries (Singapore and Japan) remain the main two markets.

The planning of tourism development at the national level invariably involves a focus on international markets because of the contribution this sector makes to export income. However, the large and increasingly wealthy population of Indonesia points to a strong domestic market which should not be overlooked. The domestic market can underpin tourism development by boosting demand for product to levels that enable economies of scale and financial viability to be achieved, while tapping this market can contribute to improving the national balance of trade situation indirectly by providing an alternative to overseas travel. The role that domestic tourism can play in regional development and counteracting the drift of the population to larger cities is also important.

The achievement of the Indonesian government's goals with respect to tourism development will largely depend upon its success in addressing several fundamental problems. First, the development of tourism on the scale envisaged will require a substantial expansion of infrastructure not only in those sectors directly concerned with servicing tourists' needs (e.g. transport systems and tourist accommodation), but also in those areas that are necessary to ensure the overall safety and comfort of visitors (e.g. waste disposal systems and health services). While the growth of Indonesia's economy may provide some of the capital for the development of this infrastructure, the scale of the investment required suggests that foreign capital will be necessary in the short to medium term. Second, while Indonesia has a large potential labour force for the tourism industry, workers in this industry need to be properly trained to ensure that a quality service is provided to visitors. Also, the education system needs to be capable of supplying graduates with the technical expertise necessary for effectively planning and managing the industry. This point is equally relevant to the third problem area confronting the Indonesian government, which concerns the vulnerability of tourism assets and the necessity of managing tourism development in a manner which ensures these assets are not destroyed. As Dieke (1993) observes, along with the required technical resources, appropriate institutional frameworks need to be put in place if sustainable tourism development is to eventuate. In the analysis that follows, we concentrate mainly on the latter problem.

The sustainable tourism development philosophy

There is a large and burgeoning literature on the concept of sustainable tourism development and its application to varying settings. The reason for referring to this literature here is not to provide a comprehensive review, but rather to draw selectively from it in order to produce an interpretation that is relevant to the Indonesian situation in general and the Sodong region of Central Java in particular.

The first formal recognition of the principles of sustainable tourism development occurred in the World Tourism Organisation's (WTO) *Manila Declaration*. In articulating the goals of modern tourism development, this statement highlighted the importance of both natural and cultural resources to the viability of a tourist destination and the equal importance of conserving these resources in the interest of both the host community and tourism (WTO, 1980). Later, in the *Hague Declaration on Tourism*, particular emphasis was given to an 'unspoilt natural, cultural and human environment [being] a fundamental condition for the development of tourism' and the need to manage tourism development in a way that preserves the physical environment and cultural *heritage* of an area, while improving the quality of life of its residents (WTO, 1989).

Drawing on these earlier prescriptions, the tourism group of the Globe '90 Conference held in Vancouver, Canada, in March 1990 concluded that 'sustainable tourism development is aimed at protecting and enhancing the environment, meeting basic human needs, promoting current and intergenerational equity and improving the quality of life of all people' (Inskeep, 1991: 459). Accordingly, sustainable tourism development is fundamentally concerned with 'meeting the needs of present tourists and host regions while protecting and enhancing opportunities for the future' and this requires resources to be managed in such a way that 'we can fulfil economic, social and aesthetic needs while maintaining cultural integrity, essential ecological processes, biological diversity and life support systems' (Inskeep, 1991: 459).

The emerging sustainable tourism development philosophy of the 1990s can be viewed as an extension of the broader realisation that a preoccupation with economic growth without due regard to its social and environmental consequences is self-defeating in the longer term. This perspective, in turn, had its roots in the 'limits of growth' movement of the 1970s (Meadows and Meadows, 1972) and has been more recently espoused by such international organisations as the World Commission on Environment and Development (WCED, 1987). As indicated above, the relevance of this philosophy to tourism has not been lost on those involved in the development of this industry. Tourism is essentially a resource-based industry, where the integrity and continuity of the tourism product in a region is inextricably linked with the preservation of its natural, social and cultural environment (Murphy, 1985) and, as observed by Murphy elsewhere (1994: 275), 'we do not inherit the earth from our forefathers but borrow it from our children'.

Options for tourism development in Indonesia

The growth in international tourism generally has been driven by the emergence of mass tourism. Mass tourism, as described by Cohen (1972), is characterised by large volumes of tourists, purchasing highly standardised, all-inclusive packaged tours and travelling in large groups. In terms of Plog's (1974) framework, mass tourists exhibit psychocentric tendencies individually. They are relatively inexperienced, unsophisticated travellers seeking traditional destinations to engage in sightseeing, relaxation and 'sun lust' activities without being challenged too much by novel and unfamiliar experiences. They aim to maximise their experience by including many destinations in their trip and they are particularly conscious of the need to display their tourism 'conquests' to friends at home.

Auliana Poon argues in *Tourism, Technology and Competitive Strategies* (1993) that mass tourism is giving way to new tourism. The so-called 'new tourists' resemble Plog's (1974) allocentric tourists in that they are more experienced and sophisticated travellers, who are more inclined to plan their trip themselves and travel independently. According to Poon, the new tourists are more spontaneous and flexible in the way they structure their travel arrangements, and they are motivated by the pursuit of special interests such as cultural tourism, nature-based tourism or adventure tourism. They are affluent, place a priority on authentic experiences and prefer short-break, mono-destination trips.

On the basis of the above classification of tourism demand patterns, a distinction between mass and new tourism oriented developments can be visualised along the lines described in Table 11.1. Here, models of development are distinguished in terms of scale, ownership and the nature of associated attractions. Thus, mass tourism is associated with large-scale, European-style, developments that are financed and controlled by external interests and involve man-made attractions. The new tourism model, on the other hand, involves small-scale,

Table 11.1 Models of tourism development

	Old (mass) tourism	New tourism
Demand	Packaged/group tourism	Independent travellers
	Psychocentric orientation	Allocentric orientation
	Sun lust/sightseeing	Seeking a variety of special interests
Supply	Large-scale	Small-scale
	European-style services/resorts	Indigenous style services/arthitecture
	Foreign ownership/linkages	Local ownership/control
	Greater dependence on	Greater dependence on pristine
	man-made attractions	culture/environment

indigenous-style development, with local ownership and control, and a greater dependence on local culture and the natural environment.

There is a parallel between the mass tourism model and what Rodenburg (1980) has earlier referred to as 'large industrial tourism' and, at the other extreme, new tourism resembles his 'craft tourism'. Rodenburg's identification of a third alternative ('small industrial tourism'), that combines elements of the other two, highlights the fact that, in reality, we have a continuum between the two extremes. Also, what has been referred to as the new tourism model corresponds to some degree with what has been referred to elsewhere as *alternative tourism* (Krippendorf, 1987) – at least to the extent that the latter refers to small-scale tourism based on the involvement of local people and the natural and cultural assets of the area.

Despite the relative recency of the emergence of Asian markets, the transition from mass to new tourism is nevertheless discernible within this context. For instance, in an article titled 'Asia's new age travellers', Baldwin and Brodess (1993) have noted a tendency among emerging markets such as the Philippines, Thailand and Korea towards low-cost inclusive tours and long multi-destination trips; while maturing markets like Hong Kong and Singapore reveal an increasing tendency towards independent travel and shorter mono-destination trips.

While mass tourism still predominates, and will do so for some time, the implications of new tourism for the industry are quite profound. It means that the market place will be more volatile and less predictable, and there will be a significant downside effect to the extent that the economies of scale previously associated with mass tourism will be less achievable. On the other hand, new tourists are seeking authentic, meaningful experiences and will therefore be less cost sensitive. They will be prepared to pay where there is value for money.

The relative immaturity of many of the emerging markets in the Asian region would seem to imply that, if Indonesia is to capitalise on these markets, it should be providing the sort of product that appeals to the mass market. However, while there are parts of Indonesia where development for mass tourism purposes may be appropriate (and, in the case of Bali, this has already occurred), the vulnerability of cultural assets in particular suggests that the mass tourism option might not be sustainable in many areas. There is, therefore, a danger of the government's enthusiasm for tourism as a centrepiece for economic development, and its ambitious targets for increased visitor numbers, resulting in an emphasis on mass tourism irrespective of the suitability of individual localities. The case of Pangandaran, a fishing village in West Java, provides an example of a situation where mass tourism appears to be inevitable despite its inappropriateness in terms of the interests of local residents and longer-term sustainability (Harris and Nelson, 1993).

The features of the old (conventional mass) tourism model of development are well known and, in the Indonesian context, are exemplified in the Bali case. References to the excesses of conventional mass tourism in general and Bali's development in particular (e.g. Hanna, 1972) might predispose many to concluding that the mass tourism model is inconsistent with the principles of sustainable tourism development. This is frequently so because the concentrations of tourist activity associated with mass tourism destinations obviously involve the risk of social and environmental carrying capacity being exceeded. Also, it is inescapable that the mass tourism option requires large (usually foreign) capital investment and is more heavily dependent upon imports to provide clients with the services they expect. As noted by Britton (1982: 355), this form of tourism development 'reinforces dependency on, and vulnerability to, developed countries' and 'results in tourists at a destination being channelled within the commercial apparatus controlled by large-scale foreign and national enterprises which dominate the industry. The greatest commercial gains, therefore, go to foreign and local elite interests.' Higher economic leakages are associated with this form of tourism owing to increased import dependency and repatriation of profits.

Conversely, the smaller-scale new tourism option has been commonly regarded as being more consistent with the interests of the local community by virtue of the higher returns associated with the higher degree of local control and a greater dependence on local technology and other inputs (Cater, 1993; Rodenburg, 1980; Weaver, 1991). Furthermore, this form of tourism is considered to be more environmentally appealing on the grounds that less tourism infrastructure is required to support it (Krippendorf, 1987: 37). This combination of features, along with feasibility of implementing local tourism development strategies largely on the basis of mobilising local resources, has encouraged many authors to advocate the adoption of the new (or alternative) tourism approach in developing countries (Cater, 1993; Craig-Smith, 1994; Hawkins, 1994; Hill, 1990; Rodenburg, 1980; Weaver, 1991). However, as Hill (1990) has noted in the case of Costa Rica, the relatively low yields associated with this form of tourism suggests that we can't rely on this approach to tourism development exclusively.

While the authors referred to above have been generally critical of the mass tourism model, others have been less scathing. In the Bali case, for instance,

McKean (1989) has commented on the apparent resilience of Bali's local culture and its adaptability to tourism, while Wall (1995) has emphasised that the erosion of traditional culture attributed to tourism is often mainly a reflection of more general social change associated with modernisation. Also, Noronha (1979: 193) highlighted some time ago that how one evaluates Bali is a matter of perspective. Referring to varying perceptions of negative developments associated with the growth of tourism, he points out that:

> For some observers these are symptoms of moral decay that mass tourism is said inevitably to bring; for others, they are mere warts that in no way detract from the greater benefits of tourism in Bali.

He goes on to suggest that 'Tourism for the Balinese has served to reinforce cultural traditions and ethnic identity.' In line with previous comments regarding economic marginalisation, however, he adds that there is a real danger of the Balinese response being increasingly determined by outside interests, with the local people being 'relegated to the role of functionaries and employees' (Noronha, 1979: 202).

Just as this is not the inevitable consequence of mass tourism, it can equally be claimed that, contrary to conventional wisdom, the new tourism alternative may not necessarily be compatible with sustainable tourism objectives. This is especially so in situations where new tourism encourages the dispersion of tourists to areas where even relatively low levels of visitation can have irreversible negative impacts on the host region. Thus, as Jarviluoma (1992: 120) suggests, 'there is good reason to suspect that *alternative tourism* spreads tourism and, at the same time, environmental, social and cultural problems to new and fragile areas'. An example of this problem has been described by Zurick (1992), who refers to the way ecotourism and adventure travel in Nepal have resulted in the intrusion of external influences in traditional communities. In another example, Weaver (1995) has drawn attention to the danger of the overly enthusiastic promotion of ecotourism resulting in environmental carrying capacities being exceeded in Montserrat. Elsewhere, Butler (1990) and Wheeler (1993) question the desirability of the new tourism model on the grounds that this form of tourism is elitist and involves incursions into formerly unspoilt areas, where even very small numbers of tourists can have a harmful effect, especially if their activities involve very close contact with locals.

With regard to the latter point, Saglio (1979) has emphasised the danger of alienating local residents in village-based tourism projects where they may become conscious of being looked at, much like zoological specimens or museum pieces. In brief, an emphasis on the new tourism model could be counter-productive in terms of sustainable development outcomes if it simply exposes vulnerable and previously undeveloped areas to the risk of adverse, irreversible impacts of tourism. Conversely, the much maligned mass tourism model could be consistent with sustainability objectives if the adoption of this approach contains tourist activity within a confined space and thus, in effect, quarantines the impacts of tourism and renders these impacts more manageable. Also, it is arguable that the mass tourism model can contribute to conservation objectives by diverting tourism traffic away from environmentally or socio-culturally fragile areas, where even low visitation levels can have detrimental impacts.

From the above discussion it is clear that, despite the intuitive appeal of the new tourism model, this approach is not necessarily more consistent with sustainable tourism objectives than the old tourism model. On this point, Butler (1992: 35) in particular has warned against making 'simplistic and idealised comparisons of hard and soft, or mass and green [new] tourism, such that one is obviously undesirable and the other close to perfect'. He adds that such an approach is 'not only inadequate, it is grossly misleading', and that 'mass tourism need not be uncontrolled, unplanned, short term, or unstable; and (new) tourism is not always and inevitably considerate, optimising, controlled, planned and under local control'. Similarly, in his reference to the relative merits of eco-tourism and mass tourism, Cater observes that:

> Ecotourism may well result in more profound and lasting changes. Unless it is properly managed, the impacts of ecotourism may be worse than those of mass tourism to clearly defined and confined resorts . . . Different types of tourism will suit different markets and destinations at different points in time . . . mass tourism fulfils an important function in certain circumstances.
>
> (Cater, 1993: 89–90)

It might be argued, therefore, that the achievement of sustainable tourism objectives in a particular region is not so much dependent upon the model of tourism development that is adopted, but rather on the establishment of a regime of planning and management systems which are effective in controlling the impacts of tourism, while ensuring that benefits to the local community are maximised both in the short and longer term. The model of tourism development that is appropriate depends on such considerations as: the range of tourism assets the region possesses and their potential appeal with respect to mass or new tourism markets; the susceptibility of the area to adverse social, cultural or environmental impacts associated with different levels of tourist activity; and the ability of local authorities to install the management systems required to control these impacts within the framework of the mass or new tourism options.

Adapting sustainable tourism development principles to the northern Central Java context

Even to the casual observer, it will soon become apparent that, like other parts of Indonesia, the northern Central Java area has a rich environmental, cultural and historical heritage which has the potential to provide the basis for a substantial tourism industry. The challenge confronting those responsible for tourism planning in the region is to develop an approach which is consistent with key elements of the sustainable tourism development principles outlined in the previous section.

Whatever particular theme or variant of sustainable tourism development principles may be favoured on the basis of an analysis of the literature, however, the successful implementation of these principles is dependent upon the extent to which they are understood and accepted by key stakeholders and the community at large in the region concerned. In other words, the general set of principles provided by the literature needs to be recast or 'contextualised' in order to be rendered more compatible with variations in the needs, attitudes, aspirations, institutional framework, circumstances and capabilities that occur at the local level. For the purposes of formulating a code of sustainable tourism development applicable to the Central Java setting, therefore, a workshop involving key representatives of the region was convened under the auspices of the Paribud-Ekoling project (Faulkner, 1995).

As outlined by Bramwell and Lane (1993: 2), the basic principles of sustainability in general involve:

- the adoption of holistic planning and a strategic approach;
- an emphasis on preserving essential ecological processes;
- recognition of the need to protect both human heritage and biodiversity;
- ensuring that productivity can be sustained over the long term for future generations; and
- equity in the distribution of the benefits derived from development.

These principles were used as a stimulus for framing the following set of guidelines for tourism development planning in Central Java.

- A holistic, planned approach is necessary. This implies:
    - effective involvement and coordination of all stakeholders who have a role to play in tourism development and/or are potentially affected by such development;
    - the integration of all facets of the destination resource base in order to achieve synergies in the delivery and management of the tourism product;
    - sensitivity to the cultural diversity of Central Java in the framing and implementation of plans.
- The importance of protecting the cultural and environmental assets of the region should be recognised within the context of the three tenets of planning currently applied to the development of individual assets – '*digali, dilaksankan* and *dikembangkan*' (to be discovered or revealed, to be implemented and to be developed).

- The forms and level of development that are promoted should ensure that:
  - productivity can be maintained indefinitely;
  - appropriate returns (defined in terms of living standards, infrastructure and recreational facilities) accrue to local communities and their interests are protected;
  - diversity and uniqueness of tourism product in specific localities is achieved.

- An approach based on these guidelines is necessary to achieve satisfaction among visitors, and thus the longevity of the product, to the extent that traditional culture and the natural landscape are important aspects of Central Java's attractiveness as a destination. The enjoyment of these features will be undermined if poor planning or inefficient management results in the deterioration of these assets or inequities that provoke a hostile community response.

As a consequence of the assessment of the region's tourism assets carried out in the Paribud-Ekoling project and the situational analysis that followed, it was concluded that the new tourism model provided the most appropriate approach to tourism development in terms of the above principles. In particular, it was noted that:

- The continuation of traditional land use practices and village lifestyle in many parts of the region represented one of its key tourism assets, and the new tourism model provided the most effective means of developing this in terms of market potential, quality of life and impact management considerations;
- Compared with other parts of Indonesia and competing international destinations, the beaches and coastline of the area offer little opportunity for the development of a competitive mass tourism oriented resort destination;
- There are significant heritage sites in the region (e.g. remnants of early colonial Dutch settlement in and around Semarang, the Ambarawa Railway Museum), but these are generally not suitable for levels of visitation similar to those that have been experienced at Borrobodur and Paramban;
- As with most other parts of Indonesia, the culture of the region (as reflected in music, dance, costume, religion, art and craft) is distinctive and potentially appealing to visitors interested in authentic cultural experiences. Given the importance of such experiences to the new tourism market, this feature of the area reinforces the orientation towards the new tourism model implied by the previous points;
- The infrastructure requirements of mass tourism cannot be satisfied in the short to medium term without distorting public sector priorities and detrimentally affecting progress towards higher living standards for the resident population; and

- In line with the Indonesian government's 'Tertinggal' (ITD) programme objectives, small-scale tourism development options that maximise local inputs to infrastructure development and service provision have the dual advantages of enhancing the involvement of small communities in the economic development process and increasing net economic gains from this process.

The Sodong agro-tourism project

While there are many small-scale operations offering services to the trickle of more adventurous backpackers passing through the area (i.e. Cohen's explorers), at this stage there is only one publicly sanctioned, high-profile project that stands out as a potential model for future development. This is the Sodong agro-tourism project, a Semarang City Council sponsored project on a 350 hectare site 20 kilometres from Semarang. A description of the Sodong project provides some insights into both the potential of this approach to tourism development and the obstacles to its successful implementation.

The area encompassed by the Sodong agro-tourism project is currently dedicated to mixed farming, with a combination of lowland padifields and annual crops, and tree plantations in the upland areas. A concept plan for the project encompasses a range of farming and related village activities which visitors can observe. These are summarised in Table 11.2.

Table 11.2 Sodong agro-tourism project activities

Category	Examples
Agriculture	Sowing, maintenance and harvesting of rice (including fish breeding in rice padis) Seasonal plantations (watermelon, corn, vegetables)
Plantation	Durian, mango, banana, coconut, nutmeg
Animal husbandry	Cattle breeding Chicken farming
Aquaculture (focused on proposed artificial lake)	Fish farming Recreational fishing and water sports
Gardening	Flowers and orchids
Forestry	Timber production in upland areas. Camping facilities to be located in forest areas
Processing and village-based activities	Processing of specific food products (salted egg, smoked fish, banana marmalade) Traditional meals Handicrafts Fruit and vegetable markets

A typical itinerary for a visitor during their trip to Sodong might therefore involve:

- observation of various farming practices (traditional padi sowing and harvest, tree plantations and produce, animal husbandry, etc.);
- witnessing of and/or participation in the processing of agricultural produce (e.g. threshing of rice);
- observing traditional fishing and fish farm techniques in the (planned) man-made lake;
- recreational fishing in the lake;
- visits to villages to meet local villagers and observe their way of life;
- observation of traditional food preparation and sample locally produced food;
- observe the production of local handicrafts; and
- overnight stay in traditional accommodation.

Visitors to the area can therefore observe traditional farming practices and examples of a village culture/lifestyle which will become increasingly rare in Indonesia as modernisation proceeds. Indeed, by placing an economic value on the traditional farming practices and way of life, tourism has the potential to provide a mechanism for conserving these important aspects of the area's history. In effect, therefore, Sodong could become a 'living museum' which is equally relevant to the increasingly urbanised domestic market as it is to the international cultural tourist.

The range of experiences Sodong can offer is likely to appeal to the more discerning international tourist with an educated interest in traditional culture/land use practices and who, as a consequence, has a particular appreciation of the authentic product. As such visitors tend to travel independently or in small groups, rather than in large groups, the area's tourism potential can be developed on the basis of small-scale, locally owned and controlled enterprises. A mass tourism approach would be inappropriate because the essential traditional features of the area would be quickly degraded if large numbers of visitors were received. Meanwhile, given the interest of this market in learning about other cultures, the success of the project may depend on the availability of guides who not only have the required language skills, but also a thorough knowledge of local farming practices and village culture.

The potential of the domestic market should not be overlooked, as it is this market that might eventually determine the longer term viability of the project. Some urban based Indonesians may be interested in visiting Sodong for similar reasons as international tourists, although in their case the primary motivation will be to 'rediscover their roots'. As mentioned previously, with the continuing economic development and urbanisation of Indonesia, the traditional way of life will be marginalised and areas like Sodong will remain as 'living museums' that will become increasingly relevant to the domestic market.

In the shorter term, however, the main domestic market for Sodong is likely to be local urban residents seeking to engage in recreational activities (e.g. fishing and camping) and purchasing fresh produce. One of the key considerations in the planning process will be the resolution of the dilemma arising from this feature of the project. While the value of the project to the local community is enhanced by the recreational opportunities it provides, the compatibility of the various markets being catered for needs to be considered. That is, will the presence of large numbers of local recreationists have a detrimental effect on the experience of international tourists and those domestic tourists who are attracted by the cultural features of the area?

There are, however, other problems to be solved and risks which need to be avoided if the Sodong project is to fulfil its potential as a model for sustainable tourism development in Indonesia.

Among the problems, the ability to attract the investment capital for infrastructure development required for the project is a major obstacle. In total, nearly 10 billion Rupiah (US $4,333,333) is being sought. Costings of some of the key requirements are summarised in Table 11.3. The dam construction for the artificial lake, road and bridge construction and tourist accommodation stand out as the major items of expenditure. Visitor facilities (including display areas, toilets, picnic areas, recreational facilities and vehicle parking areas), and the upgrading of water supply and waste disposal systems are also major items.

For the project to succeed, the Semarang authorities also need to establish a management structure regime which is capable of:

- ensuring that the infrastructure development process is properly coordinated;
- protecting the interests of the host community;
- facilitating the development of the Sodong tourism product itself; and
- coordinating the efforts of local tourism agencies and providers to the degree necessary to develop and market viable tourism packages and create synergies with destinations elsewhere in Central Java.

Table 11.3 Major capital investment requirements of the Sodong project

Item of expenditure	Estimated cost	
	Millions of Rupiah	000s US $
Dam construction	1,500	650.0
Road and bridge construction	1,522	659.5
Tourist accommodation	2,500	1,083.3
Other visitor facilities	1,223	530.0
Water and waste disposal systems	110	47.7
Total of major items (cf. grand total of 10 billion Rupiah)	6,855	2,970.5

Source: Government of Semarang Municipality, 1995

Following on from the latter point, it is important to reiterate the earlier observation that, relative to other areas of Indonesia (e.g. Bali and the southern part of Central Java around Yogykarta), the Semarang region does not have a critical mass of attractions. Without the parallel development and marketing of other attractions, within and beyond the region, therefore, Sodong may experience difficulty attracting the volume of visitors required for economic viability.

The fact that the risks or threats associated with the project are numerous re-inforces earlier observations regarding the central importance of the management factor as the lowest common denominator in sustainable tourism development. Key threats include:

- The potential for negative attitudes towards tourism and tourists developing if the activities of visitors become unduly intrusive or disruptive to the normal daily lives of villagers;
- The risk of the disruptive effects referred to above being accentuated by ignorance of local social and religious sensitivities among international visitors in particular;
- The risk of income from tourism being diverted to tourism enterprises or government agencies, rather than the local community. Mechanisms need to be established to ensure that the Sodong community receives an equitable return;
- The risk of attempting to appeal to a diversity of markets and, in the process, failing to satisfy any of them owing to incompatibilities in demand. For example, success in attracting local recreationists could undermine the presentation of an authentic traditional setting for cultural tourists;
- The temptation to increase income by increasing the volume of tourists will affect the sustainability of the product by reducing the quality of the visitors' experience and placing pressure on local residents' adaptive capabilities, and thus risking an adverse reaction to tourism; and
- Increased levels of visitation will be accompanied by a greater risk of environmental degradation as a consequence of waste disposal problems, encroachments on cultivated and/or natural area, traffic noise and associated air pollution problems.

The prospects of the project proceeding would appear to have received a major boost recently, with the announcement by a local (i.e. Central Javan) business man that he is prepared to provide the investment capital required for capital works. Acceptance of this proposal, however, may be at the expense of compromising the original concept of the project and, more importantly, exposing it to several of the risks referred to above. One of the reasons behind the investor's interest is his desire to 'integrate' the agro-tourism project with a racecourse and theme park development. The mass tourism orientation of these plans is clearly inconsistent with the agro-tourism proposal. The racecourse/theme park option will not only

217

appeal to a very different market, but also the volume of visitors attracted will detract from the agro-tourism experience, produce negative social and environmental impacts and, therefore, ultimately undermine the sustainability of the product.

This predicament of the Sodong project highlights the scale of the development dilemma that could be conceivably confronting tourism planners in both the developing and developed worlds. On the one hand, destinations where the sustainability of tourism development hinges on the adoption of the 'new' or 'alternative' tourism formula often need a certain level of capital investment in order for them to become competitive and viable. On the other hand, as potential investors are understandably seeking to maximise the returns on their investment, they have a tendency to advocate a scale of development (and an associated volume of visitors) which exceeds the level that is appropriate in terms of sustainability.

In the case of Sodong, it appears that this dilemma might be resolved in two ways. The worst case scenario would see development proceeding in line with the investor's plan. If it is assumed that the racecourse/theme park option succeeds in attracting visitors, this will result in the value and sustainability of the agro-tourism product being undermined by the area being inundated by an excessive number of tourists, many of whom will only have a cursory interest in what the agro-tourism component of the destination has to offer. Meanwhile, those who have a legitimate interest in agro-tourism and authentic cultural experiences will be discouraged from visiting the area and this, combined with the previous effect, will result in the incentive for maintaining authenticity being diluted.

In the best case scenario, we would have the original concept being adhered to. However, as it is unlikely that the investor would commit himself to the project under these conditions, it may be necessary to review infrastructure requirements in order to scale down the capital investment to a level which is both more commensurate with the reduced volume of visitors that will be received and more realistic in terms of the availability of funds. Under these circumstances, an incremental approach may be appropriate.

Conclusion

With the growth of tourism, both globally and within the Asia/Pacific region in particular, this industry has the potential to make a significant contribution to the economic development of countries such as Indonesia. Apart from the opportunities provided by the emerging economies of its neighbours, Indonesia's substantial population base will provide a domestic market which will itself fuel the growth of tourism as the modernisation process continues.

There are, however, challenges that need to be addressed at both the macro and micro levels if the country's tourism potential is to be realised. Essential infrastructure needs to be expanded to support the growth in visitor numbers and, as a corollary of this, investment capital is necessary. While, as in the case of the Sodong project, some investment capital will be generated by economic growth, the scale of investment required for major infrastructure suggests that foreign sources will need to be tapped. A substantial investment in education and training will also be necessary if a satisfactory standard of services is to be provided to visitors, and if the technical skills required for the effective planning and management of tourism development are to be available.

The vulnerability of many of Indonesia's tourism assets highlights the necessity of a planning and management regime that is based on the sustainable tourism development principles. Apart from the equally formidable challenges of developing the technical capabilities and systems for implementing such an approach, however, a major challenge will be in determining the volume of visitors at particular destinations which balances the demands of economic viability with the necessity of ensuring visitation levels do not exceed social and environmental carrying capacities. Meanwhile, reconciling the demands of tourists with those of local recreationists will become an increasingly vexing problem as Indonesia's large population becomes more affluent and therefore more demanding in terms of recreational opportunities.

References

Special sources

Baldwin, P. and Brodess, D. (1993) 'Asia's new age travellers' *Asia Travel Trade* 12–17.
Faulkner, H. W. (1995) 'Paribud-Ekoling: sustainable tourism development in Central Java', *Tourism Management* 16, 7: 545–8.
Government of Semarang Municipality (1995) *Planning Agro Tourism Development of Sodong, Semarang*.
Repelita VI (1993) *Repelita VI, 1994/5–1989/9*, Keputusan Presiden, Jakarta.
Statistik Pengeluaran dan Pandangan Wisatawan Mancanegara di Jawa Tengah Tahan (1993).

General references

Bramwell, B. and Lane, B. (1993) 'Sustainable tourism: an evolving global approach', *Journal of Sustainable Tourism* 1, 1: 1–5.
Britton, S. G. (1982) 'The political economy of tourism in the third world', *Annals of Tourism Research* 9: 331–58.
Butler, R. W. (1990) 'Alternative tourism: pious hope or Trojan Horse?' *Journal of Travel Research* 28, 3: 40–44.
Butler, R. W. (1992) 'Alternative tourism: the thin edge of the wedge', in V. L. Smith and W. R. Eadington (eds) *Tourism Alternatives: Potentials and Problems in the Development of Tourism*, Philadelphia: University of Pennsylvania Press: 31–46.
Cater, E. (1993) 'Ecotourism in the third world: problems for sustainable tourism development', *Tourism Management* 14 , 2: 85–90.
Cohen, E. (1972) 'Towards a sociology of international tourism', *Social Research* 39, 1: 164–82.

Craig-Smith, S. J. (1994) 'Pacific island tourism: a new direction or a palimpsest of western tradition?' *Australian Journal of Hospitality Management* 1, 1: 23–6.

Dieke, P. U. C. (1993) 'Cross-national comparison of tourism development: lessons from Kenya and The Gambia', *Journal of Tourism Studies* 4, 1: 2–18.

Hanna, W. (1972) 'Bali in the seventies. Part 1: Cultural tourism', *American Universities Field Staff Reports*, Southwest Asia Series 20, 2.

Harris, J. E. and Nelson, J. G. (1993) 'Monitoring tourism from a whole-economy perspective: a case from Indonesia', in J. G. Nelson (ed.) *Tourism and Sustainable Development: Monitoring, Planning, Managing*, Waterloo: Department of Geography, University of Waterloo: 179–200.

Hawkins, D. E. (1994) 'Ecotourism: opportunities for developing countries', in W. Theobald (ed.) *Global Tourism: The Next Decade*, Oxford: Butterworth-Heinemann: 261–73.

Hill, C. (1990) 'The paradox of tourism in Costa Rica', *Cultural Survival Quarterly* 14, 1: 14–19.

Inskeep, E. (1991) *Tourism Planning: An Integrated and Sustainable Development Approach*, New York: Van Nostrand Reinhold.

Jarviluoma, J. (1992) 'Alternative tourism and the evolution of tourist areas', *Tourism Management* 13, 1: 118–20.

Krippendorf, J. (1987) *The Holiday Makers*, London: Heinemann.

McKean, P. F. (1989) 'Towards a theoretical analysis of tourism: economic dualism and cultural involution in Bali', in V. I. Smith (ed.) *Hosts and Guests: The Anthropology of Tourism*, Philadelphia: University of Pennsylvania Press: 119–38.

Meadows, D. and Meadows, D. (1972) *Limits to Growth*, New York: Universe Books.

Murphy, P. E. (1985) *Tourism: A Community Approach*, London: Methuen.

Murphy, P. E. (1994) 'Tourism and sustainable development', in W. Theobald (ed.) *Global Tourism: The Next Decade*, Oxford: Butterworth-Heinemann: 274–90.

Noronha, R. (1979) 'Paradise reviewed: tourism in Bali', in E. de Kadt (ed.) *Tourism: Passport to Development? Perspectives on the Social and Cultural Effects of Tourism in Developing Countries*, Oxford: Oxford University Press: 177–204.

Plog, S. C. (1974) 'Why destination areas rise and fall in popularity', *Cornell Hotel and Restaurant Administration Quarterly* 14, 4: 55–8.

Poon, A. (1993) *Tourism, Technology and Competitive Strategies*, Wallingford: CAB International.

Rodenburg, E. E. (1980) 'The effects of scale in economic development; tourism in Bali', *Annals of Tourism Research* 7, 2: 177–96.

Saglio, C. (1979) 'Tourism for discovery: a project in lower Casamance, Senegal', in E. de Kadt (ed.) *Tourism: Passport to Development? Perspectives on the Social and Cultural Effects of Tourism in Developing Countries*, Oxford: Oxford University Press: 321–5.

Wall, G. (1995) 'Rethinking impacts of tourism', *International Academy for the Study of Tourism*, Cairo, Egypt.

Weaver, D. B. (1991) "Alternative to mass tourism in Dominica', *Annals of Tourism Research* 18, 3: 414–32.

Weaver, D. B. (1995) 'Alternative tourism in Montserrat', *Tourism Management* 16, 8: 593–604.

Wheeler, B. (1993) 'Sustaining the ego', *Journal of Sustainable Tourism* 1, 2: 121–9.

World Commission on Environment and Development (1987) *Our Common Future*, Oxford: Oxford University Press.

World Tourism Organisation (1980) *Manila Declaration*, Madrid: World Tourism Organisation.

World Tourism Organisation (1989) *The Hague Declaration on Tourism*, Madrid: World Tourism Organisation.

Zurick, D. (1992) 'Adventure travel and sustainable tourism in the peripheral economy of Nepal', *Annals of the Association of American Geographers* 83, 4: 608–28.

12

LOCAL PARTICIPATION IN TOURISM IN THE WEST INDIAN ISLANDS

Lesley France

For many years tourism has been identified as a method for economic development by governments of potential and actual destination areas (Wolfson 1967, Mathieson and Wall, 1982) and especially those in developing countries such as the Caribbean islands (Ferguson 1990). Yet writers like Naipaul (1962) suggest that tourism has a flawed record in this role, because it has led to a perpetuation of economic dependency on the metropolitan core countries and thereby replicates the plantation system of an earlier colonial era. One of the questions posed here is whether tourism is a dependent, or necessarily a dependent activity. A range of theories has been evolved by academics concerned with dependency (see Forbes 1984 for a detailed analysis) and, with little adaptation, have been considered useful frameworks for the description and explanation of emerging patterns of tourism in developing countries (Britton 1980, Harrison 1992), including those in the Caribbean (Woodcock and France 1994). New approaches addressing this problem, in particular, notions of sustainability introduced by the Brundtland Report (WCED 1987) and reformulated by Agenda 21 at the Rio Conference (Kirkby *et al.* 1995), were relatively slow to spread to tourism. This chapter focuses on one aspect of sustainability, which is central to modern development thinking – the concept of participation. Participation is concerned with stakeholders and therefore has clear links with dependency. The Caribbean provides an appropriate backdrop for exploring the relationship between dependency and participation since tourism is an important economic activity in the majority of the island nation-states, but varies in its nature, intensity and history among the islands.

Towards a more sustainable approach

The Brundtland Report (WCED 1987) popularised notions of sustainable development that were slow to be adopted by the tourism industry and by many host communities. The main barriers to adoption have been the pressures of rising demand and vested interests, including the profit-motivated attitude of many of those involved in tourism, such as the multinational companies that dominate the industry. There has been an acknowledgement of the importance of economic health at a destination, together with optimum satisfaction of guest requirements (Muller 1994) and an emphasis on the need to protect natural resources and host cultures by academics (Mathieson and Wall 1982, Lea 1988, Burns and Holden 1995). However, many of the principles of sustainable development have been difficult to apply at destinations where traditional mass tourism structures have been established in which benefits are reaped from economies of scale. Unlike Brundtland, Agenda 21 touched specifically on tourism (Stancliffe 1995), which was mentioned as offering sustainable development potential to some communities, particularly in fragile environments. Tourism was also implicitly affected by the action programme, since it is an activity whose impacts are clearly influenced by the policy and management under which it operates (Stancliffe 1995).

Among the various approaches to tourism management that have been enunciated over the years, the concept of sustainability has become the most pervasive. Based largely on the dual principles of intergenerational and intra-generational equity, this approach led to an increasing belief that there should be no exploitation at the expense of ecology, implemented through the application of environmental impact assessment and, further, no exploitation of ecology at the expense of opportunity for our children's children. This becomes a political debate involving all sections of the community and consequently one in which participation is highlighted. The approach to tourism management has therefore changed, theoretically, with the role of the state as a planning agent diminished. As five-year plans, and especially physical planning, have declined, so the state has become a coordinator rather than a controller. The outcome has been a search for stakeholders (owners, workers, secondary workers) beyond government. An increasing community emphasis is also encouraged by government, by aid agencies (in developing countries), and by pressure groups and public opinion in both host and tourist originating countries that influence the policies of governments and multinational companies. This leads to participation and the state adopting the role of enabler. Some of these elements can be clearly identified within a West Indian context and provide a new management frame for tourism planning.

A framework for the study of participation in tourism

For the purposes of this study, participation is defined as the active involvement of people in decision making (Slocum and Thomas-Slayter 1995). It is also extended to include empowerment, through which individuals, households, local groups, communities, nations and/or regions shape their own lives and the kind of society in which they live (Nelson and Wright 1995, Slocum and

Thomas-Slayter 1995). Indeed, participation is a process of empowerment that helps to involve local people in the identification of problems, decision-making and implementation which can contribute to sustainable development.

When linked to tourism, participation relates to a number of arguments that will be examined in this case study. These include:

- The extent to which tourism can be seen as 'a good thing', which in turn involves the identification of those who 'gain' and those who 'lose'. This obviously links with stakeholders and therefore with dependency, mentioned earlier, and highlights the neo colonial role of multinational companies and elites.

- A review of the nature of participation and, in particular, of participation in relation to tourism. The development model developed by Pretty (1995) is a useful starting point, identifying a spectrum from the more common passive and incentive-driven forms of participation towards those which are more interactive. This accords well with the superimposed nature of tourism activity, that is frequently grafted on to an economy and society in a 'top-down' manner. However, the inevitability of such a model should be questioned and a search made for alternative models. One such alternative approach might be to begin with the concept of local decision-making and the capacity of local people to create their own enterprises. The increase and spread of new skills can broaden the effects of participation outside tourism. Within this overall picture there will be spatial and temporal variations dependent upon a range of factors including gender, age, level of education and ethnicity.

- Scale is an element worthy of separate consideration since it affects the intensity and nature of participation within tourism, especially in small, open economies such as those of the Caribbean islands. What is possible at an individual or local scale may not be feasible at national or regional level.

Table 12.1 shows Pretty's (1995) model applied to tourism and modified by extending it backwards to include an earlier phase, common in the Caribbean islands, in which the exploitative plantation management system is seen as laying the foundation for tourism development before it emerged from a colonial into a neo-colonial situation. Since Pretty clearly describes the characteristics of each type of participation he identifies in some detail, there is no attempt to do other than summarise these features here. Instead, the focus is upon applications specifically within the tourism industry. The complexity of tourism, with its many and varied actors, each with a multiplicity of motivations, both within and outside the industry itself is, in a Caribbean context, grafted on to former slave societies that have become small, independent nations. Generally, the open economies of these countries contain relatively few natural resources, other than those ideally suited to meet the current demands of large tourist-generating areas located principally in the metropolitan countries. At a macro-level, the

Table 12.1 Types of participation in tourism in relation to the West Indian islands

Typology	Characteristics	Characteristics related to tourism
Plantation	'Plantation' management system. Exploitative rather than developmental. Payment in kind. Possibly paternalist	Hotels owned by individuals or companies with socio-cultural relations associated with the management system of a plantation economy. Purely for material gain of owners. No attempt at participation by workers who are often racially and culturally distinct and separate from 'management' and certainly from owners
Manipulative and passive participation	Pretence at participation. Local employees are told what is decided	Multilateral or bi-lateral aid/loans and military investments used for tourism purposes. Some highly centralised MNCs, e.g. hotel chains, tour operators and airlines based in developed countries. Neo-colonial attitudes prevail through the use of expatriate labour, capital, technology. Those involved in tourism in all but menial capacities are likely to be expatriate or non-black West Indians who are involved to gain access to local power structures
Consultation	Locals consulted but external definition of problems and control. No local share in decision-making	Some MNCs, e.g. locally based airlines, hotel chains, where management of day-to-day affairs is devolved from metropolitan centres to local elites – both black and non-black
Material incentives	Local people contribute resources but have no stakeholding	Local employment in accommodation where there is increasing use of local expertise, e.g. locals appointed at management level. Employment in agriculture – producing for the tourism industry
Functional participation	Participation seen by externals as a way of achieving goals. Major decisions are external	Increasing use of local expertise, capital and technology/culture. Some, though often small, locally owned hotels. In larger hotels decisions often locally made but are dependent upon external factors – often on demand from overseas markets. Minority elites often most likely to be participants
Interactive Participation	Local people contribute to planning. Groups take control of local decisions	Hotels owned by local individuals or consortia of business people. Local taxi firms; locally run and owned excursion tours; locally owned restaurants and attractions. Maintenance/revitalisation of cultural events for the benefit of tourists and locals
Self-mobilisation	Independent initiatives	Local people who have amassed capital from tourism strengthen and extend their activities and also use this capital to move into a wider range of economic activities

Source: Pretty (1995).

Caribbean nation states have different levels of participation in tourism, while sectors of the industry vary, along with individual establishments within each sector. The mix of types of participation highlights the complex evolutionary situation in which level 7 (self-mobilisation) is seen as more developmental with greater interactive participation than level 1 (plantation). A range of processes, including intervention by government and outside agencies, can lead to movement along the participation spectrum towards higher, more interactive levels.

Table 12.2 shows the explanatory model illustrated by means of a range of examples drawn from the West Indian islands.

Processes that change participation in tourism

The processes that move the model on from one stage to the next are varied and complex. They do not always combine neatly to form a clear threshold at the start of each new phase. An investigation of these indicators is not yet complete, but a number of significant factors can already be identified and suggestions made for their position along the spectrum of the model.

- Ownership patterns vary according to the scale of the establishments, but clearly affect the search for stakeholders.
- Small hotels and restaurants have often been under local ownership throughout the evolution of tourism, from its early beginnings to the more sophisticated modern era. However, these control only a very tiny share of the international tourism market.
- Medium to large hotels in the Caribbean islands often began under aristocratic ownership by a local minority elite or by expatriates, falling within the plantation or manipulative/passive phases of the model. While some maintained this structure, many were taken over by multinational companies based in the metropolitan countries of North America or North West Europe. In terms of the model, this has resulted in a move towards the consultation or even material incentive phase, depending upon the policies pursued by each multinational company. Local business syndicates and/or elites have increasingly acquired some of these accommodation establishments and, among this group, West Indian minority elites have begun to act in a multinational, and almost a neo-colonial capacity. When local businesses control aspects of the industry, functional and interactive types of participation can be recognised. However, if elites begin to act in a neo-colonial capacity on other islands, it could be argued that a retrograde step has been taken and a return to the manipulative/passive or consultation phase has occurred as far as the indigenous population is concerned. But it can also be postulated that the mobilisation of capital and the entry of West Indians themselves into the international economy takes participation in tourism in

the Caribbean islands a step further along the spectrum towards the most interactive phase.

- Management patterns frequently replicate those of ownership, although local involvement in middle and senior management positions occurs earlier in the spectrum. Local, especially minority, elites are likely to be involved in management in the earliest phases of the model, while by the consultation stage the local black majority have usually achieved positions in middle and lower management.

- Autonomy is a critical factor. Colonial and neo-colonial attitudes dominate in the first stages of the model. Day-to-day decisions are likely to be local when the consultation phase is reached, but there is no stakeholding until functional participation is achieved and even then many major decisions are made externally. It is only in the most interactive phases that groups and individuals take control, and independent initiatives become feasible. Even then the highly competitive global nature of the tourism industry constrains the full autonomy of any destination area.

- Food is a good metaphor for local culture and local culture reflects the local environment, including attitudes towards management. There is often a tension between local food – whether offered in indigenous restaurants or in the creole nights scheduled weekly within major hotels – and international fare. The degree of penetration of local food, both ingredients and menus, into the tourist market is a reflection of the level and nature of participation. This is especially the case in terms of gender and indigenous/local partici-pation in the food preparation labour market. Figure 12.1 shows that local food is often produced by unskilled, female labour that is unorganised and operates on a piece-work basis. When multinational hotels and restaurants are constructed to serve the tourism market, they create enclaves that are usually dominated by skilled, male chefs who tend to be relatively highly paid and unionised. Some of these chefs may be expatriate workers, especially in the early stages of development when manipulative/passive participation occurs. However, as visitors become more sophisticated in their tastes and fashion favours 'ethnic' foods, indigenous involvement is facilitated. Participation by local male chefs and by more females in tourism establishments occurs alongside the introduction of a greater range of local ingredients and creole dishes. A higher level of secondary employment is also evident. Such involvement pushes the model on to the material incentive and functional participation stages and can be encouraged by government employment policies, especially those relating to the provision of work permits for overseas labour.

- Types of tourist and types of tourism can influence the type and extent of participation by local people as the industry evolves. According to a range of models (quoted in Pearce 1989, 20–22), including that of Thurot which was based on the Caribbean, class is an important factor, as are the person-alities of a range of travellers (Cohen 1972, Smith 1977). It is suggested by

Table 12.2 Examples drawn from the West Indian islands

Typology	Examples from the West Indies
Plantation	Ownership of early hotels and tourist attractions by the aristocracy of companies, e.g. Myrtle Bank Hotel, Port Antonio, Jamaica, was built for the benefit of passengers and executives of the United Fruit Company and run by them; St Nicholas Abbey, Barbados, an Elizabethan great house was seen by some of the early visitors. Expatriate 'aristocracy' included true aristocrats, e.g. the Lascelles family who had shares in the Caribee Hotel in Barbados in the 1960s, and leaders of society like Noel Coward and Ian Fleming, both with properties along the north coast of Jamaica that became tourist attractions
Manipulative and passive participation	Airport built on Grenada after the 1983 US invasion; ownership of hotels like Sam Lord's Castle and the Colony Club by MNCs or expatriates on Barbados. British Airways, Britannia, America Airlines along with tour operators like Thomson and Kuoni are metropolitan-based MNCs. Luxury hotels like Jalousie Plantation on St Lucia failed to bring as much local employment as hoped and obtruded into the local environment and social scene. Tourists often restricted themselves to familiar foods hence imports are significant, e.g. St Lucia. Atlantis submarine on Barbados is run and owned by white expatriates. Familiar 'international' foods imported for tourists although they can often be grown locally – few links with local farmers, e.g. Jamaica in 1950s, St Lucia in 1980s
Consultation	International car rental agencies exist, like Avis on Grenada and Hertz on St Lucia. The Coral Reef Club on Barbados has been run by an English expatriate family since 1950
Material incentives	MNC-owned hotels still important but often have locals in middle and lower management posts, e.g. Kuoni-owned Discovery Bay Hotel and the Hilton Hotel on Barbados and Ramada Renaissance Hotel on Grenada. The Nutmeg Bar on Grenada in the 1960s was owned by an American and was air conditioned, but had local employees (even at management level) and local food. Local food on Dominica. Increasing local food on Barbados, where locally grown flowers are used in hotels and supplied by shops owned by minority elites. Links begin between food in hotel and restaurants and local agriculture leading to some import substitution – variations exist among islands depending on climatic factors, marketing structures and farmers' willingness to accept risk. Local chefs begin to be appointed and pseudo-local menus appear occasionally in hotels
Functional participation	Creole landladies in 18th- and 19th-century Barbados highlight the role of owners and managers catering for overseas and local visitors at an early stage. Ocho Rios, Jamaica, in 1992, typified the large hotels since 95 per cent employed in

228

Table 12.2 continued

	accommodation were Jamaican and 84 per cent of rooms were Jamaican owned, but through management mechanisms like franchises and management contracts, 53 per cent of rooms were linked to MNCs; however most decisions were made locally. Links with local agriculture improve still further and local food is used where possible. Amount of imported food begins to decline. Increasingly more genuine local dishes, adapted for tourists' tastes, feature on menus. Ecotourists integrate, consume local food and stay in small, indigenously owned and run establishments, e.g. Dominica. Mass tourists to islands are often well-travelled and willing to try local food and local dishes and/or self-cater, buying local ingredients and eating in restaurants where local dishes can be found e.g. Grenada, Barbados
Interactive participation	Ecotourism Conferences run by the Carribean Tourism Organisation promote interactive tourism. Local businesses and syndicates own many tourism enterprises, e.g. local ownership and management of: small, budget hotels along the south coast of Barbados; 70 per cent of units in Dominica; Green Parrot and Anse la Raye Hotels on St Lucia; Morne Fendue Guesthouse on Grenada. Local operators provide land tours by bus and taxi, e.g. Dominica, Barbados. On Barbados only local car hire firms exist. West Indian elites act in a neo-colonial capacity – perhaps retrograde but mobilises capital and gains West Indian entry to international economy, e.g. Sandals Hotels based in Jamaica. Conservation and restoration of attractions for tourism, e.g. Old San Juan in Puerto Rico; Brimstone Hill on St Kitts; upgrading neglected botanical gardens on Nevis; cleaning the sea off Grenada; creation of marine park Montego Bay, Jamaica.
Self-mobilisation	Ownership of taxi firms on Barbados and Jamaica. Craft industries on Dominica and Barbados. Owners may use profits from tourism to extend tourism businesses or move into more widely available services, e.g. leisure facilities like sports, fashion, beauty; or to buy property, as on Barbados. Increase of tourists from within Caribbean, e.g. in Dominica over half of stayover visitors are from other islands, so the gap between hosts and guests narrows. Guests more prepared to follow local lifestyles in terms of accommodation and food

Sources: Aspinall (1930); Apa (1988); Bendure and Friary (1994); Bousquet (1990); France and Wheeller (1995); Graves (1965); Honeychurch (1991); Momsem (1994); Panser (1990); Pretty (1995); Shapiro (1991); Weaver (1991); Wilder (1986); Woodcock and France (1994).

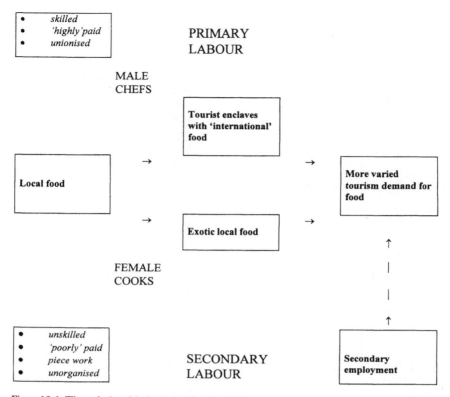

Figure 12.1 The relationship between food and labour

these authors that destinations are frequently opened up by adventurous extroverts (drifters and explorers [Cohen 1972] or elite tourists [Smith 1977]) who may be wealthy. As these destinations become known they develop more facilities specifically aimed at the tastes and demands of the market. The provision of an environment of familiarity, together with cost advantages consequent upon economies of scale, encourages more tourists, who are less adventurous and self-sufficient and who tend to be drawn from the less wealthy middle classes. Cohen's (1972) individual mass tourists or Smith's (1977) incipient mass tourists therefore dominate at this stage. This marks the beginnings of mass tourism as the inexorable movement across the spectrum of types of visitors towards the anxious, inhibited individuals organised mass (Cohen 1972) or charter (Smith 1977) tourists occurs. This group demand a risk-free environmental 'bubble' at their destination, on the journey and in the organisational process, in which familiarity dominates and the exotic must be tamed and circumscribed.

The explorers/elite tourists frequently favour alternative forms of tourism, such as ecotourism, to remoter areas and are prepared to sample local lifestyles in accommodation and food. They are therefore more attuned psychologically to the interactive end of the participation spectrum, whether this is on remoter islands like Dominica, where physical and economic circumstances have been reinforced by government policy to produce this form of development, or on the more popular islands, where visitors can stay in smaller locally owned and run guesthouses and small hotels, and local ingredients are presented in authentic dishes at meals. In contrast organised mass or charter tourists, who may be wealthy (drawn from the upper middle classes) or may be budget-conscious (from the middle and lower classes), choose the security of enclave environments either in villas or in multinationally owned establishments, where those facilities found in the metropolitan countries are demanded, e.g. air conditioning, insect-free rooms, 'international' food, the services of a courier or representative of the multinational company that frequently organises their visit to act as a buffer between local people and the tourists. Such mass tourism occurs towards the passive end of the participation spectrum and typifies a relative lack of state planning restrictions or incentives to encourage indigenous developments.

Individual or incipient mass tourists and the modified packages on which they travel are prepared to countenance contact with some local people and are therefore willing to enter an environment in which they feel less threatened by limited amounts of local participation than do the psycho-centrics (Cohen 1972; Smith 1977). They therefore occupy the middle ground of the spectrum covered by the participation model.

Conclusion

The application of Pretty's modified model to tourism in the Caribbean clearly demonstrates the extent to which participation and empowerment have emerged within the islands. Spatial and sectoral variations are apparent but, above all, the analysis stresses the dependence of tourism, and thus the majority of the island economies, upon the tourism-generating countries of the metropolitan core. The global nature of tourism and the power exerted by the multinational companies inevitably constrain the opportunities available to the indigenous population in their efforts to achieve greater participation in an industry which is, for most, their only possible route to development. In spite of this caveat, much has been achieved in the last half century. If local people and multinational companies are to work together for the benefit of both, then constructive attempts should be made to move participation towards the more interactive end of the spectrum, through investment in the training and education of local people, who should then be provided with the incentive of participation to improve effectiveness and efficiency.

The emergence of more instances of the interactive phases of participation are beginning to generate greater linkages among different sectors of the economy and need to be encouraged. In agriculture, farmers are diversifying from a reliance on highly vulnerable export-oriented production to import-substituting commodities like vegetables, fruit and flowers. Marketing systems are the biggest problem, but appropriate mechanisms are being created through joint action by farmers and government. Farmers' cooperatives on St Lucia provide examples of this. The manufacturing sector is only slightly developed. As yet many products are imported. Nevertheless, traditional handicrafts and small-scale local industries are being expanded to provide tourists with souvenirs, e.g. pottery, T-shirts, basketwork; and some basic necessities, e.g. insect repellent, soap. Locally organised schemes to enable tourists to 'meet-the-people', and the encouragement by local tourist organisations and government for visitors to take part in local festivals, bring greater contact between hosts and guests and have led to improved understanding between different societies. In consequence tourists may be more willing to accept a higher level of participation by local people in a range of skilled and responsible positions in the tourism industry, thus widening opportunities for the indigenous population. In this respect a common cultural and linguistic inheritance between hosts and guests, that has been derived from the colonial past, can be used to advantage. A number of familiar customs and traditions, and a common language within an exotic context, is precisely the controlled environment that many tourists seek (Cohen 1985, Burns and Holden 1995). As such, it is exploited by both the multinational companies and the host nations as a marketing mechanism and can be used to involve greater numbers of local people who thereby benefit in socio-economic terms.

As advantages accrue to local residents, so attitudes towards tourism and its effective management will improve. The precise emphasis on each aspect of approaches like those mentioned above varies among the multinational companies, according to the market sector they serve, and it is these variations that create different levels of opportunity for participation and empowerment on the part of local people. Ultimately then, in a demand-led industry where fashion and the marketing campaigns of multinational companies dictate the nature and direction of tourism movements, it is the opportunities granted by the market that determine the final level of participation and empowerment achieved in host areas like those of the Caribbean.

Stakeholders beyond government have historically shaped the development of the tourism industry and continue to do so. While state planning and policy could influence patterns of tourism in a manner aimed at increasing levels of participation, most island governments have in general not endeavoured to exert such an influence.

References

Aspinall, A. (1930) *A Wayfarer in the West Indies*, London: Methuen.
Bendure, G. and Friary, N. (1994) *Eastern Caribbean*, Hawthorn, Australia: Lonely Planet.

Bousquet, E. (1990) 'Hands off St Lucia's pitons', Equations Newsletter (mimeo), Tourism Concern.

Britton, S.G.(1980) 'The spatial organisation of tourism in a neo-colonial economy: a Fiji case study', *Pacific Viewpoint* 21, 2: 144–65.

Burns, P.M. and Holden, A. (1995) *Tourism. A New Perspective*, Hemel Hempstead: Prentice-Hall.

Cohen, E. (1972) 'Towards a sociology of international tourism', *Social Research* 39: 164–82.

Cohen, E. (1985) 'The tourist guide origins, structure and dynamics of a role', *Annals of Tourism Research* 12, 1.

Ferguson, J. (1990) *Far from Paradise*, London: Latin America Bureau.

Forbes, D.R. (1984) *The Geography of Underdevelopment*, London: Croom Helm.

France, L.A. and Wheeller, B. (1994) 'Sustainable tourism in the Caribbean', in D. Barker and D.F.M. McGregor (eds) *Environment and Development in the Caribbean. Geographical Perspectives*, Kingston, Jamaica: University of the West Indies Press.

Graves, C. (1965) *Fourteen Islands in the Sun*, London: Leslie Frewin.

Harrison, D. (ed.) (1992) *Tourism and The Less Developed Countries*, London: Belhaven.

Honeychurch, L. (1991) *Dominica. Isle of Adventure*, London: Macmillan.

Kirkby, S.J., O'Keefe, P. and Timberlake, L. (eds) (1995) *The Earthscan Reader in Sustainable Development*, London: Earthscan.

Lea, J. (1988) *Tourism and Development in the Third World*, London: Routledge.

Mathieson, A. and Wall, G. (1982) *Tourism. Economic, Physical and Social Impacts*, London: Longman.

Momsen, J. (1994) 'Tourism, gender and development in the Caribbean', in V. Kinnaird and D. Hall (eds) *Tourism. A Gender Analysis*, Chichester: Wiley.

Muller, H. (1994) 'The thorny path to sustainable tourism', *Journal of Sustainable Tourism* 2, 3: 131–6.

Naipaul, V.S. (1962) *The Middle Passage*, London: Penguin.

Nelson, N. and Wright, (1995) 'Participation and power', in N. Nelson and S. Wright (eds), *Power and Participatory Development Theory and Practice*, London: Intermediate Technology Publications Limited.

Panser, H.S. (1990) *The Adventure Guide to Barbados*, Edison, NJ: Hunter Publishing.

Pearce, D. (1989) *Tourist Development*, 2nd edn, London: Longman.

Pretty, J. (1995) 'The many interpretations of participation', *In Focus* 16: 4–5.

Shapiro, M.J. (ed.) (1991) *Time Out in Barbados*, Dublin: Caribbean Magazine Group, Consultants Overseas Ltd.

Slocum, R. and Thomas-Slayter, B. (1995) 'Participation, empowerment and sustainable development', in R. Slocum, L. Wichhart, D. Rocheleau and B. Thomas-Slayter, *Power, Process and Participation – Tools for Change*, London: Intermediate Technology Publications Limited.

Smith, V. (1977) *Hosts and Guests: The Anthropology of Tourism*, Philadelphia: University of Pennsylvania Press.

Stancliffe, A. (1995) 'Agenda 21 and tourism', unpublished mimeo.

Weaver, D. (1991) 'Alternative to mass tourism in Dominica', *Annals of Tourism Research* 18: 414–32.

WCED (1987) *Our Common Future*, Oxford: Oxford University Press.

Wilder, R. (ed.) (1986) *Barbados*, Insight Guide. Singapore: Apa Productions Limited.

Wolfson, M. (1967) 'Government's role in tourism development', *Development Digest* 5, 2: 50–56.

Woodcock, K. and France, L.A. (1994) 'Development theory applied to tourism in Jamaica', in A.V. Seaton *et al.* (ed.) *Tourism. The State of the Art*, Chichester: Wiley.

13

THE ECONOMIC EVALUATION OF THE ROLE OF CONSERVATION AND TOURISM IN THE REGENERATION OF HISTORIC URBAN DESTINATIONS

Mike Stabler

The issue of the regeneration of tourism destinations has often been embodied in studies of the life-cycle process (Butler, 1980; Cooper, 1992) in an analytical approach closely resembling the economic concept of technical innovation and diffusion (Ferguson, 1988). In these studies attention has concentrated on destinations whose principal function is tourism, such as seaside resorts and spas. More recently it has been recognised that major Old World cities, hitherto considered immune from cyclical effects, for example Athens, Amsterdam, Rome and Venice, and New World cities, for instance New York, Sydney and Vancouver, are showing signs of suffering in the same way as specific tourism destinations, notwithstanding that they possess a multifunctional economic base and therefore are not so heavily reliant on tourism. Even flagship tourist historic cities like Bath, Copenhagen, Edinburgh and Paris acknowledge a need to maintain their core attractions and introduce new ones and undertake refurbishment and/or renewal. Similarly, other large cities, in common with these of the first rank, contain commercial, industrial and residential areas which have declined as a result of economic, social and cultural change. However, it is increasingly being recognised that, rather than undertaking complete redevelopment, especially of central business districts, as in the past, there are many structurally sound buildings of historic interest capable of performing a number of new functions to meet the needs of the urban residents and support activities, such as tourism. This recognition has, of late, coincided with fears concerning the drift of populations from urban centres to the suburbs and the burgeoning of out-of-town retail malls and commercial and industrial development which respectively threaten the viability of high street shops and office and factory employment, which in turn undermines the local tax and services base. It is now realised that there is a need to reverse such trends in order to create a vibrant and buoyant urban economy. The initiation of schemes to revitalise and diversify the economic, social and cultural foundations of cities and towns has been reinforced by calls to pursue sustainable development. While the concept is subject to much confusion both in its definition and its implementation, having been interpreted in many ways (Pearce et al., 1989; Turner et al., 1994), a consensus is evolving as to what its implications are for achieving sustainable cities (Hunter and Haughton, 1994). An important element of sustainability is the regeneration of run-down areas in which conservation schemes make an important contribution. Such schemes in many premier league historic and modern tourist cities can act as benchmarks for other potentially attractive destinations.

The attributes of cities and towns vary so that there is a need to consider the different circumstances prevailing in each. For example, it is instructive to distinguish those which are post-industrial as opposed to being service-based, some of which may be tourism specific. It might be possible to identify common features in a review of representative cases which determine the success or failure of regeneration schemes and their impact. Although there are many examples of urban regeneration projects involving the conservation of historic buildings and areas, there are few studies of either the process or the effect. There are even fewer evaluations in monetary terms and virtually none of an economic nature in which social benefits and costs, which do not manifest themselves in the market, are included (Stabler, 1995; Allison et al., 1996). This is surprising given that in times of economic recession and endeavours to cut public sector expenditure as a long-term policy, governments need to justify support of what is apparently spending which yields no explicit market return. Moreover, organisations with responsibility for the listing of historic buildings and designation of conservation areas, such as English Heritage in the UK, ought to be concerned to show that conservation gives value for money, if only to support their application for funds to fulfil their remit.

The aim of this chapter is to examine, from a largely economic perspective, the role of conservation and the extent of its complementarity with tourism in triggering destination regeneration. Regeneration is perceived as a generic term encompassing the physical and economic improvement of an area and the revitalisation of its social and cultural life. The emphasis is not so much on explaining what has and is occurring but on demonstrating the contribution which economics can make to measuring the value of conservation and whether it confers a net benefit or cost in both monetary and non-monetary terms. The political dimension of the economic analysis of conservation is that it might be possible to show that public sector grants and subsidies are justified because they are more than outweighed by the ultimate benefits. In this sense expenditure on conservation is akin to the rationale for urban and regional policies to revive areas which have suffered structural changes and have inadequate and inappropriate social capital. Urban and regional problems and policies have been a long-standing branch of economics (Armstrong and Taylor, 1993) and concepts and methods applied in this field are equally relevant to considering the role of conservation and tourism.

The main thrust of the chapter is the examination of a case study, particular attention being paid to how conservation might act as a catalyst facilitating urban regeneration and the processes it sets in train. At appropriate points an economic interpretation and appraisal of the case study is given, including as necessary explanations of relevant concepts, primarily where it can be shown that a number of benefits and costs associated with it have not been identified and evaluated. The implications of such omissions are examined, especially where substantial and continuing public sector financial support is required. In the concluding section the practical difficulties of initiating conservation schemes which contribute to regeneration and tourism development are touched on, especially the need for cooperation between land and property owners and occupiers, developers, the commercial sector, voluntary bodies, local authorities, land use planners and central government. Some final observations are offered on the way forward with regard to the need for further research.

Conservation processes and tourism in urban regeneration

The contribution of the listing of buildings and designation of conservation areas to the regeneration of urban economics and how this interacts with tourism has not been extensively researched. It should also be noted that urban conservation is not necessarily the only engine of regeneration and that its role is uncertain. Therefore, the connection between the two essentially remains a hypothesis. Nevertheless, there are theoretical arguments, resting largely on economic urban and regional concepts, particularly relating to property, that conservation plays a positive role. Here, emphasis is placed on the dynamic rather than the static value of conservation and accordingly attention concentrates on the processes set in motion by initiatives, whether relating simply to the action of listing and designation or the implementation of schemes.

Urban dynamics

There is a strong interdependence between the value of one property and another in a neighbourhood. While expenditure on maintenance and upkeep tends to increase the value of a specific property, it also increases the value of adjacent ones. The owners of these, if they are strictly rational, will maximise their welfare or returns per unit expenditure on maintenance by adopting a strategy of minimising such spending. In economic terms the owners are 'free riders' until everyone in the locality adopts the same strategy. In short, there are at first what economists refer to as beneficial externalities, defined below in the context of the case study, conferred on neighbours by the first persons undertaking improvement. Conversely the externalities become detrimental or negative should all neglect their properties and the area deteriorates. The first movers face what is known as the 'prisoners' dilemma, a well-known example in game theory. This idea is equally applicable to residential and commercial properties and is the key to the effect of the conservation of architecturally or historically significant buildings in a designated area. Public authorities can intervene to encourage the upgrading of properties with a consequent 'knock-on' effect as can be demonstrated by multiplier analysis, explained later in the chapter. Conservation initiatives, especially if accompanied by funding, if only of a pump-priming nature, can be the trigger by which people with new skills can be encouraged to move into those areas, or skilled people persuaded to remain there. This would very likely lead to new businesses, including tourism-related ones, being located in the conserved buildings/areas and in turn facilitate the emergence of new business opportunities, additional spending and engender yet more new business creation. In effect the implication is that public spending does not simply fill a gap that the market, left to itself, would not, but that the designation and any initial public investment will act as the leverage for further private investment, both in upgrading other buildings in the area, and in the creation of new businesses/retention of businesses that would otherwise have failed or left. A longer-term process of cumulative upgrading occurs as the import of new skills reinforces the base of local entrepreneurial activity, and only improves the image of the area. Thus, by this process, cities and towns may be reverting to their former functions; for example they may have been industrial centres, but are now returning to administrative, business services and commercial activity. They also become generators of cultural services and, increasingly, as the providers of urban amenities, both enhance the quality of life of their residents and provide the economic base for urban tourism and other leisure-oriented service industries. The key difference, however, from the pre-industrial era is that the activities identified above now contribute both the major, and a growing part of urban economic output and welfare.

The cumulative effect is reinforced as there is evidence (Keeble, 1989) that the rate of new and successful business formation tends to be higher amongst immigrants and the more educated groups attracted by the enhanced living environment urban conservation can generate. Additionally, these inhabitants spend their income on urban cultural and recreational activities and personal investment in their houses. Therefore, the conservation and improvement of a group of buildings will result in higher rents and prices being obtained. Moreover, this will have beneficial spill-over effects on property values and rents in locations surrounding the initial area of environmental improvement, perhaps also altering the social character of the area. Indeed, this gentrification itself may be part of the process generating the environmental improvement.

Paradoxically, however, while the listing of an individual building has the effect of reducing its value because the way in which it can be altered is constrained, the designation as a conservation area of the surrounding location may increase its value. This occurs because the freedom of manoeuvre of the other owners is also now reduced. They cannot do things to their properties which would devalue other properties through allowing or causing the environment to deteriorate. Morover, greater certainty is introduced so that owners feel more secure and therefore confident about undertaking further investment. Thus, conservation might initiate the dynamic process to generate benefits beyond the initial public, and subsequently personal and private, investment.

The empirical context

Surprisingly, despite the many schemes, there are few case studies where full records have been kept of their cost and impact; virtually none have been continuously monitored. Exceptions are the Recite Urban Pilot Projects emanating from the European Union, of which the study considered below was one. Thus, empirically, it is very difficult to ascertain precisely what creates the conditions for successful regeneration to occur as opposed to an environmental improvement in the physical fabric. This is simple to observe and carries the danger of a kind of 'architectural determinism' in the sense that provision of an attractive built environment is alone sufficient to give rise to economic, cultural and social revitalisation. There is much anecdotal evidence but few concrete documentary records which corroborate this intuitive assessment. There are numerous examples in the literature with respect to British, French, German, Italian and Spanish cities in Europe and others in Australia and North America, all describing how regeneration occurred and that conservation was a contributory factor. However, the reporting of these schemes tends to be relatively uncritical so that it is difficult to ascertain the principal economic and policy factors. With respect to the role of tourism in complementing the conservation process, there is a substantial body of work on both its positive and detrimental aspects which are admirably reviewed in tourism texts, for example Mathieson and Wall (1982) and Pearce (1989). The literature on historic cities and heritage tourism is less relevant for it is more concerned with established environments and the marketing of these (Ashworth and Tunbridge, 1990) rather than the process of regeneration.

In a rare analysis, Pearce (1994) has examined the evidence relating to a number of these urban regeneration initiatives. The part played in them by successful urban conservation and the adaptation of the specific heritage of the local built environment has yet to be investigated, particularly the role of key factors. Although conservation would appear to be significant in a number of successful urban regenerative schemes, it is certainly not a necessary condition for success, since some schemes have not included conservation areas, nor a sufficient condition, as some urban conservation areas have not resulted in further urban regeneration. As shown in examining the case study, economic mechanisms can be elucidated which suggest how conservation may lead to regeneration, but in the examination of any successful regeneration schemes it is difficult to distinguish the variables which have resulted in that success, particularly when public sector authorities have often targeted an area and adopted several policies specifically to assist and encourage regeneration.

This then is the current position so that it is obvious much research needs to be done. While economic concepts and methods can inform the manner in which systematic and rigorous investigations should be undertaken, there is much initial 'ground-clearing' that needs to be undertaken before this stage is reached. For example, the thresholds which should be exceeded before successful regeneration occurs and the critical mass, in terms of the size of the conservation area, that is able to sustain this are as yet unknown. Furthermore, the requirements by the property sector and the role it can play have hardly been established. In addition the criteria for successful partnership arrangements between the public and private, and even voluntary, sectors have not been identified. It is therefore imperative that suitable case studies should be selected and their performance monitored over time to identify the key factors. The research conducted by Allison *et al.* (1996) has yielded a preliminary checklist of apparently successful regeneration in which a number of significant variables can be discerned, such as Covent Garden in London (architecture and tourism), the Jewellery Quarter in Birmingham (local community skills), Temple Bar in Dublin (community, culture and tourism), Chinatown in Victoria, Vancouver Island (ethnic cohesion) or Gas Town in Vancouver (historic and tourism).

The case study

The perspectives on the current state of understanding of the urban dynamics of conservation, tourism and regeneration underline the close interrelationship between the economic, cultural and social activities, local community, built environment and the wider urban economy, of which tourism is part. Accordingly, an excellent case study of urban regeneration which both illustrates and offers insights on the role of urban conservation and its interrelationship with a number of key variables, in particular tourism, is Temple Bar, Dublin. This case is also instructive in that it is characterised by not being a high-profile, flagship project. Rather it is more typical of many quarters in cities which have an important role in the local context but may have the potential to contribute to the wider urban, regional or national one. Also it is characterised by factors which may or may not be applicable in other contexts, the identification of which is important in order to avoid making inappropriate generalisations. Moreover, some factors which triggered its success may also be those which could lead to its possible decline in the future and therefore offer pointers as to dangers in following a specific line of development and management. Finally, Temple Bar has the merit of being a scheme which has been closely and continuously monitored since it was initiated so that it is possible to discern the principal factors which have influenced its development.

As indicated in Figure 13.1, Temple Bar is situated immediately south of the Liffey River between O'Connell Bridge to the east and Dame Street to the south, is 97 hectares in size, consisting of tightly clustered business premises and houses built in the seventeenth and eighteenth centuries. It was mostly derelict for many years as it had been earmarked as the site of a central transport station for the national bus company, which bought up properties as they became available in readiness for redevelopment. However, the seeds of its present revival and growth were sown when artists, musicians, clothes designers and makers, bar, cafe and shop proprietors took short leases on low rents, to join those small businesses already in the area. Impetus was given to regeneration as opposed to redevelopment by lobbying by the Temple Bar Development Council with the result that local and central government collaborated to adopt the Temple Bar Area Renewal and Development Act. By 1991, which coincided with Dublin's year as European City of Culture, 225 businesses employing nearly 1,600 people were actively trading, four-fifths in rented premises of which half paid less than £5,000 per annum in rent and taxes. By the end of 1995 a further 1,200 jobs had been created, including 150 trained in the cultural and environmental fields, 72 new businesses had started, and from a baseline of 250 residents there were 750. The pedestrian count had trebled and tourism visits were estimated to be the third best in Dublin (Bord Fáilte, 1995).

The potential for the revitalisation of the area originated when enterprising business people, at a time when no financial or delivery mechanisms were in place, decided to bid for European Union funding for its renewal involving rehabilitation of the buildings and limited redevelopment to create open spaces and cultural centres as opposed to merely conserving the existing fabric of buildings. The aim was to combine the renewal with the needs of the local community for a strong socio-cultural base.

A scheme, one of thirty-two mounted between 1989 and 1993 under the Recite programme for urban pilot projects (UPPs), was initiated in January 1991 with an initial budget of 9.4 million ECU (£6.7 million), half of which came from the European Regional Development Fund, leading to the establishment of Temple Bar Properties.

The purpose of the Temple Bar UPP was to improve pedestrian access and flows, develop marketing and communication programmes and human resources. What emerged from its implementation was the stimulation of private sector interest and innovative features and examples of good practice. The definition of clear objectives in a flexible framework plan, the establishment of a single-purpose agency with full executive responsibilities, simultaneous 'top-down/bottom-up' initiatives, mixed use (cultural, residential, retail/commercial and recreational) developed concurrently with complementary human resources development and environmental programmes, were all considered inspirational in building confidence that the scheme would succeed.

The commitment of the local residents and entrepreneurs and the Recite funding attracted additional funds and generated wider schemes. For example,

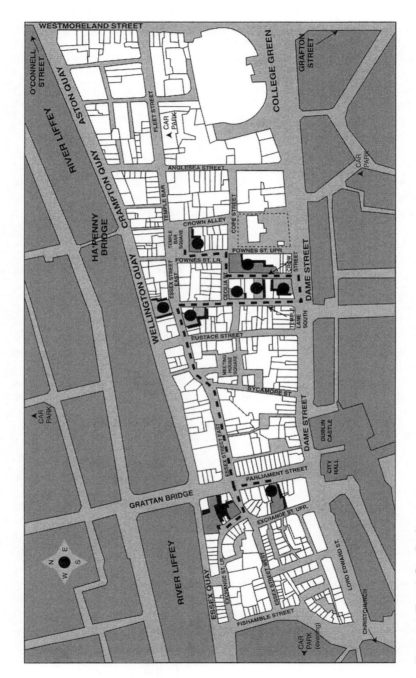

Figure 13.1 Temple Bar district

an architectural competition in late 1991 led in 1992 to the adoption of a cultural, environmental, residential and retail development. By the end of 1995 when the Recite programme was completed, total investment from national and European Union sources totalled around 120 million ECU (£86 million) and this has levered further investment from the private sector of the order of 70 million ECU (£50 million). Additionally, under the administration of Temple Bar Renewal Ltd and Temple Bar Properties Ltd, more generous tax incentives than elsewhere in the City of Dublin were given including relief from property taxes and investment allowances. Also, powers of compulsory purchase and restrictions on the assignment of leases and subletting by tenants were enforced to facilitate the attainment of UPP objectives.

In Temple Bar, special attention has been paid to accessibility to the area and within it, traffic control and protection of pedestrians, the encouragement of residence by those working in the area, and extension of open spaces. The idea of a master plan was not pursued, it being felt to be more appropriate, within what was regarded as a framework plan, to allow development of individual sites but in a structured way. The Development Programme of 1992 confirmed the intention of creating small-scale mixed use where commercial activities would cross-subsidise those which were not likely to be fully commercial.

The main features of the pilot project proposed five actions: new public squares; new pedestrian links; hard landscaping; lighting and signing; creation of a film centre and research marketing and communication. Economic growth, the magnitude of which was indicated earlier, in response to these infrastructural improvements, stimulated further public sector investment.

An assessment by the organisations involved in the Temple Bar project concluded that the primary effect was that it had acted as an excellent pump-priming scheme. It was also felt that the urban pilot project enabled good practice to be explored before embarking on a course of action, especially as initiatives were responsive to the needs of the local community and helped to build wider support for the larger programme. The pilot project increased private sector confidence at a time when a likely successful outcome was uncertain. Basic infrastructural works on roads and lighting and the carefully thought out marketing programme also encouraged private sector investment. The setting up of an information centre on the project was perceived as a key factor in this, particularly as it created a distinct image and contributed to informing not only the inhabitants but the general public, interesting them in patronising the area. Tangible signs of the beneficial effect of renewal spreading into the surrounding area were observed. Also, Temple Bar had a significant national impact in terms of changing the direction of urban development policy elsewhere. It was accepted that area-based schemes are an efficient delivery mechanism and several other programmes have been initiated. One feature accentuated in the project was its cultural base, particularly relating to the film industry which has had a stimulative effect at a national level. Indeed, investment in the arts was seen as an important trigger of revitalisation sustaining considerable employment and generating income. The Irish Tourist Board (Bord Fáilte, 1994) attributes the sizeable increase in tourists to Dublin between 1989 and 1994 to the attractions of Temple Bar.

Appraisal of the Temple Bar project

The evidence of the increase in the economic and tourism activity generated by the Temple Bar project over four years or so of the early 1990s, tends to corroborate the supposition that the very act of listing and designation gives rise to static effects and that conservation schemes, in conjunction with other initiatives, do engender continuing and increasing dynamic benefits which multiplier analysis predicts would occur. Also, the interrelationships between the public, private business and personal sectors, with consequent spill-over effects within the city of Dublin, the region and nationally, squares with the linkages which input–output analysis identifies. Furthermore, the account given above of the Temple Bar programme, reveals that many other more intangible effects occurred, such as an enhanced image, a sense of belonging, social cohesion and cultural enrichment.

Some work on the more intangible impacts of the Temple Bar programme is given in studies of it by Coopers and Lybrand (1994), Montgomery (1995a), Nexus Europe (1991) and Urban Cultures (1991) which focused on the role of the local community, especially its actions in retaining the essential character of the area to meet local needs and demand arising from the inhabitants of Dublin.

What is of interest in the studies referred to is the emphasis on the cultural dimension, defined in its widest sense to embrace the business environment, commercial activities and the social mores, as well as a very strong arts-based foundation. Montgomery (1995b) discusses how culture can promote revitalisation and an enhanced quality of life by extending economic activity into the evening, animate a place, underpin commercial enterprise, change and build a distinctive image, give a sense of place and provide a critical level and pattern of activities to draw visitors. This, as argued by Bianchini (1991), can also knit with the physical attributes brought about by the original nature of the buildings in terms of their scale, style, age and cohesiveness. Imaginative regeneration can further improve the area by sympathetic rehabilitation, street and open space design and layout, and the accompanying infrastructural upgrading, such as pedestrianisation and lighting.

However, the observations by the commentators cited above and the claims made by official utterances on the Temple Bar project are both incomplete and imbalanced. They are incomplete as neither all the benefits accruing to the immediate area nor those occurring in the City of Dublin and the country at large were included. There is an imbalance because the direct and indirect costs of the scheme are almost entirely ignored in what are essentially uncritical reports. Therefore, in order to give a more rounded appraisal of the case study, it is of value to consider it in a wider context. First, any available evidence drawn from other illustrative studies, which are relevant and indicative of the impact of conservation on urban regeneration, are briefly reviewed. Next, the limited extent of the evidence of the benefits and costs generated by the Temple Bar and other studies is examined. Then, to highlight the gaps in the current assessment of the case study, the scope and nature of the approach advocated by economists is introduced. This perspective considers the identification of benefits and costs, the methods which can be employed to measure them and the overall decision-making process which can be applied to show how a full appraisal of such projects should be conducted.

Effects are experienced in the form of both indirect monetary and indirect non-monetary ones by those who live or work in historic buildings and conservation areas. Although the evidence is conflicting or somewhat intuitive with regard to direct monetary impacts, the capital values, rental incomes and yields of buildings are certainly affected by listing. In the UK, a study commissioned by the Department of National Heritage, English Heritage and the Royal Institution of Chartered Surveyors (Scanlon; et al., 1994) indicated that the rental income of historic properties used commercially was depressed because redevelopment opportunities were denied; it was also suggested that adverse effects might spill over into surrounding areas. On the other hand research by the Investment Property Databank (IPD) (1993, 1995) found that listed commercial buildings outperformed unlisted ones. In the residential market casual observation reveals that the capital values of listed buildings are higher than comparable non-listed property. In the United States Asabere et al. (1989) and Moorhouse and Smith (1994) and Ford (1989) offer evidence of enhanced value differentials of up to 20 per cent for period properties and historic district designation. Countering this, though again hard evidence is lacking, the need for public sector support to assist occupiers of listed buildings indicates that costs are heavier than for non-listed premises. For example, not only are redevelopment options narrowed, but obligations regarding maintenance involving expensive materials, such as hand-made brick or stonework, and higher running costs, for instance heating, are imposed.

Although these studies identify that indirect effects on occupiers and/or owners of listed buildings occur, they offer no hard measurable evidence. It is acknowledged that they may derive welfare from living or working in attractive and elegant surroundings which are more likely to be on a human scale and reduce the stress often associated with modern high-rise properties. Additionally, inhabitants in, and visitors to, old cities and towns or quarters within them, are likely to experience a welfare gain from looking at beautiful heritage artefacts. On the other hand, both inhabitants and visitors can also suffer a welfare loss where buildings are neglected and become dilapidated or are ugly or out of character with those which are adjacent. Paradoxically, however, property in blighted areas can confer advantages, as occurred in Temple Bar.

The moves by those who decided to reside and set up businesses in the area were not motivated by a desire to preserve key buildings of significant historic interest, although the area consisted of properties representative of the seventeenth and eighteenth centuries, which were also largely homogeneous in character. Indirect benefits arose because of the poor state of properties. For occupiers, capital values and rental payments were relatively low so that it was possible to run viable businesses on modest sales revenue. A further benefit was that the high-density, low-rise nature of the area attracted a particular kind of entrepreneur and resident so that its social structure was uniform and cohesive. The businesses which developed also appealed to the inhabitants of Dublin outside Temple Bar so that impetus was given to the potential value of saving the area in its entirety rather than redeveloping it. Offsetting these indirect benefits were the static costs of the scheme. Before its regeneration was instituted, because of uncertainty over its future, buildings fell into disrepair blighting the area, therefore rendering renovation more expensive and extensive.

In the context of urban conservation and regeneration and the role tourism plays in it, it is the dynamic effects which are most significant. The major impact of tourism has been implicitly recognised by the Bord Fáilte (1995) in its estimates of increases in tourist numbers, which in turn leads to the generation of income and employment through direct, indirect and induced effects of their expenditure. Other kinds of more diffuse impacts have not been explicitly identified and certainly not measured. For example, consequent improvements in the infrastructure – transport, local authority services, cultural activities – which enhance the quality of life of inhabitants and the experiences of visitors. Conversely, overlooked in multiplier analysis and in appraisals of the Temple Bar scheme, costs are often associated with income and employment growth, especially where activities such as tourism are stimulated. For instance, there is likely to be an increase in more far-reaching and intangible costs, such as visual intrusion, congestion, pollution, crime and delinquency, and perhaps over-use and degradation of built and natural environments, which reduce the quality of life for residents and the experience of visitors.

The official publications (Bord Fáilte, 1995; Recite, 1995) are silent on such ongoing monetary and non-monetary costs of the Temple Bar scheme. Furthermore, while the existence is acknowledged of indirect monetary costs, such as tax breaks and investment allowances, these have not been quantified. There are also other recurring indirect non-monetary costs which are more difficult to isolate and quantify, for example the requirement for increased local authority services concerning refuse collection, sewerage and street cleaning which fall on all the inhabitants of Dublin, not just those living and working in Temple Bar and its visitors.

There are even more subtle costs associated with the project. In a recent study of the area (Duncan et al., 1996) some possible problems in the future were identified, arising from the very success of the scheme. The multiple stores are becoming increasingly interested in locating there, being attracted by the high turnover and profits potential. This will almost certainly increase the pressure to enlarge the floor area of retail units so threatening the character and scale of the buildings and in turn leading to higher property values and rents. This will tend both to push out the smaller proprietors and tenants and change the ambience of the area. However, a key factor is the extent to which continued private and public sector funding will be required to maintain employment levels and the viability of certain activities which are non-commercial. Although it is rather a crude measure, the statistics presented earlier indicate that employment increased by 1,200 in the four years following funding of the scheme in 1991, which amounted to £136 million, suggest that each job created cost £0.113 million. This cost would be even greater if the tax reliefs were to be included. However, it must be acknowledged that this expenditure may be continuing to increase employment which is as yet unrecorded. Moreover, a proportion was undoubtedly devoted to housing projects which increased the number of residents by 500. Nevertheless, regarded in this way the level of expenditure has been substantial. It is also possible that financial support might need to be maintained for non-commercial activities. Currently, cultural activities are cross-subsidised. If support is withdrawn, the cultural base of the area will almost certainly be impaired. Furthermore, at present the facilities and services offered appeal to a young local clientele whose expenditure offsets the net costs. There is no certainty that such patronage will be sustained as it grows older. Likewise, tourists attracted to the area may not necessarily retain a taste for it in the long run.

An economic perspective on the case study

A number of concepts and methods in economics can be applied to appraise the Temple Bar case study more comprehensively and rigorously to show the gaps and discrepancies in current assessments of the impact of the programme. In introducing these concepts and methods, attention is concentrated on how such an appraisal would be conducted rather than attempting to estimate actual benefits and costs to arrive at an overall net effect.

Static and dynamic effects

A distinction is made between static and dynamic effects because different methods need to be adopted to identify and evaluate them. Static benefits and costs arise as a consequence of an event at one point in time with a largely once and for all effect. For example, listing, as indicated by some of the evidence given earlier, imparts an upward or downward shift in property and perhaps land values. Dynamic impacts have an ongoing effect, possibly getting greater with the passage of time. Thus, in the Temple Bar case, its designation as a Recite-funded project had an immediate static impact on the capital and rental value of properties and land. However, no attempt was made to measure this in a 'before' and 'after' monitoring. The dynamic impact has been easily discernible as demonstrated by the increasing numbers of Irish and overseas visitors attracted to it who, as indicated earlier, generate both beneficial and detrimental effects.

However, despite being so visible, its measurement has not been proposed and certainly evaluating the costs associated with greatly increased visitor rates has not been contemplated. The benefits and costs respectively accrue to or are suffered by owners and/or occupiers of buildings and the inhabitants and employees of, and visitors to, the area.

The concept of value

In conventional analysis, under the limiting conditions of a highly competitive market structure in which goods and services are traded, economics argues that the exchange prices determined by the interaction of supply and demand are a reflection of value. In practice this theoretical ideal is never met, so that it is accepted that goods and services possess use value which is greater than their exchange value. Relatively recently in the amenity and environmental field, the concept of total economic value has been invoked as a basis for attempting to measure both user and non-user non-priced (indirect) benefits and costs in addition to priced (direct) ones. Thus in the context of urban conservation and regeneration it is possible to refer to:

Use value
- direct – the value derived from the occupation and use of an historic building
- indirect – the value conferred by a building's or area's appearance, or design, or historic associations on occupiers, the local community and visitors

Non-use value
- option value – the value of heritage resources which is the expression of the willingness to pay for their conservation so that the possibility of using them in the future is retained
- existence or intrinsic value – the value of retaining heritage resources irrespective of their instrumental use to human beings
- bequest value – the value placed by the present generation on retaining resources for the benefit of future generations

With respect to Temple Bar, the direct value has been shown to be lower because of the state of the buildings which lowered market capital and rental values. However, the largely homogeneous character of the area and the development programme which has enhanced its attractiveness, has more than compensated occupiers and owners. This is not necessarily in indirect monetary terms in the form of increased business activity but rather more with regard to intangible benefits concerning a sense of place, image and community spirit. Non-use values are harder to establish because currently mechanisms for the expressions of option, existence and bequest value have not been applied. It could be argued that the funding of £136 million is partially justified on the grounds of a willingness to pay for non-use benefits. Nevertheless, the examination of the case study indicates that only the direct value has been taken into account. Passing reference only has been made to indirect values while non-use values have not been overtly acknowledged. Therefore, it would appear that the total value of the Temple Bar scheme is much greater than existing assessments suggest.

Collective consumption goods and externalities

The notions of static and dynamic effects and total economic value are closely bound up with the long-standing and widely accepted concepts of collective consumption goods (often referred to as 'public' goods) and externalities which singly or together are features of most forms of economic activity. Markets fail to operate efficiently to take account of the total value to consumers or the full costs of production of goods and services. Collective consumption goods are essentially characterised by being non-excludable. This means that once provided everyone can gain access to them and derive benefit without having to pay for them. Accordingly, because the private sector will not supply them if no revenue and profit can be gained, it is necessary to provide them through the public sector, funded through the tax system or some kind of collective charge. Closely associated with collective consumption goods are what are termed 'externalities'. The market certainly operates but it fails to embody social benefits and costs arising from consumption of such goods. These two concepts can be illustrated by reference to the Temple Bar project. Owners and occupiers and visitors enjoy such benefits as the improved appearance of renovated and improved buildings, provision of open spaces, creation of traffic-free areas. Conversely they suffer the costs of, for instance, congestion, pollution and perhaps increased levels of crime which come with greater numbers of tourists. These are respectively beneficial and detrimental externalities. The interrelationship of externalities with collective consumption goods occurs because free access to the latter can lead to their overuse and degradation. All users are responsible as, for example, excessive use of vehicles causing air pollution can damage the fabric of buildings, or too many pedestrian visitors can, by walking over them, destroy grassed open areas or damage delicate artefacts by handling them. Tourism authorities and businesses contribute to the problem in promoting and marketing Temple Bar and thus generating demand. Thus, free access and the absence of market mechanisms for consumers and suppliers to pay the full costs of goods and services adversely affect listed buildings and conservation areas. Equally, owners and occupiers confer benefits for which they receive no compensation, for instance where the additional costs of maintaining an attractive historic building enhances visitors' experiences.

Neither the benefits nor costs of providing collective consumption goods and the beneficial and detrimental effects of externalities have been accounted for in the Temple Bar study. Therefore, together with the failure to recognise the existence of, let alone measure, all the elements of total economic value, there are many discrepancies and omissions in the appraisal of the programme which this economic perspective points up. This perspective can be widened by considering how the omitted benefits and costs might be measured and by proposing an analytical framework in which the net effect of the scheme might be determined.

Measuring the value of conservation and regeneration

Static effects

Methods of measuring the non-user value, i.e. non-priced components, of historic buildings and conservation areas have been extensively applied in the natural environmental and recreation fields, but there is little indication as to their relevance to the built environment (Owen and Hendon, 1985).

Three methods: the hedonic pricing method (HPM), the travel cost method (TCM) and contingent valuation method (CVM) are widely accepted as feasible, both singly and in combination to measure static values. A fourth, the Delphi technique, has been advocated. These have been reviewed by Allison *et al.* (1996) and Stabler (1995) with respect to the value of conservation, and at a higher theoretical level in Hanley and Spash (1993). Lack of space precludes even a brief exposition here but their main conceptual and technical features and an evaluation of their relevance is given in Appendix I. It is necessary to point out at this juncture that only the third, the CVM, is comprehensive in endeavouring to capture all static user and non-user values, covering residents, employers, employees and visitors.

Dynamic effects

The measurement of dynamic effects in urban and regional economics (the spatial aspect of the subject) and tourism, has relied on multiplier or input–output analysis, both of which are outlined in Appendix I, with some examples of their applications. However, these two methods have their shortcomings. As currently developed, while they measure the value of inputs (including employment) and outputs or income flows, they are not suitable for identifying or measuring all benefits and costs, certainly not those of an intangible nature, such as externalities and non-user values.

A project appraisal framework

In discussing the case study and evidence of its impact, it has been apparent that attention has concentrated almost entirely on the benefits of the scheme. It was argued that this yields an imbalanced assessment. To redress this by considering the associated costs, it is necessary to conduct an appraisal which identifies and evaluates both the monetary and non-monetary effects. Given the nature and magnitude of urban conservation and regeneration schemes like Temple Bar, the analytical framework needs to take account of not only capital factors but recurrent ones also. Moreover, there are implications for resource allocation and consumers' welfare, especially where displacement may occur. For example, with regard to resources allocation, if development potential is foregone, property investment may take place outside the area and blight occur within it. This is indeed what happened in the Temple Bar area before the initiation of the scheme.

It was a relatively large conservation area and although there is no firm evidence, its designation may have forced up the prices of developable sites in the surrounding urban area and caused delays, particularly as inventories of the listed building and plans on the phasing of conservation had to be drawn up. With respect to social welfare, an enhancement of what was an historic residential area which subsequently attracted higher income groups displaced lower income groups, or if they were not forced out of an area, imposed higher housing and living costs on them. Furthermore, financial resources to fund the projects were partly drawn from property or indirect tax revenues gathered elsewhere. Such taxes are regressive, meaning that lower income groups met proportionately more of the costs. Yet another crucial factor is the time horizon which is likely to be a very long one. Also, as already shown, the social benefits and costs of conservation give rise to different attitudes by consumers and producers in comparison with their behaviour in the market for normal goods and services. These are the principal reasons why it is not only necessary to adapt a broadly encompassing project appraisal framework but also why it is inappropriate to use appraisal methods applied in the commercial sector where resource allocation and consumer welfare effects are normally ignored, even where a discounted cash flow approach is adopted. The elements of cost–benefit analysis (CBA), which is a project appraisal method advocated by economists, are explained in Appendix II. Thus, since the Temple Bar scheme is characterised by many of the issues identified here and encompassed in a full CBA appraisal, it is imperative, in order to establish its net benefit or cost, that all aspects of its impact are incorporated. No such appraisal has been conducted so no clear conclusion can be drawn on the efficacy of the scheme.

Factors facilitating regeneration

The value of studying the Temple Bar case lies in a number of its features which offer pointers as to the appropriate actions which can be taken where circumstances are similar.

Underpinning the success of the project has undoubtedly been the importance of local involvement and control in which activities, facilities and businesses primarily focus on the local market rather than specifically the tourism one. Such a management structure is likely to prevent the detrimental trends which have been identified by restricting development to a scale in keeping with the character of the area and creating conditions conducive to innovative entre-preneurship. Temple Bar also demonstrates how the cultural base facilitates the

251

development of the evening economy in conjunction with the design and layout of an area and an appropriate mix of residential and business properties. This clearly links with conservation of the physical fabric so that the regeneration process has been an integrated one from the outset, not purely initiated by the renovation of buildings or rehabilitation of designated areas in the hope that economic and socio-cultural factors will revitalise an area deterministically. The Temple Bar example in fact suggests the opposite in that a cultural initiative preceded the physical one. Businesses were already operating multifunctionally, there was an active street life consisting of a variety of activities and people both lived and worked in the area. All these factors gave the area vitality which acted as a springboard for attracting European Union funding and encouraging the central and local government to become involved to support the scheme. The careful planning of the development, keeping the programme flexible and at a human scale, due regard being paid to the interrelationship between physical environment and the aspirations and activities of the community, has given a firm foundation to the revival of the area. Nevertheless, it must be recognised that some features of the success of Temple Bar are peculiar to that location alone, and the circumstances which existed at the time the regeneration occurred, and to an extent still exist. It is in an area of the city south of the River Liffey towards which the commercial centre of gravity is moving anyway. The entrepreneurial characteristics of the people who opened businesses in the area, the EU underwriting of funding, the property slump which created the opportunity for short leases on low rents, the young age structure of Dublin (70 per cent are aged under 40), the Irish lifestyle, the stewardship undertaken by local businesses and the commitment to the project by the Irish government all contributed to the impetus of the rejuvenation.

From the foregoing analysis, taking a largely economic standpoint, and from the illustrative case study, it is possible to give some insights into what are perceived as factors promoting or retarding the fortunes of local economies. Tackling urban and regional problems has had some impact on the regenerative role of conservation, especially where funding is made available for specific heritage projects. While central government stands behind national bodies charged with responsibility for heritage matters, it ultimately determines, through the funds allocated, the importance it attaches to the use of conservation in urban regeneration. This is certainly the case in the impetus given at local level, where the public sector plays a key role in creating conditions conducive to the initiation of regeneration schemes, which in essence are not radically different from other large development schemes. However, it is participation by those living and running businesses in listed buildings and conservation areas who ultimately determine successful regeneration. Thus the three major sectors – public, business and personal, including the voluntary movement – mindful of the multifunctional nature of urban existence, need to work cooperatively.

The contribution each of these sectors can make has been examined in some detail elsewhere (Stabler, 1996) and an indication has been given in discussing the case study above. However, it is of value to summarise their roles at this point. The public sector, largely through the local authority and planning system, can contribute by means of strategy documents concerning development and planning applications, but above all by leading and showing commitment to conservation and regeneration schemes. The private sector's role certainly lies in its specialist expertise, such as in property development, funding and marketing. The personal and voluntary sectors can also provide expertise at low cost, through local societies and amenity groups with knowledge of local materials, building design and historical information. Also they are independent of market pressures and offer a different perspective on schemes reflecting local opinions and attitudes.

An increasing number of examples globally reveal that there are several prerequisites for success, an important one being the recognition that there are different categories of conservation schemes which require different approaches. The benchmark cities with large historic cores and established cultural and tourism activities, such as Bath, Barcelona, Bologna, Lille and Rouen do not necessarily offer a pattern for former industrial cities and towns, for example Birmingham, Bradford and Halifax, while particular quarters, irrespective of the overall character of the urban area, may be so diverse as to warrant very specific schemes, for instance the harbour area in Baltimore, Ghiredeli Square in San Francisco, the castle and Arab quarter in Lisbon and The Rocks in Sydney.

Relating to the physical factors rather than reiterating the economic and socio-cultural aspects already identified, but at the same time acknowledging the inappropriateness of solely property-led conservation and regeneration, it is important that there is a clearly defined spatial boundary, a mixture of residential and business premises on a human scale at affordable rents. There should be a minimum critical mass in terms of size of the conservation area and a number of distinctive listed buildings as 'anchor' attractions. Care should be taken in creating an environment where people can interact conveniently and in safety, through the provision of, for example, open spaces, adequate lighting, traffic-free areas, diversity of facilities and services.

Postscript

Considerable attention has been paid to expounding economic principles and methods with respect to measuring the static and dynamic benefits and costs of conservation, respectively. Much can be learned about the former quite quickly as the valuation of static benefits can be undertaken by applying well developed economic methods used elsewhere. However, little is known about the latter so consequently what research should be undertaken is more difficult to specify. While there are many examples of conservation initiatives, a number of which have been identified above, there have been no full economic studies to ascertain and evaluate them and certainly, until the advent of the Recite projects in Europe, no attempt to undertake continuous monitoring to assess their 'true' costs and long-run viability commercially. Thus there is a crucial need to undertake longitudinal case studies to try to understand how the process works and conservation's roles in it, requiring the following initial steps.

- The derivation of a taxonomy of cases, the prime objective being to identify what differentiates successful from unsuccessful conservation and regeneration schemes.
- The establishment of the economic impact of conservation schemes in terms of: changes in the physical, cultural and social environment; character of economic activities; and movements of the population both within and outside the designated area.
- Monitor the changes in property values, rents and yields over time within and outside the designated area.
- Conduct an initial assessment of the benefits and costs to establish the net effect of conservation in selected conservation areas and the adjacent urban areas; periodic repetition of such an assessment will be necessary.
- Identify the economic functions that both complement and are competitive with conservation and regeneration.
- Ascertain the policy mechanisms and instruments which enhance or deter conservation and regeneration.

Underpinning this research should be the objective of establishing whether public sector expenditure on conservation is justified in terms of value for money, by which is meant meeting the objectives (which may be social as well as economic) at minimum cost to the public purse rather than insisting that direct monetary returns exceed direct monetary costs.

Appendix I

Methods of measuring static value

Indirect or revealed preference methods

HEDONIC PRICING METHOD (HPM)

The hedonic pricing methodology (HPM) was developed by Rosen (1974), based on earlier consumer theory work of Lancaster (1966). It aims to determine the relationship between the attributes of a good and its price, and is arguably the most theoretically rigorous of the valuation methods. It is strongly rooted in microeconomic consumer theory, and takes as its starting point that any differentiated product unit can be viewed as a bundle of characteristics, each with their own implicit or shadow price. For example, in the case of housing, the characteristics may be structural, such as the number of bedrooms, size of plot, presence or absence of a garage, or of an environmental benefit, for instance air quality, the presence of views, noise levels, crime rates, proximity to shops or schools. In the context of the listing of historic buildings, or the designation of conservation areas, it is possible to attribute the impact of such listing or designation for properties by observing the difference in value between two identical buildings, one of which is in a conservation area, and the other of which is not. Thus the price of a given property can be viewed as the sum of the shadow prices of its characteristics. Recent applications of the method relevant to amenity and conservation valuation have been undertaken by Asabere *et al.* (1989), Garrod and Willis (1991) and Willis and Garrod (1993).

TRAVEL COST METHOD (TCM)

The travel cost method (TCM) was developed by Clawson and Knetsch (1966). The method is based on the premise that the cost of travel to recreational sites can be used as a measure of visitors' willingness to pay and thus their valuation of those sites. The real costs of travelling to a site are taken as a proxy for the price of the product. Thus, even if visitors do not pay to use a site, they may have incurred expenditure either implicitly or explicitly in travelling to it, which could be used as a measure of (or at least as a lower bound to) their valuation of that site. Time can be perceived as an implicit cost while explicit costs are petrol or public transport fares. Whether on-site and travelling time should be incorporated into the estimate of total cost is a point of debate in the literature. For sites to which the majority of visitors walk, valuing the time they take is the only measure which can feasibly be used (Harrison and Stabler, 1981). This might suggest that on-site time should be the true focus of the debate. If it is decided to include on-site or travelling time, it may be difficult to assign a value to it. The opportunity cost in terms of foregone earnings or leisure time which could have been spent doing something else might be taken. Garrod *et al.* (1991) is representative of applications of the method to public gardens in cities while Englin and Mendelsohn (1991) is an example of combined use of the HDM and TCM methods concerning the compensation required if access to a site is foregone but the empirical application is not strictly relevant in considering urban conservation.

Direct or expressed preference methods

CONTINGENT VALUATION METHOD (CVM)

In contrast with hedonic pricing and the travel cost method, which are indirect methods of eliciting valuations from consumers by considering their revealed consumption in related markets which exist, the contingent valuation method (CVM) directly questions consumers on their expressed or stated willingness to pay (WTP) for, say, an environmental improvement or their willingness to accept (WTA) compensation for a fall in the quality of the environment. Since respondents are questioned directly, it is possible to ask them whether they would be willing to pay, for example, to preserve a recreational site, a tropical rainforest or an historic building of which they are not users. Thus an advantage of the method over others is that it is possible to obtain, at least in principle, option and existence valuations as well as user values. Few studies in an amenity or built environment context have been undertaken; reference was made in the text to Willis *et al.* (1993) and Hanley and Ruffell (1993) is a useful illustrative example concerning recreation.

DELPHI TECHNIQUE

Instead of attempting to ascertain individuals' – who may be irrational, impulsive or poorly informed – WTP for an environmental improvement or WTA a degradation, it is possible to employ a panel of 'experts' and to elicit their views on the valuation of environmental changes. The technique was developed by the RAND Corporation in the 1950s (Dalkey and Helmer, 1963), and has been found to be particularly useful in cases where historical data are unavailable or where significant levels of subjective judgement would be necessary (Smith, 1989).

An evaluation of static measures

The hedonic pricing method (HPM) embodies the value that consumers place on environmental attributes only in so far as they are capitalised into property prices; i.e. it captures internalised benefits. Thus the technique is not suitable for the consideration of option, existence or bequest values, nor does it incorporate valuations by visitors. It is the most acceptable from an economic standpoint because of its rigour, reliability and robustness and is eminently suitable for identifying the characteristics of historic buildings or properties in conservation areas, the values of which are enhanced or reduced when listing or designation occurs.

The travel cost method (TCM) works best when the majority of the visitors live a long way from the site. Hence TCM models may be inadequate for the valuation of urban recreational sites where the distance travelled is likely to be very short and so tends to nullify the technique. It may be resurrected in these cases if the value of travel time is incorporated so that even those who walk to the recreational site incur an implicit cost. The travel cost method is appropriate for estimating the value of tourism relating to historic artefacts by both residents and visitors, as long as the former actually do use the sites. However, if residents have specifically moved nearer to sites in order to visit them more often the TCM will undervalue benefits because travel costs will be lower. From an economic view-point, it is perhaps the least preferred as, although it is conceptually acceptable, it does not have a strong theoretical underpinning and is difficult to implement. Moreover, it does not measure wider non-use values and it is difficult to aggregate values obtained for individual sites because of overlapping catchment areas of competing destinations. Virtually no studies of applications of the method to visits to historic cities and towns have been conducted.

The contingent valuation method (CVM) is possibly the least rigorous, but arguably now the most popular of the three main valuation techniques. It has become widely accepted, certainly in the political arena, as it is compatible with the concept of democracy and valuation by the population at large; for example, it has already been used as evidence in compensation claims in the US. It has the merit of not only being simple, flexible and easily understood but it is the broadest in covering use and non-use values and both residents and visitors. Potentially, therefore, in the context of urban conservation with the many associated elements of collective consumption goods and externalities, CVM can make a key contribution because of its comprehensive embodiment of all aspects of total economic value. Its main drawback is that to date it has been used to investigate virtually only hypothetical situations and few studies have fully investigated the relevance of its application to urban conservation. For example, one by Willis *et al.* (1993) did not explore non-use values in CVM estimates of the heritage benefits of Durham cathedral.

The Delphi technique has not been applied in the urban conservation field although it is also a potentially fruitful approach which can be used on a comparative basis with the other methods referred to, particularly the CVM technique. It is equally simple, flexible and wide-ranging, giving the possibility of offering insights into the stance of experts on the value of conservation. However, it is inherently subjective and as a qualitative approach does not allow value to be expressed in objective monetary terms.

Methods of measuring dynamic value

MULTIPLIER ANALYSIS

Multiplier analysis has been widely applied since its advocacy by Keynes (1936) to reflate the economy in an era of acute recession in the UK and the same approach can be used in smaller and local areas. Considerable analysis to measure the effectiveness of such a policy at a lower level has been conducted with respect to urban and regional economies and tourism expenditure, for example there are useful summary reviews in Armstrong and Taylor (1993) concerning regional policies and Pearce (1989) covering the impact of tourism, but the methods are equally applicable to other injections, such as cultural, recreational and sporting activities. The basis of multiplier analysis is that new money coming into an urban area or region – investments or central government grants or tourist expenditures – stimulate the local economy. The impact does not end with the first expenditure round – the direct effect – as further rounds and indirect and induced effects are generated so the ultimate effect on income is greater than the magnitude of the initial injection. It follows that increased income flows in one area in turn stimulate employment, the direct, indirect and induced impact of which can also be analysed.

INPUT–OUTPUT ANALYSIS

Input–output analysis is a rather more all-embracing analysis, for it traces the interrelationships of different industrial sectors (for example, accommodation, facilities, transport and services), specifically the origin and destination of inputs and outputs and can incorporate the contribution of built and natural environments and the impact of economic activity on them (Pearce and Turner, 1990). As a more comprehensive analytical framework, especially spatially, the input–output approach is ideally suited to assessing the economic impact of tourism which may be stimulated as a result of conservation schemes in cities of sub-regional and regional importance. Moreover, income and employment multipliers can be derived from it.

An evaluation of dynamic measures

The value of multiplier and input–output analysis respectively, in which the principal injections or inflows and leakages or outflows of expenditure of an area are indicated, is that they can identify a significant aspect of listing and the designation of conservation areas, namely spending in historic tourism destinations. The main point to comprehend is that the objective of calculating tourism multipliers is to ascertain the net effect of an initiative on a destination's income and employment by identifying not only the direct, indirect and induced income (Pearce, 1989) but the ultimate beneficiaries. A crucially important flow is the value of leakages or 'imports'. If a destination has to import a high proportion of its goods and labour then the benefits are diminished. In general, the larger the area the more self-sufficient it is. Thus, for instance, in the early 1990s the UK had a multiplier of 1.73 whereas in Western Samoa it was 0.39 (Pearce, 1989). This means, in essence, that the ultimate economic impact of inward tourism in the UK was nearly three-quarters greater than the initial round whereas in Western Samoa barely two-fifths of the total earnings from tourism remained in the island. Similar values for the multiplier to that in Western Samoa have been found for individual urban areas in the UK (Archer, 1982). With respect to employment, tourism is favourably considered as an activity likely to be beneficial in regenerating an area. Archer (1976), for example, found that every £10,000 spent then created nearly five jobs in tourism compared with almost three jobs, on average, in other economic activities.

Although multiplier and input–output analysis have yet to be applied to ascertain the impact of specific conservation schemes, they have considerable potential to contribute to their evaluation within holistic analytical frameworks in addition to their value for measuring dynamic benefits.

Appendix II

An explanation and evaluation of Cost–Benefit Analysis

The derivation of cost–benefit analysis (CBA) as a suitable framework for the assessment of projects in the public domain which specifically accommodates non-monetary benefits and costs have been admirably reviewed in standard texts by Hanley and Spash (1993), Little and Mirlees (1974) and Pearce and Nash (1981). Urban conservation, however, does pose specific problems because of the characteristics of the built environment in terms of its function, physical structure, location, surrounding environment and legal status, as well as the many and multi-dimensional – cultural, ecological, economic, environmental and social – externalities of an intangible nature which are generated.

Two important features should be an integral part of any appraisal of urban conservation. The first is the need to derive what are known as shadow prices where market prices are inappropriate or non-existent. For example, the size of the project may be significant enough to affect the market for inputs: thus a considerable increase in the demand for specialised labour or building materials in the face of an acute shortage of both might lead to a rise in their prices. The second factor is the choice of the discount rate by which the benefits and costs are adjusted to express them in terms of their present or future value. It is widely accepted (Pearce and Nash, 1981) that the social opportunity cost rate should be applied, reflecting the cost of funds which might be used in alternative projects in the public domain rather than the private or commercial one, where time horizons are shorter and expected rates of return higher, so that higher discount rates are applied, tending to reduce the present value of more distant benefits over the projected time-span of the projects. This tends to discriminate against projects, such as urban conservation schemes, where the pay-offs may not be significant until well into the future.

The need for an all-embracing version of CBA has been advocated by Lichfield. His development of the method has been instrumental in widening not only the scope of studies concerned with public sector project appraisal but also in focusing on key issues. Indeed little work apart from his has been done on the overall framework most appropriate for evaluating urban conservation. His early contribution (1956) was to suggest the use of a planning balance sheet approach (PBSA), originally put forward as a means of incorporating non-monetary benefits and costs. The method was further refined by him (Lichfield, 1988) in his deriva-tion of community impact analysis (CIA) or community impact evaluation (CIE). The objective of CIE is to indicate which sections of the community are likely to gain or lose from decisions relating to planning, urban renewal or regeneration programmes, i.e. the method takes account of the redistributional effects. This was a radical departure from conventional social cost benefit analysis (SCBA) because, in arguing for the necessity to recognise welfare transfers between different sectors of society, the danger of double-counting is introduced. Such a notion is at odds with the tenets of economics which are concerned with aggregate or overall net increases or decreases in welfare rather than their distribution. The need to broaden CBA has recently been developed even more by Lichfield *et al.* (1993), when applied to conservation and heritage, in a document published by the International Council of Monuments and Sites (ICOMOS). An associated aspect of this view is that appraisal should aim to ascertain the true costs of conservation in capital and operating terms on a 'with' or 'without' basis, i.e. what the benefits and costs and distributional and allocational consequences are of undertaking conservation as opposed to not doing so. Accordingly this suggests that the direct benefits and costs of conservation should be assessed by estimating the private opportunity cost while for the indirect, or social, ones the social opportunity cost should be estimated. Against these opportunity costs should be set the benefits of conservation. Clearly such benefits are lost if no action is taken and the situation remains as before.

Most of the cost of urban conservation is almost certainly direct. The incidence of this is in both the private and public sector. Individual property owners undertake expenditures and, as a result of listing or other regulations, have additional costs imposed on them. Public funds are used directly in the conservation of historic buildings and the required administration consumes real resources. The indirect costs of conservation, most obviously the foregone values of redevelopment already mentioned, are also substantial.

The point at issue is therefore when does one adopt the Lichfield-type methodology as opposed to the strict CBA one? The choice probably may depend on the purpose of the investigation and the significance of gains and losses experienced by different sections of society. If the purpose is merely to establish the net benefits or costs of conservation then, at an aggregate level, the impacts of a distributional nature, because they are transfers and do not have a net effect, can be ignored. Lichfield's CIA methods are probably most suitable with respect to specific projects. If the question is more one of principle then the more general form of PBSA or SCBA is appropriate. The same conclusion applies to other methods such as Zahedi's (1986) analytic hierarchy process which was proposed as a means of formalising the choice procedure of alternative schemes. The more conventional CBA approach continues to be the favoured analytical framework as demonstrated by Greffe (1990), for example, and it seems to be the most appropriate in the context of whether simply assessing whether listing and designation are worthwhile.

References

Special sources

Bord Fáilte (1994) *Report and Accounts*. Dublin.
Bord Fáilte (1995) *Report and Accounts*. Dublin.
Coopers & Lybrand (1994) *Employment and Economic Significance of the Cultural Industries in Ireland*. Dublin: Temple Bar Properties Ltd.

Duncan, S., Gurdip, G. and Kodis, A. (1996) *Project Report on Temple Bar*. University of Reading, Department of Land Management and Development. Unpublished.

Montgomery, J.R. (1995a) 'Urban vitality and the culture of cities', *Planning Practice and Research* 10(2): 101–9.

Montgomery, J.R. (1995b) 'The story of Temple Bar: creating Dublin's cultural quarter', *Planning Practice and Research* 10(2): 135–72.

Nexus Europe (1991) *The Impact of the Temple Bar Development on Local Enterprises*. Dublin: Temple Bar Properties Ltd.

Recite (1995) *Dublin*, Urban Pilot Projects No. 11. Brussels: ECOTEC.

Urban Cultures (1991) *Creating Dublin's Cultural Quarter*. Dublin: Temple Bar Properties Ltd.

General references

Allison, G., Ball, S., Cheshire, P.C., Evans, A.W. and Stabler, M.J. (1996) *The Value of Conservation?* London: English Heritage, the Department of National Heritage, the Royal Institution of Chartered Surveyors.

Archer, B.H. (1976) 'Uses and abuses of multipliers', in G.E. Gearing and W.W. Swart (eds) *Planning for Tourism Development: Quantitative Approaches*. New York: Praeger.

Archer, B.H. (1982) 'The value of multipliers and their policy implications', *Tourism Management* 3(4): 236–41.

Armstrong, H. and Taylor, J. (1993) *Regional Economics and Policy*, 2nd edn. London: Harvester Wheatsheaf.

Asabere, P.K., Hachey, G. and Grubaugh, S. (1989) 'Architecture, historic zoning, and the value of homes', *Journal of Real Estate Finance and Economics* 2: 181–95.

Ashworth, G.J. and Tunbridge, J.E. (1990) *The Tourist-Historic City*. London: Belhaven.

Bianchini, F. (1991) *Urban Cultural Policy*, National Arts and Media Strategy Discussion Document. London: Arts Council.

Butler, R.W. (1980) 'The concept of a tourist-area cycle of evolution and implications for management', *Canadian Geographer* 24: 5–12.

Clawson, M. and Knetsch, J.L. (1966) *The Economics of Outdoor Recreation*. Baltimore: Johns Hopkins University Press.

Cooper, C.P. (1992) 'The life cycle concept in tourism', in P. Johnson and B. Thomas (eds) *Choice and Demand in Tourism*. London: Mansell.

Dalkey, N. and Helmer, O. (1963) 'An experimental application of the Delphi method of the use of experts', *Management Sciences* 9(3): 458–67.

Englin, J. and Mendelsohn, R. (1991) 'A hedonic travel cost analysis for valuation of multiple components of site quality: the recreational value of forest management', *Journal of Environmental Economics and Management* 21: 275–90.

Ferguson, P.R. (1988) *Industrial Economics: Issues and Perspectives*. Basingstoke: Macmillan.

Ford, D.A. (1989) 'The effect of historic district designation on single-family home prices', *AREUEA Journal* 17(3): 353–62.

Garrod, G. and Willis, K. (1991) 'The environmental economic impact of woodland: a two-stage hedonic price model of the amenity value of forestry in Britain', *Applied Economics* 24: 715–28.

Garrod, G., Pickering, A. and Willis, K. (1991) 'An economic estimation of the recreational benefit of four Botanic Gardens', *Countryside Change Unit Working Paper* No. 25. Newcastle: University of Newcastle.

Greffe, X. (1990) *La Valeur economique du patrimoine*. Paris: Anthropos, Economica.

Hanley, N. and Ruffell, R. (1993) 'The contingent valuation of forest characteristics: two experiments', *Journal of Agricultural Economics* 44: 218–29.

Hanley, N. and Spash, C.L. (1993) *Cost–Benefit Analysis and the Environment*. Aldershot: Edward Elgar.

Harrison, A.J.M. and Stabler, M.J. (1981) 'An analysis of journeys for canal-based recreation', *Regional Studies* 15(5): 345–58.

Hunter, C. and Haughton, G. (1994) *Sustainable Cities*. London: Jessica Kingsley.

Investment Property Databank (1993) *The Investment Performance of Listed Buildings*. London: Investment Property Databank.

Investment Property Databank (1995) *The Investment Performance of Listed Buildings: Update*. London: Investment Property Databank.

Keeble, D. (1989) 'The dynamics of European counterurbanisation in the 1980s: corporate restructuring or indigenous growth?' *Geographic Journal* 155(1): 70–74.

Keynes, J.M. (1936) *The General Theory of Employment, Interest and Money*. London: Macmillan.

Lancaster, K.J. (1966) 'A new approach to consumer theory', *Journal of Political Economy* 84: 132–57.

Lichfield, N. (1956) *Economics of Planned Development*. London: Estates Gazette Ltd.

Lichfield, N. (1988) *Economics in Urban Conservation*. Cambridge: Cambridge University Press.

Lichfield, N., Hendon, W., Nijkamp, P., Ost, C., Realfonzo, A. and Rostirolla, P. (1993) *Conservation Economics*. Sri Lanka: International Council on Monuments and Sites.

Little, I.M.D. and Mirlees, J.A. (1974) *Project Appraisal and Planning for Developing Countries*. London: Heinemann.

Mathieson, A. and Wall, G. (1982) *Tourism: Economic, Physical and Social Impacts*. Harlow: Longman.

Moorhouse, J.C. and Smith, M.S. (1994) 'The market for residential architecture: 19th-century row houses in Boston's South End', *Journal of Urban Economics* 35: 267–77.

Owen, V.L. and Hendon, W.S. (1985) *Managerial Economics for the Arts*. Ohio: University of Akron.

Pearce, D.G. (1989) *Tourism Development*, 2nd edn. Harlow: Longman.

Pearce, G. (1994) 'Conservation as a component of urban regeneration', *Regional Studies* 28(1): 88–93.

Pearce, D.W. and Nash, C.A. (1981) *The Social Appraisal of Projects*. London: Macmillan.

Pearce, D.W. and Turner, R.K. (1990) *Economics of Natural Resources and the Environment*. Hemel Hempstead: Harvester Wheatsheaf.

Pearce, D.W., Markandaya, A. and Barbier, E.B. (1989) *Blueprint for a Green Economy*. London: Earthscan.

Rosen, S. (1974) 'Hedonic prices and implicit markets: product differentiation in pure competition', *Journal of Political Economy* 82(1): 34–55.

Scanlon, K., Edge, A. and Willmott, T. (1994) *The Listing of Buildings: The Effect on Value*. London: English Heritage, the Department of National Heritage and the Royal Institution of Chartered Surveyors.

Smith, S.L.J. (1989) *Tourism Analysis: A Handbook*. Harlow: Longman.

Stabler, M.J. (1995) 'Research in progress on the economic and social value of conservation', in *The Economics of Architectural Conservation*, Institute of Advanced Architectural Studies, York: University of York.

Stabler, M.J. (1996) 'The role of planning, heritage conservation and tourism in urban regeneration: an economic and real property perspective', paper prepared for the ACSP – AESOP Joint International Congress, Toronto, 25–28 July.

Turner, R.K., Pearce, D.W. and Bateman, I. (1994) *Environmental Economics: An Elementary Introduction*. London: Harvester Wheatsheaf.

Willis, K. and Garrod, G. (1993) 'The value of waterside properties: estimating the impact of waterways and canals on property values through hedonic price models and contingent valuation methods', *Countryside Change Unit Working Paper* No. 44. Newcastle: University of Newcastle.

Willis, K., Beale, N., Calder, N. and Freer, D. (1993) 'Paying for heritage: what price for Durham Cathedral?', *Countryside Change Unit Working Paper* No. 43. Newcastle: University of Newcastle.

Zahedi, F. (1986) 'The analytic hierarchy process: a survey of the method and its application', *Interfaces* 16: 96–108.

14

CONSERVATION AND REGENERATION: TWO CASE STUDIES IN THE ARAB WORLD

Martin Crookston

This chapter examines the approach to regeneration and conservation adopted in two historic areas in the Arab world. The two localities are the Ottoman city of Salt in Jordan, and the merchant quarter of Dubai known as the Bastakia. In both, tourism played an important part in the strategy for renewal and recovery.

The two projects share many common features. Both have an Arab Islamic cultural heritage and setting, a recent history of neglect and decline, a growing local recognition of the potential resource that each area represents, and considerable Western aid and encouragement for the conservation and tourism development initiatives.

However, there are marked differences too. First, culturally: behind Salt's former glory lies the Turkish Ottoman tradition, whilst Bastakia's roots lie across the Persian Gulf in southern Iran. Second, financially: in a Gulf Emirate like Dubai, funding is inevitably much less of a problem than in Jordan with its fast-growing population, new burden of returned emigrants, and lack of natural resources. Third, in scale: Bastakia is a unified little clump of forty-odd buildings alongside Dubai Creek, whilst the historic core of Salt is a collection of hundreds of old houses, schools, mosques, churches, mixed in with much newer development along three wadi floors and their hillsides.

In this chapter, the two projects are described in turn. The regional and social settings, the physical built heritage and the tourism potential of each project is described, along with the strategies adopted. The third part of the chapter then considers some of the lessons that might be drawn for approaches to area rejuvenation and transformation, in the developing world in particular.

Looking at the two case studies in detail, at their genesis, evolution and realisation, highlights some interesting themes about how tourism, conservation and economic regeneration can work together in less developed countries. The limits of the possible appear quite clearly. The two cities' experiences suggest that Western advisers need to be cautious and modest about the possible benefits, and potential threats, which may be entrained by a tourism closely associated with the cultural heritage.

From the perspective of tourism management and development, these two Arab-world initiatives show how crucial the institutional background is. Dubai Municipality, wealthy and used to intervention to achieve its purposes, is almost too eager to bring about solutions. The Jordanian state, acutely conscious of the heritage value of its historic city but with many other calls on its very limited resources, can scarcely afford even to pump-prime the recovery effort. The organic and the planned are always in tension, and rarely more so than in settings like these.

Salt: regenerating an historic Jordanian town centre

The first case study, that of Salt, is the story of a developing country's rediscovery of a part of its history and built heritage. A heritage that was hitherto rather undervalued when placed alongside the archaeological and classical heritage, but with real potential to link to economic regeneration at the city-specific level of a real local economy.

In 1988 the Salt Development Corporation, a non-government organisation (NGO) set up by concerned Jordanians, commissioned a study intended to lead to projects which would help to conserve the historic city of Salt in the Kingdom of Jordan. Tourism is a key element in the strategy, both in supporting and justifying conservation, and in stimulating local economic development. The Corporation appointed Jordan's Royal Scientific Society (RSS), with technical assistance from Llewelyn-Davies Planning (LDP), and backed by financial support from USAID.

Jordan, despite being a small country (4 million people) and quite poor (GDP/ head US $1,750) has quite a high profile. Internationally, this is because of King Hussein and the peace process. In tourism terms, it arises mainly from its biblical connections and the presence of the incomparable Petra. Jordan proper is the land between the River Jordan and Iraq. That is to say, its western boundary is the Rift Valley of the River Jordan, its spine is the uplands of Gilead and Moab, and its wide eastern extent starts as wheatland near Amman and turns from semi-desert to true desert by the time it reaches the Iraqi and Saudi frontiers laid down by the Mandate settlement of 1919–20.

The people of Jordan are predominantly Arab. They are predominantly Muslim though with a large Christian minority. They are a young, fast-growing population, relatively highly-educated and often going abroad to work; a mixture of Jordanians themselves, Palestinians of the 1948 and 1967 diasporas, and immigrant labour from the very low-wage countries (Egypt, Philippines, Sri Lanka). The people are housed in everything from refugee camps to luxury villas, but mainly in the sprawling suburbs of ever extending Amman. Economic development and urban regeneration are live, indeed acute, issues throughout the country, and tourism is recognised as having a role to play in this regard.

265

The economy is fragile. It is dependent on remittances from Jordanians abroad and on foreign aid, as much as on natural resources like phosphates and the small amount of Jordanian oil. Tourism is an important part of the economy, and one of the few real 'export' sectors. It was identified by the World Bank (1990) as almost the only sector with clear growth potential.

Jordanian tourism, even so, is actually quite a small industry. Even an attraction of world renown like Petra brought in fewer than 100,000 visitors per year in 1990. Although numbers are now growing quickly, principally because of the great interest of the Israeli tourist industry since the 'peace settlement', the figures are still very low by international standards, and the number two attraction, Roman Jerash, gets only about 60,000–70,000 paid entries per annum.

The tourism markets are of two separate kinds, as reflected in the image emphasised in national advertising. These are the beach and adventure tourism to Aqaba, Wadi Rumm and the south, and the cultural/heritage tourism to the rest of the country, which is often tied in to Jerusalem, Bethlehem and Palestine. The two markets really only overlap at Petra. Key marketing images were identified in a Pannell Kerr Forster study (1986) as being ancient history and archaeology, biblical connections, and cultural, geographical and historical diversity.

The city of Salt

The city of Salt itself, midway between Amman and the Jordan Valley, is thus located more or less on one of the main tourism axes. Figure 14.1 shows the location and potential tourism circuits. It is only 35 km from Amman, and actually only 80 km from Jerusalem, though even today and despite the peace process, that can mean a full half-day of journey and queuing. It is a hill town, built essentially in the late Ottoman era and experiencing its Golden Age surprisingly late – up to and including the period of the First World War. Yet to a European eye it looks as if it was completed shortly before the Wars of the Roses, with massive stone arches opening into dark yards where caravanserais used to form up for the journey across the desert to Baghdad or Saudi Arabia. Two steep-sided wadis form its core. Above and behind them, a rim of newer development around the town spreads on to the surrounding plateau, to the extent that the Old City at the centre contains now only a quarter or so of Salt Municipality's population of 50,000.

The Old City has a texture and quality which make it unique in Jordan, and possibly in the whole of the Near East. The golden-stone houses clustered on the slopes of the hills, and the unity and historic significance of the Ottoman architecture, make it an unusually intact and integrated survival.

So the first stage of the conservation and tourism study assessed its potential and its problems as follows:

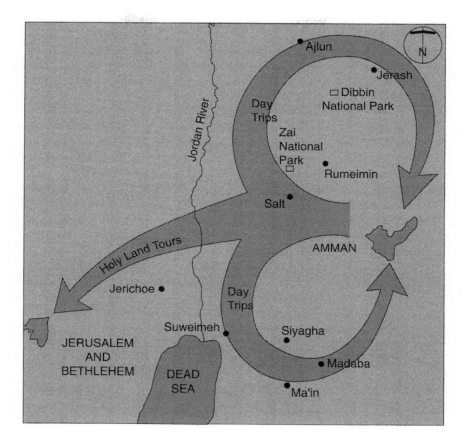

Figure 14.1 Salt: location and tourism circuits
Source: Llewelyn-Davies.

- it has conservation value, with dozens of buildings in the 'Grade 1' class, far more than any other Jordanian city, as well as hundreds in the lower grades, and intrinsic quality arising from the townscape value of the building clusters;
- it has tourism potential because of that character, but there is as yet no existing tourism trade to speak of;
- the Old City's economic role is declining. It contains Jordan's last traditional souk, but its shops and craftsmen suffer from the fact that too many of their fellow-Saltis want to shop in more modern showrooms on the Amman Road; and
- the historic fabric is crumbling year by year, so rescue conservation is an urgent need.

267

Policies for renewal

The appropriate policies and proposals in such circumstances are clearly not only tourism development policies. The study by the Royal Scientific Society and Llewelyn-Davies (1990) sought to set out an integrated approach containing four main strands:

- On conservation, the team surveyed every historic building in the Old City, in order to define a town-wide strategy for buildings, groups of buildings and the wider urban landscape. This allowed the study simultaneously to identify the priority actions (whether for 'rescue' or in order to provide attractive visitor features), and to offer a way of tackling the problems in bite-sized chunks, an important consideration with such limited resources;
- The approach to town planning stressed the need for complete scrapping of the zoning approach that actively encouraged redevelopment, insensitive scale, and demolition for new roads;
- The economic strategy identified measures for small business development related partly to the start-up of tourism, and also focused on ways of channelling money to housing renewal;
- The tourism strategy and proposals combined a historic building attraction with restoration of the Souk, a small hotel, and a tourist trail. A central aim was to provide uses for, and draw attention to, the most important old buildings.

Tourism is very much seen, then, as integrated with the wider development aims for Salt. It is valued partly for its direct contribution, as visitors increasingly discover the town and spend money in new attractions. Similarly, there is value in the spin-off effects of tourism in helping support existing, and perhaps new, small enterprises (cafes, crafts, market stalls). There is a third very important contribution associated with tourism's value in image terms. By validating the Old City and its often shabby old buildings, tourism encourages resident – Jordanian people – to value the old at least as much as the new, and to consider it as a valid option for living and working. Figure 14.2 shows the proposal for the Souk, much of it now implemented.

This kind of 'organic' urban/cultural tourism seemed to the team to have several advantages. It fits well with the existing local economy, and supports it, as it is, without demanding major change to ways of living. It does not impose, as so much tourism product development does, an inevitable cost in terms of imported materials demanded by the visitors. It is welcomed by Saltis, who are proud of their city, happy to receive visitors, and unthreatened by a tourism which is relatively small scale compared with the size and bustle of their town. It also helps in the revaluing, by Jordanians, of the old and the traditional: a hearts-and-minds job, without which all theory about sustainability and best use of resources is, sadly, irrelevant.

Figure 14.2 Restoration of the old souk, Hammam Street, Salt
Source: Llewelyn-Davies.

The regeneration has already started in Salt. One of the mini-merchant palaces, the Beit Touqaan, has been, as the study recommended, rehabilitated to become the formal town hall, by one of a group of very talented Jordanian conservation architects. Others are being improved and brought back into use, as family houses, a 'Salt Zamaan' cafe-restaurant, and as workshop/training centres. Some of the recommended policy and institutional changes are under way as well. The peace process, however fitful, offers some hope that the tourism circuits from Jordan to Jerusalem will become more easy and varied, and can start to include Salt as its attractiveness develops.

Bastakia (Dubai, United Arab Emirates): a unique survival

The second case study again explores a location which has been rediscovered only relatively recently, by its people and government. Even more than in Jordan, what is left of the Bastakia is a very important proportion of what is left of Dubai's built heritage.

Around 2,000 km south-east of Jordan, almost at the other end of Arabia, Dubai is the great entrepot of the Gulf. So it was a hundred years ago, when the little trading quarter of the Bastakia began to grow. Now Bastakia is Dubai's only intact survival from its early days. Recognition of its significance is quite recent, and indeed demolition of the half that had managed to survive into the 1990s was still a possibility as recently as 1994, when the Dubai Municipality (DM) appointed the Llewelyn-Davies consultancy to report on the area's conservation and renewal. In a city-state like those of the Gulf, there is no doubt that personal relationships can play a vital role in policy shifts, and it may well be no coincidence that the shift in emphasis toward conservation came about not long after a visit to Dubai by the Prince of Wales, who visited houses in Bastakia – literally adjoining the palace of the Ruler, Sheikh Maktoum.

Dubai is the most sophisticated of the Gulf cities, the most culturally varied, socially relaxed and economically open. It is less oil-dependent than the other Emirates, having built on its historic merchant tradition to create a current role as the transhipment and forwarding centre of the region's sea and air trade. That merchant tradition is at the heart of Bastakia's *raison d'être* as the district was founded and inhabited by Persian merchant families, from the city of Bastak across the Gulf in the late nineteenth century. They followed the opportunities of trade with the mainly nomad Arab peoples of the peninsula, settled on land alongside Dubai Creek granted by the Sheikh whose fort and palace stood nearby and brought with them skills and traditions from the Persian mainland. Most notably – because these still are the most obvious defining characteristic of Bastakia – they built their houses with the great 'Wind Towers' of coral and sea-stone, the traditional yet effective form of air-conditioning developed over hundreds of years on the Iranian plateau.

The Emirate's population, some 600,000 people, of whom over three-quarters are expatriates from India, Pakistan, Egypt and Palestine, now shows little sign of that heritage. A few families still speak the 'Bastaki' melange of Farsi and Arabic, but they are fully assimilated United Arab Emirate (UAE) nationals of the rich merchant class. Some may still own property in the district, but hardly any now live there. Those who now live in Bastakia are mainly there because it is the cheapest, most run-down and crowded area of the city. They are often Indian labourers or taxi drivers using the historic merchant houses as bachelor 'chummeries' for fifteen or twenty young men. Dotted about among them are a few longer-established families and even one or two Westerners who have rented and restored houses to a very high standard.

Economically, Dubai is extremely prosperous, with one of the highest GDP per head figures in the world. It combines oil revenue (about a fifth of the scale of Abu Dhabi's annual oil production), shipping and distribution as the UAE's principal port, manufacturing and assembly mainly in the Jebel Ali Free Zone, business and financial services and, increasingly, tourism.

Dubai's tourism

Tourism in Dubai is now quite large scale, and still growing, with over a million hotel guests in 1993, nearly 5,000 rooms in deluxe class alone, and a highly organised package-tourism sector attracting European visitors to the beach hotels in particular. The growth has been very rapid with the 2.8 million bed-nights for 1993, 15 per cent up on 1992, double the 1987 figure, and nearly three times that for 1984. The study team produced their own estimate of the holiday tourism market in order to assess the potential for visits to attractions in Bastakia. This required making allowance for business visitors, and for the very noticeable (but probably transitional) phenomenon of ex-Soviet-bloc trips, mostly to buy and sell a bizarre mixture of goods from Red Army binoculars to blankets made in Kerala. With those exclusions, the study team's estimate was of a holiday tourism market probably running at some 150,000–160,000 visitors per year, with perhaps 20,000 per month in the peak of the winter season. Northern and Central Europe account for over 42 per cent (Germany 16 per cent, UK 13 per cent), and there is also an important market segment of visiting friends and relations (VFR) associated with the extensive British business presence in the Emirate. A total annual market for heritage tourism attractions has been estimated at 200,000. The trade is more seasonal than it used to be, with the peak month now being December (10 per cent of annual total), indicating that it is the leisure tourism that is growing fastest.

The Bastakia district is already on the tourism map of Dubai in a small way. One of the brochures mentions 'the Bastakia district of old wind-towered and mud-walled houses' in offering a City Tour as part of the additional excursion package that can be pre-booked. All but one of the City Tours includes at least a brief call at or drive past the district. Perhaps 10,000–15,000 people per year are shown the area, essentially the art gallery in one of the houses on the edge, plus something between 5 and 30 minutes looking at the buildings and taking photographs. Figure 14.3 shows the city context and linkages to other attractions.

In developing the strategy for the area, the study team had to consider viable and sustainable future uses for the historic buildings, as well as specifying the detailed works of restoration that were needed. The strategy report explained that heritage and cultural uses are closely bound up with tourism. As attractors of visitor activity, all three can bring life and economic value to an area. But it also warned the Municipality that such activities can risk ruining an area. If they are handled carefully, tourist attractions can contribute to the appeal of Bastakia. They can contribute to displaying the area to everyone (including, as in Jordan,

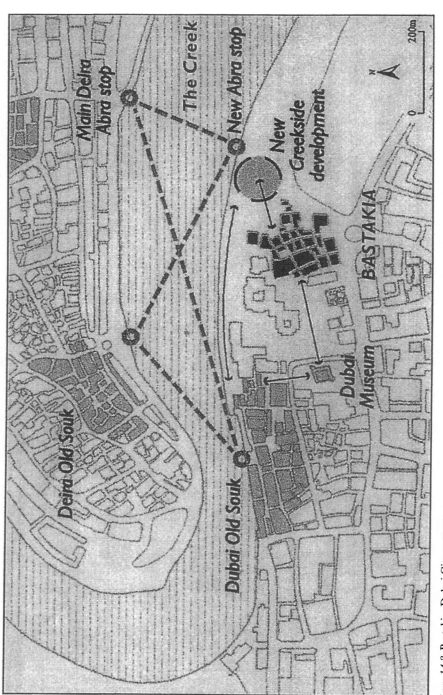

Figure 14.3 Bastakia: Dubai City context
Source: Llewelyn-Davies.

the nationals, and not just overseas visitors), and they will support the cultural and heritage buildings financially.

The market segments to be served, and thus the attractions which were considered, could be drawn from a distinct range of different requirements, each with differing implications for Bastakia. From cultural tourists, who might wish to spend a whole day or even longer exploring the old houses, to coachloads of beach hotel visitors passing half an hour there on their way to the Bur Dubai and Deira Souks.

The role of the built heritage

The starting-point was heritage uses, because it is Bastakia's built heritage which is so special. Bastakia is a unique survival of old Dubai. Clustered near the Creek, the historic buildings and the pattern of 'sikka' pathways that criss-cross between them offer a vital link back to the city as it was lived in only a generation or two ago. The predominant, and best-known, feature is the Wind Towers, of which twenty-five survive in Bastakia, more than anywhere else in Dubai or indeed the Emirates. Around the Wind Towers, many other valuable features add to the architectural significance of the old houses: balconies, verandahs, courtyards and elaborate detailing in coral, mangrove and gypsum mortar. For the people of Dubai, it is of great cultural and historic significance, and steadily coming to be recognised as such. No other comparable grouping survives, as opposed to single individual preserved buildings, nowhere else shows the city as it used to be. Its location, too, is full of history, close to the Palace, the Creek and the Souk.

There is no specific heritage use of the Bastakia buildings at present, though the preserved Al-Fahidi Fort occupied by the Dubai Museum is only 300 metres away, and the adjoining Ruler's Diwan and Quayside are of great historical importance to Dubai. While the Bastakia area and houses are recognised as being of heritage value, they have no formal status as monuments, and there is as yet no interpretation or explanation of the buildings for visitors.

The heritage theme was seen as very promising for future uses. The study team proposed a Wind Tower Museum, to explain and interpret the Wind Towers and the history of Bastakia. An appropriate and very attractive use of one of the buildings, it is now being worked up in detail as a 'Key Project' in the restoration phase of the overall project. A second quite similar use proposed was a Restored House Museum, focusing on lifestyle in order to show how these buildings were used sixty years ago, and how people lived, with artefacts, fabrics and furnishings. Figure 14.4, a page from the study report (Llewelyn-Davies Planning 1994a), shows the concept applied to one of the wind-tower buildings.

Also central to the idea of Bastakia's future is its cultural significance. The area has a special meaning for the people of Dubai, as the last district which recalls the former way of life of the long-established families, many of whose members can still remember it in great detail. In future, it can play a lively part in the cultural and social life of Dubai, as the city is lived in by locals and expatriates

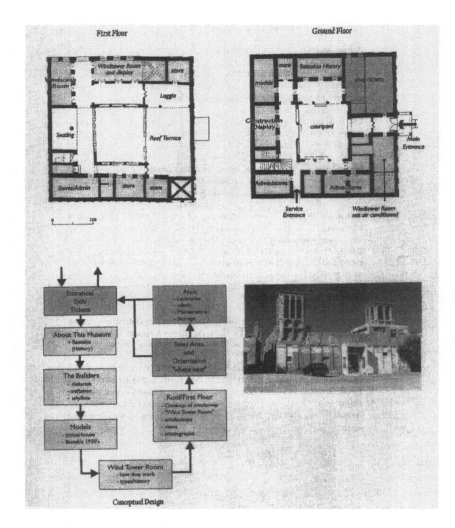

Figure 14.4 Bastakia Windtower Museum concept
Source: Llewelyn-Davies

alike. At present, Bastakia is just about on the cultural map of the city and the Emirates. As noted, there is one gallery, the Majlis Gallery, which attracts perhaps 10,000 visitors per year, at least some of whom buy art works. It also acts in a small way as a cultural-social space, with at least one major exhibition per winter season. Possible future uses in this sector were seen as including a Cultural Centre for artists and the Arts Society, and further art galleries. There is already interest in opening one or more galleries, and it was suggested to the study team that another two would be both beneficial and acceptable.

The single main conclusion was that a mix of uses was both feasible and desirable. There was evidence of current interest in several uses and it was recognised that potential value in the heritage, culture and tourism sectors; and residential occupation would give the area continued life.

The recommended strategy therefore sat tourism squarely at the heart of a mixed-use approach. It had two basic elements:

- A balanced mix of future activities, including housing, tourism, cultural and possibly some commercial use. Such a mix would reflect current interest in occupying Bastakia houses, exploit the tourism and cultural potential of the heritage and variety. The unstated recommendation was that the area should not become a 'Tourist Village';
- Careful, subtle physical change to the buildings and their surroundings in order to restore them and adapt them to modern use. The aim would be a cleaner, safer, better cared-for version of what is there now, not a radical change in layout or appearance. Here the unstated recommendations were that the area was not to become a pastiche, and it was not to involve a clearance of all but one representative fine building.

The Municipality has now moved on to the implementation phase. Paths and public spaces have already been resurfaced, basic restoration work is under way, and the buildings suffering the worst misuse have been emptied and secured. In 1998, the first new uses should be opening up, and starting to add to the variety and interest of Dubai's urban tourism product.

Themes

There are perhaps three themes that can be drawn out of these two similar, but distinct projects in Jordan and the Gulf. They are to do with the role of tourism in regeneration; the 'fit' between tourism markets and the possible product; and the implementation and institutional issues of who should do what.

Tourism as part of the regeneration effort

Tourism is a very familiar component of the regeneration effort in Britain and the USA. Urban tourism, the heritage industry and economic revival are more and more closely intertwined. In the less developed countries, this link has until recently been much less explicit or evident. Tourism has been a force for economic change, without doubt. It has not, however, tended to be harnessed to the rejuvenation of local economies and the built fabric, rather being seen as a new and powerful engine to be coupled on to the train of regional economic development from a low initial base.

Salt and Bastakia show this more locally focused effort starting to be applied, though they also show its limitations. At the national and regional scale, Jordan's tourism can no doubt benefit in a small way from Salt's contribution to a more varied and stronger product, with the city offering its unique mix of living 'mediaeval' market town to the historical/Biblical/cultural package that is already attracting visitors to Jordan and the Holy Land. But the contribution in that direction is probably not as significant as the contribution that Salt itself can get from joining the tourism product in this way. The city can benefit in several ways: a broader base to the local economy, a way of funding its infrastructure and rehabilitation needs, a new vitality, role and belief in the historic central area.

The 'fit' between markets and the heritage product

The issue of how well tourism demand can be made to fit with a heritage product that may be inherently fragile and vulnerable is also central to the role that urban tourism can play in this setting. In the Jordanian example, it seemed to the study team that the fit was potentially very good. What the city of Salt has to offer can, with careful upgrading and management, match very well what Jordan's existing and likely future visitors will be looking for, on an acceptable scale, without worrying residents, and with likely direct benefit to the weakest part of the local economy.

Dubai's example is more difficult to judge. The Emirate's holiday market is predominantly, though by no means exclusively, package beach-and-adventure tourism, with the potential of Bastakia being 'consumed' by a visitor very quickly and swamped in volume terms all too easily. The tourism monoculture is difficult to avoid – very many small towns and delightful villages have ended up like Rocamadour, Sidi Bou Said or Haworth, on the way to becoming little more than theme parks. This is a real danger for Bastakia. The lesson for places like this in the developing world is surely to balance such dangers (of loss of authenticity, or of day-to-day liveability) against a clear assessment of the real benefits of tourism development. The question must always be: what actual objectives is it fulfilling? There is no point in tourism development for its own sake. If the benefits to Dubai's tourism are marginal, and the threat to Bastakia's authenticity and character significant, then the 'fit' is not a good one. In those circumstances, host governments are right to be sceptical of a tourism-led regeneration strategy and they may want to consider diverting the main demand to a manufactured 'traditional holiday village' on a new site. This is a serious possibility in Dubai.

Who does what: institutional and implementation themes

One of the sets of lessons we can try and learn from each tourism/conservation/ regeneration project is to do with the mechanisms at local level. Institutional arrangements and financial requirement will obviously vary from place to place and milieu to milieu. The Royal Scientific Society and Llewelyn-Davies Planning report (1990) set out a series of recommendations tailored to Salt and to Jordan, and it would be fruitless to recite them here. Essentially, though, the challenge is to get the renewal of the Old City to become self-sustaining, and self-financing by the private sector. But this usually requires a massive and sustained injection of public sector support. Of course the private sector must be involved, but this will not happen without evidence of sustained commitment from the public side, and without an investment curve being initiated by public money. We just have to hope that the private and public investment curves intersect sooner rather than later – and stick with it until they do.

Curiously, in Dubai, the threat to this balance is from the other end. The Municipality is inclined to try to do everything and endangers the project's long-term sustainability if it does so. The buildings of Bastakia require constant attention and maintenance, because of their construction and materials. They used to get that attention when they were owned by the families who had built and expanded them. They have not had it since the start of the blight associated with the threat of demolition, and decay has been serious. Although the Dubai Municipality could, and will, implement the Key Projects and tourism invest-ments, the study team argued that the Municipality could not really expect to intervene in a way which required them to keep up 40–50 buildings for the next hundred years. The need is to find uses that are comfortable in the area, viable, lasting, and capable of paying for their own upkeep. The same could, in principle, go for the initial investment in restoration. If a market response were to be guided carefully, it could produce an organic response that would be more sustainable and flexible than a completely Municipality-led and planned solution. Such thinking is relatively recent and difficult for administrations to handle even in the West. In the Gulf, with its combination of boundless resources, paternalist sheikhly rule, and an administrative tradition imported from Paris via Cairo, there will be resistance to this approach and its acceptance will rely on the new thinking percolating through the ranks of younger concerned UAE nationals who are gradually coming to exert a more subtle and relaxed approach to municipal interventions.

Conclusion

While the two case studies examined in this chapter differ in certain respects, they both illustrate the potential role historic cities of the developing world can play as important tourism assets. The restoration of historic buildings within these cities can therefore contribute to economic rejuvenation by supporting tourism development, while the income from tourism activity can, in turn, contribute to the conservation of aspects of the urban environment that have significant heritage value to the local population.

However, the Bastakia and Salt examples also illustrate how the potential of heritage tourism will not be realised without a thoroughly considered and balanced management approach. In particular, it is important that tourism markets and the heritage product are matched in a manner that ensures that fragile historical assets are preserved. Also, although income from tourism might eventually be sufficient to cover the cost of restoring and maintaining these assets, seed funding from the public sector is generally necessary as a primer for these activities. Viable commercial linkages are necessary in the longer term in order to ensure that the burden of maintaining heritage buildings does not fall on the public sector and, ultimately, the resident population.

Bibliography

Beazley, E. and Harverson, M. (1985) *Living with the Desert – Working Buildings of the Iranian Plateau*. Warminster: Avis and Phillips.

Economist, The (1996) 'The World in 1997', London: Economist Publications.

Llewelyn-Davies Planning (1994a) *Bastakia Conservation*. Dubai: for Dubai Municipality.

Llewelyn-Davies Planning (1994b) *Bastakia Conservation Technical Note 1: Tourism Background*. Dubai: for Dubai Municipality.

Pannell Kerr Forster (1986) *The Future Course of Tourism in the Hashemite Kingdom of Jordan*. Washington: USAID.

Roberts, J. and Zandi (n.d.) 'Dubai Town Walk', Dubai, Majlis Gallery, available 1994.

Royal Scientific Society, with Llewelyn-Davies Planning (1990) *Salt – A Plan for Action*. Amman: for Salt Development Corporation.

World Bank (1990) *Country Report: Jordan*. Washington: IBRD.

Part III

CHANGE MANAGEMENT STRATEGIES IN TOURISM OPERATIONS

INTRODUCTION
TO PART III

Gianna Moscardo

The introductory chapter and Parts I and II have all been centred on the theme of change and its relevance and importance to tourism. Part III takes this theme of embracing change and examines it at the scale of specific tourism operations and markets. The chapters in this section demonstrate the need for each tourism operation to develop appropriate management strategies for the changing circumstances it faces. Tourism managers and operators need to be able to understand the nature of the changes that will be important to their operations and to respond to these changes. For some the initial imperative is to create the conditions for change, while for others the prime need is to evaluate the success of the management strategies put in place. The chapters presented here examine a variety of responses to change and provide examples from a range of different types of tourism operations. In addition to the major theme of change, three other important themes can be found in the cases presented in this section. These are the search for sustainability, a growing recognition of cross-cultural issues, and the need to develop and use more sophisticated approaches to gathering information about vital aspects of tourism business, especially about markets.

The search for sustainability

Recent years have seen the publication of numerous studies and reports concerned with examining changes in technology, political systems and socio-demographics and how these might influence tourism. While a large number of specific trends have been described, one major theme has emerged. This is the importance of sustainability as a principle to guide the development of tourism. Sustainability is clearly the catchphrase of the decade and currently dominates discussions of tourism development, planning and management. Sustainability is about quality, balance and continuity. For tourist operators sustainability means maintaining the quality of the experience or product offered to visitors, of the environment that supports tourism and of the lives of those who live in or near tourist destinations. It means finding a balance between the needs of residents and visitors and between the costs and benefits of tourism for host communities. It also means ensuring continuity of the resources required by tourism, of the support of communities and of the interest of tourists.

One challenge faced by tourism operators is that of minimising the negative and enhancing the positive impacts of tourism. Several chapters in this section, including Ryan, Moscardo and Woods, Buhalis and Cooper, and Pearce, Kim and Lussa, provide case studies describing and/or evaluating management strategies which are meant to enhance the benefits of tourism and/or manage its costs. The first two chapters focus on the physical environment, while the chapters by Buhalis and Cooper, and Pearce, Kim and Lussa concentrate more on host community issues. In the Buhalis and Cooper chapter economic returns for host communities is the major topic addressed, while in the Pearce, Kim and Lussa chapter cross-cultural impacts are highlighted.

Many authors have both predicted and proposed the expansion of ecotourism as an option for developing a more sustainable tourism industry. Chris Ryan's chapter provides an evaluation of the success of three different types of ecotourism operations in New Zealand. The evidence presented suggests that it may not be as easy as some suggest to develop and manage economically viable ecotourism operations. Ryan provides some insights into the difficulties faced by operators seeking to minimise environmental and cultural impacts.

Moscardo and Woods also directly address the issue of sustainable tourism operations in their chapter on a rainforest cableway in North Queensland, Australia. The chapter evaluates the role of interpretation as both an important tool for minimising tourist impacts and a mechanism for creating quality experiences for visitors. The results of this evaluation indicated that an effective interpretation programme can be successful in creating a better understanding of the environment amongst tourists and in encouraging more sustainable behaviour. This case study also demonstrated the value of interpretation for enhancing visitor satisfaction.

The third chapter in this section, by Buhalis and Cooper, is concerned with the economic aspects of sustainable tourism operations. A common criticism made of tourism is that the economic benefits often flow to large companies which are usually owned by interests outside of the destination area. One of the options for increasing the sustainability of tourism is to increase the economic benefits that flow to locally owned businesses. This case study involves a SWOT analysis of small and medium tourism enterprises in the Greek Islands and uses the results of this analysis to suggest a mechanism for increasing the competitiveness of such businesses. The authors describe a networking process designed to assist these tourism operators to maximise returns from marketing and to gain access to technology and training. An important feature of such a network is the potential to encourage better management responses to change.

The growing recognition of cross-cultural issues

The chapter on cross-cultural contact by Pearce, Kim and Lussa provides an evaluation of another proposed strategy to minimise the negative impacts which can result from tourism. Here the emphasis is on methods for ensuring successful interaction between hosts and guests from different cultures and two major issues are addressed. The first deals with the nature of the problems which can arise when people of different cultures interact and it suggests a set of principles which can be used to predict likely problems which may need intervention. In this instance examples of problems faced by Korean tourists in Australia are used to highlight the value of the approach. The second section evaluates a strategy designed to improve the communication between Australian visitors and Indonesian hosts. As one of the trends identified for future tourism patterns is increasing travel to and from the Asia-Pacific region these two cases also provide interesting insights into two Asian cultures.

This chapter shares a common thread with the next one, by Richards and Richards, in that both emphasise the importance of visitor management for the ongoing success of tourism operations. The Richards and Richards chapter, which evaluates the development of a Disney theme park in France, also highlights the need for tourism managers to be aware of cross-cultural differences. Within this case study there is substantial discussion of the negative consequences of not being aware of the cultural differences that exist between different groups of tourists, and specific examples of such differences and their implications for theme park management are described.

Seeking and using marketing information

The problems of lack of consideration of cultural differences is one aspect of failing to effectively manage information about markets. The Richards and Richards chapter on Disney in France provides more general insights into the importance of a market orientation in tourism management. The case study describes the problems that can arise when managers become too complacent and removed from the expectations and needs of their customers. This chapter provides an excellent example of Langer's concept of 'mindlessness' (1989) in business with the North American Disney managers attempting to run an operation in another location without changing their approach. The solution to the resulting problems are a major feature of this case study and a number of factors that contribute to successful theme park management are identified.

The next three chapters in this section by Edgar, Hsieh and O'Leary, and Burns and Murphy, all concentrate on the importance of successful marketing for tourism operations. As previously noted, sustainability is not just about managing the environmental, cultural and economic impacts of tourism. Sustainable tourism must also provide a quality experience for its customers. Many analysts have proposed that tourism operations must become more market oriented or customer focused if they are going to successfully compete in a world with greater market differentiation and rapidly changing lifestyles (Poon, 1993).

The chapter by David Edgar concentrates on evaluating marketing strategies in tourism management. In this case the analysis is of the marketing strategies used by Scottish hotel operators. The study examines the success of different types of marketing strategies for attracting tourists for short break holidays and matches the strategies to different types of hotel operations. Change is an important theme in two features of this case study. First, the author argues for the use of more sophisticated models for evaluating the success of different strategies and a substantial part of the chapter describes such a model. Second, the author uses the marketing strategies of different types of Scottish hotels to attract short break customers, as a specific example of how the model can provide suggestions to guide changes in marketing practice.

The next chapter by Hsieh and O'Leary concentrates on understanding the factors which influence destination choice amongst Japanese pleasure travellers. The rise of Japan as a major source of tourism is one of the most noted changes in tourism in recent times and this case offers some interesting insights into the nature of travel decisions made by these tourists. In particular it is argued that travel motivation and activity preferences are important factors in travel decisions which are not usually investigated. The reported results provide support for this argument and implications for tourism marketing can be drawn from the information provided.

Burns and Murphy provide a case study of the promotional strategies of marine tour operations. They argue that it is important to match the content of promotional activities to both the destination and products being offered and the characteristics of the desired markets. In the first case dissatisfaction can result if operators promise more than they can deliver in their promotional campaigns, while in the second case it is important that promotional campaigns be designed with intended markets in mind. For example, if the major market is a family-oriented one seeking passive experiences, a promotional campaign showing young singles engaged in adventure activities is unlikely to be successful at attracting customers. This case study examines the content of promotional brochures used by reef tour operators and matches the images used with the characteristics of reef tourist markets. The chapter offers some insights into the value of developing promotional campaigns based on market information. Again a major message in this chapter is that tourism operators cannot be complacent and must use information to generate change in their approaches to tourist markets if they are to maintain their position in a very competitive environment.

References

Langer, E.J. (1989) *Mindfulness*, Reading, MA: Addison-Wesley.

Poon, A. (1993) *Tourism, Technology and Competitive Strategies*, Wallingford: CAB International.

284

15

DOLPHINS, CANOES AND *MARAE*

Ecotourism products in New Zealand

Chris Ryan

This chapter considers issues pertaining to ecotourism in New Zealand from two perspectives. For the most part the definition of ecotourism utilised in this chapter is tourism that is primarily motivated by a wish 'to study, admire, or appreciate the scenery and its wild plants and animals, as well as any existing cultural manifestations (both past and present) found' (Blamey, 1995: 14). Generally, discussions of ecotourism incorporate an educational component allied with a wish for activities to be sustainable, that is having the minimum disturbance possible upon physical and cultural environments. The first issue is, what are the implications of seeking to develop ecotourism for a country like New Zealand in terms of its impacts upon the National Parks within a market-led governmental policy? Second, what are the problems of sustaining small businesses based upon seasonal demand? Both issues are considered through three case studies. Each will review the operator's practices and the beliefs underlying those operations. Subsequently, the question is asked, is financial viability for small ecotourism operators only sustainable through high pricing which limits the market to the comparatively affluent, or alternatively by the operator sustaining high personal costs in both financial and psychological terms?

New Zealand has, for a long time, promoted itself as a tourist destination which offers open spaces, a clean environment and unique, fresh and green scenery. There is validity in these promotional images used by the New Zealand Tourism Board, although as Dennis Marshall (Minister of Conservation in 1996) commented in an address to members of the New Zealand Tourism Research Association, 'Unfortunately we have no great secret, other than a small population to account for our "green" condition.' The Department of Conservation (DoC) is intimately involved with tourism in New Zealand – many photographs used by the New Zealand Tourism Board in promotional literature show scenery from the conservation estate. Further, in 1996, the Department deliberately began to use its scarce resources to upgrade facilities in the more easily accessible parts of its estate for tourist use, while permitting huts in other areas like the 'High

Country' to perish over time. While this is partly a zoning exercise, it is also a means of obtaining more revenue from tourism. Moreover, New Zealand is not so unique as to be protected from the ills of a modern world, whether these problems are social, economic, political or environmental. New Zealand's tourism industry is very aware of the problems caused by tourism, as evidenced by the DoC's brief to the incoming National Government in 1993. It noted that:

> overseas visitors may not find the predicted level of use acceptable from a social perspective because of crowding, particularly at huts and camp-sites. There is certainly a great deal of concern among traditional New Zealand users of the back-country over the level of contact with other users which will occur and the diminishment of the sense of isolation . . . We must ensure that the clean, green image is matched by reality, or New Zealand's credibility with potential visitors will be destroyed
> (Department of Conservation, 1993: 21–2)

It might be argued that these concerns are bound to be expressed by those involved in conservation given the 'user pays' market-driven philosophy espoused by governments in New Zealand since 1987. Hence the DoC is required by the Treasury to use its assets where possible to fund its conservation efforts. Such policies are a direct challenge to the conventional conservation ethic of the Department. Yet, even as the DoC goes through a painful process of adaptation, so too the private sector is beginning to realise that it must change some of its practices. In both 1995 and 1996 the New Zealand Tourist Industry Association Annual Conference spent a day considering sustainability issues in tourism, and the Association has been an important body in setting up the New Zealand Tourism Research Association and in lobbying the Public Good Science Fund for monies to research the economic, social and environmental impacts of tourism within the country.

It is far too soon to argue that tourist businesses in New Zealand promote forms of tourism that are economically, socially and environmentally sustainable (however this might be defined), but undeniably there is an increased awareness of these issues. Indeed, the Resource Management Act of 1991 requires tourist developments to occur solely within frameworks that seek to protect the environment. There remains, however, debate about the forms of tourism development that can be tolerated. One example is the argument over the noise intrusion of scenic flights. Braun-Elwert, Director of Alpine Recreation Canterbury has written:

> Nobody doubts the importance of scenic flights to the tourist industry. All are aware that much pleasure is derived by many thousands of people being able to see wild places from the air. It is also acknowledged that flying people over such areas can protect fragile environments . . . due to the fact that people aren't intruding on the natural infrastructures in the

wilderness below . . . [but] Who wants to climb Mount Cook accompanied by the thunder of aircraft under no particular flight restrictions? Who wants to hear the drone of an aircraft drown out the sounds of nature on a Milford Track walk? No one. So let's address the issue before it's too late.

(1995: 10)

Booth (1996) also reported evidence that visitors to Mount Cook National Park were perceiving such flights as an intrusion into 'natural quiet'. Consequently there are demands by environmental lobby groups that flight paths must comply to minimum flying distances as in US National Parks, and that landing rights issued by the DoC should be linked to clearly defined flight paths subject to annual review and monitoring. It is of interest to note the findings by Higham and Kearsley (1994) and Kearsley and O'Neill (1994) in their surveys of overseas visitors to New Zealand's South Island and its National Parks. Over 85 per cent of such visitors regarded the purpose of the parks as being the preservation of scenery and wilderness, and only 22 per cent thought that helicopter flights were appropriate.

Another example of the pressure from conservation groups is the demand to cease recreational fishing in the Poor Knights Marine Reserve off Northland, the only marine reserve in New Zealand where fishing is permitted, with a resultant depletion of fish stocks. Indeed, divers who visit from other parts of the world are among those who seek a total ban (Forest and Bird, 1995). However, when the Minister of Conservation sought to impose a total fishing ban in 1994, recreational fishing groups sought, and obtained, an injunction to stop the order. Throughout 1994 and the first half of 1995, various legal actions were commenced, and withdrawn by the Minister of Conservation, and in 1996 a new public submission phase was initiated.

Thus, like other places around the world, New Zealand is subject to concerns and debate about tourism and its impacts. However, the debate is all the more critical to New Zealand because of its reputation as a destination for outdoor-oriented tourism. The interests of overseas visitors to New Zealand are well documented through international visitor surveys. Table 15.1 provides data derived from the International Visitor Surveys for 1993 and 1995 undertaken by the New Zealand Tourism Board. During this period the Board notes that the numbers of overseas visitors to New Zealand increased by more than 10 per cent, but the table clearly shows that the popularity of outdoor activities climbed more rapidly than that. It is noticeable that use of tramping trails and longer bush walks appears to have doubled. Furthermore, for both 1993 and 1995 more than 50 per cent of all overseas visitors went to a national, forest or maritime park while on their stay in New Zealand. The National Parks most visited were Fjordland, Mount Cook and Westland – all South Island parks. However what Table 15.1 very clearly shows is the rapid increase in usage. In particular, Westland experienced an increase of almost 50 per cent in visitors to about 284,000 in just

Table 15.1 Outdoor and ecotourism activities by overseas visitors in New Zealand, selected activities 1993 and 1995

Activity	1993	1995
Bush walks (<half a day)	310,000	422,755
Bush walks (> half a day)	70,000	153,746
Tramping/hiking	40,000	89,679
Whale watching	45,000	102,491
Rafting	80,000	89,679
Visits to National Parks		
Fjordland	275,000	303,819
Mount Cook	220,000	288,634
Westland	190,000	284,500

two years. That the increase in Fjordland has been held to 'only' a 10 per cent increase in numbers (from 275,000 in 1993 to about 303,000 in 1995) has been due to the use of booking systems related to hut capacities on the main tracks.

The market for outdoor activities can be segmented by nationality. For example, as shown in Table 15.2, in 1995, of those going tramping, five nationalities account for 56 per cent of the total. What Table 15.2 also shows very clearly is that the growth of interest in outdoor activities has climbed significantly, and in many cases faster than the rate of growth of overseas visitors to New Zealand, which in the same period grew by approximately one-third. However, when other active outdoor sporting activities are analysed, nationality apparently becomes more of a determining factor. For example, the data show that, for 1995, 10.7 per

Table 15.2 Numbers of overseas visitors undertaking selected outdoor activities – analysis by nationality

Activity/Nationality	Numbers 1993	% of group	Numbers 1995	% of group	% growth 1993–5
Tramping					
Australians	6,000	1.8	9,300	2.6	55
Britons	6,000	6.6	10,500	9.5	75
Americans	5,000	3.8	9,500	3.8	90
Germans	3,000	6.2	6,800	10.9	126
Japanese	2,500	2.0	8,500	5.6	240
Climbing, caving, moutain biking					
Australians	17,500	5.4	14,000	3.9	−17
Britons	11,000	12.0	13,700	12.4	24
Americans	10,000	7.6	14,000	10.7	40
Germans	10,000	20.8	12,200	19.6	22
Japanese	5,000	4.0	20,100	8.1	302

cent of Americans visiting New Zealand go climbing, caving or mountain biking, as do 8.1 per cent of Japanese visitors, 19.6 per cent of Germans, and 12 per cent of Britons. However, specific research into backpackers shows that this group accounts for a third of all such outdoor activities by overseas visitors, and that the common characteristic of this group is one of comparative youth (50 per cent are between the ages of 25 to 34). Table 15.2 also provides further evidence that the Japanese market is increasingly taking on the characteristics of the longer established European and North American markets. In interpreting the data presented it should be noted that the percentages relate to *all* overseas visitors by nationality regardless of purpose of visit.

It is therefore not surprising to record that the numbers of operators providing outdoor and ecotourism activities is high given the small size of New Zealand's population. Table 15.3 provides an estimate of those numbers derived from the *Annual Directories* of 1996 and 1997 of such operators (Hobbs, 1995, 1996). This is an incomplete listing, but represents the majority of outdoor and ecotourism operators which have been operating for at least three years. The distinction between outdoor operator and ecotourism operator is blurred. In a number of cases the outdoor operator is offering an adventure in a natural environment, but the product is the thrill, and any appreciation of the setting and its significance is incidental to the main product. In other situations operators have classified

Table 15.3 Numbers of outdoor and ecotourism businesses in New Zealand 1996 and 1997

Location	Outdoor 1996	Outdoor 1997	Eco-tour 1996	Eco-tour 1997
Northland	22	15	4	3
Auckland	34	29	1	2
Waikato and Waitomo	7	5	1	0
Coromandel	22	16	5	4
Bay of Plenty and Rotorua	20	20	2	0
Eastland and Hawks Bay	10	7	3	2
Taupo, Tongariro and Ruapehu	29	28	1	1
Taranaki, Rangitikei, Wanganui, Manawatu	13	11	0	1
Wellington and Wairarapa	3	6	0	0
Nelson, Marlborough and Kaikoura	37	35	12	7
West Coast	22	22	3	4
Canterbury	36	32	5	1
South Canterbury and Mount Cook	21	20	2	2
Queenstown and Wanaka	76	62	3	5
Fiordland	17	22	4	4
Otago	6	2	5	3
Southland and Stewart Island	4	3	4	3
Total	379	335	55	42

themselves as 'outdoor products' but a significant part of their appeal is the chance provided for relaxation in quiet, natural surroundings, with an opportunity for learning, while simultaneously having as little impact as possible upon the environment. Table 15.3 suggests there are approximately 350 adventure and 50 ecotourism operators (Hobbs, 1995, 1996) compared with 598 for Australia (Econsult, 1995). Given the respective sizes of population and numbers of overseas visitors between the two countries, this suggests that 'the outdoors' might be a more important component of New Zealand's tourism product than is the case of its Tasman neighbour. Blamey (1995) attempted to assess the growth of ecotourism in Australia, and to test the hypothesis that ecotourism is growing faster than tourism as a whole. It is interesting to note his conclusion that:

> Little evidence currently exists to support the claim that ecotourism growth is, *overall* [original emphasis], greater than that of inbound tourism as a whole. In part this reflects the lack of evidence. One might expect, however, that particularly high levels of ecotourism growth might be reflected in IVS (International Visitor Surveys) results pertaining to bushwalking, outback safari tours and scuba diving/ snorkelling. The average annual growth rate in visitation of bushwalkers, for example, was 11 per cent over the period 1989–1994, only slightly higher than the rate of 10 per cent for all international tourists, and less than the equivalent rate of 13 per cent for those indicating holidays as main purpose of visit.
>
> (Blamey, 1995: 145)

Arguably Blamey's findings are applicable to New Zealand. It has been noted that tramping has increased faster than the rate of growth of overseas visitors to New Zealand. Yet more detailed analysis is required. For example, do older visitors have different preferences, is their actual pattern of tramping different to that of younger groups, and to what extent do they require family-oriented outdoor pursuits? Research being conducted in New Zealand has only recently been concerned with such questions (e.g. Bassett, 1996).

Another means of examining ecotourism is through market segmentation based on psychographics, and here the evidence is complex. A large number of motivational studies of different tourist groups exist, and tourists have been categorised based on factors like their needs for relaxation, excitement, social interaction, friendship, status, ego enhancement, and a wish to explore and discover, or to revisit familiar locations. (One such study is that of Eagles using a 55-item scale developed for the Canadian Tourism Attitude and Motivation Study, 1992). Unfortunately limitations exist with many of these studies. Some, like those of Yiannakis and Gibson (1992) and Pearce (1982) report tourist typologies and their characteristics, but not the respective importance of each segment, while studies of eco-tourists may isolate such tourists *per se* from other categories, and hence interrelationships between the typologies are not discussed.

In part the problem is definitional, and might be encapsulated as one where it is asked, are eco-tourists simply tourists with an interest in nature, or are they 'amateur' botanists; that is people with high levels of interest and knowledge in nature who are prepared to experience discomfort in pursuit of their interest? If the former, then as Blamey (1995: 61) notes 'The results of the three nation-wide studies (Newspoll, 1994; Blamey and Braithwaite, 1995; US Travel Data Center, 1992) suggest that the potential ecotourism market is broad in terms of demographic and lifestyle variables.' There is also a significant problem that tourists are not uni-behavioural; that is, they can shift from one form of behaviour to another, and thus may move from one psychographic profile to another depending upon the needs of a particular moment, and the structure and opportunities presented by specific holiday locations. Pearce has noted, 'the comparative approach has yet to emerge as a distinctive, readily recognizable methodology in tourism research' (1993: 20). As a result, the literature has many motivational studies using techniques like factor and cluster analysis, but unfortunately researchers tend to use differing items in their questionnaires. The consequence is that no consistent finding has emerged as to the size of the eco-tourist market. One small-scale New Zealand study of 203 respondents noted the tautology that international tourists visiting ecotourism ventures were motivated by wanting 'to seek, learn about and capture nature and a unique outdoor environment' (Juni et al., 1996: 219), but it also found that such visitors also had an interest in a warm climate, inexpensive meals, trying new foods and having fun, with nationality also being an important variable. The authors support Bottrill's conclusion that 'while it may have been the "ecotourists" that created the demand for ecotourism, in reality they are unlikely to constitute more than a portion of the business for . . . new "ecotourism" ventures' (1992: 28).

For planning purposes, in New Zealand, the Department of Conservation's 1995 *Visitor Strategy* identified seven visitor groups who visited the National Parks and other areas that form New Zealand's 'Conservation Estate'. These were the 'short-stop travellers', 'day visitors', 'overnighters', 'back-country comfort seekers', 'back-country adventurers', 'remoteness seekers', and 'thrill-seekers'. Of these, the 'thrill-seekers' might more properly be called adventure tourists rather than eco-tourists because they engage in sporting activities like downhill skiing, paraponting, rafting, bungy jumping and snowboarding. From other studies in New Zealand (Ryan, 1995), it would appear that this group is primarily young and wanting a good social life based on bars and clubs in the evenings after their 'thrill-seeking', hence the natural setting of their activities has value primarily as a resource for adventure rather than as a subject for enquiry into the being of nature. For 'short-stop travellers' and 'day visitors' the natural environment has interest, but their participation is limited. The DoC describes the short-stop traveller as requiring 'high vehicle accessibility with visits of a short duration of up to one hour's length or associated with lunch/cup of tea/toilet stop/stretch the legs or natural attraction visit' (1995: 18). For this group, like the 'thrill-seeker', the natural environment is primarily an attractive background that helps to make more pleasant the prime activity in which they engage. 'Day visitors' however, seek 'an experience in a natural (or rural) setting with a sense of space and freedom. The degree of challenge and risk taken will vary considerably depending on the activity undertaken' (DoC, 1995: 19). While seeking the space of the countryside, their use of that space can vary considerably. Some are families enjoying walks and picnicking, others might be wanting opportunities for learning about conservation values, while others may be substituting for more 'high country' experiences because of time constraints. Like the previously discussed groups, this segment requires comparatively easy access and high road accessibility.

The remaining clusters share a want for remoteness, but vary as to length of stay and degree of comfort wanted for accommodation, and, in the degree of sociability wanted. 'Overnighters' are described by the DoC as representing 'the New Zealand family holiday experience' based on camping activities and campervans. The comfort required by the 'back-country comfort seekers' is of a type consistent with remoteness; that is DoC huts (possibly with a warden) with cooking facilities, running water, and long-drop toilet facilities. They tend to stay 2 to 5 days. 'Back-country adventurers' tend to avoid contact with people, requiring only a few basic facilities, while 'remoteness seekers' are similar in wanting very few facilities and being highly self-reliant. As a group these last two sections form but a small part of the total tourist population. The DoC notes that only 4 per cent of the New Zealand population goes tramping, while 7 per cent of New Zealand's overseas visitors actually went tramping (NZTB, 1994, 1996). These classifications are supported by other evidence emanating from New Zealand researchers. Higham and Kearsley (1994) report high levels of concerns from overseas visitors to Fiordland National Park that it was 'essential' that wilderness areas should be accessible only by foot. Concerns about over-crowding have been expressed about some parts of the conservation estate, and Kearsley and O'Neill (1994) report that Greenstone Caples in particular, and Milford Sound to a lesser extent, are already being perceived as being over-crowded to moderate or extreme degrees by over a third of all visitors. Warren and Taylor (1994) also point to pressures at Taiaroa Head (based on albatross watching) and Kaikoura (marine mammals watching), with a mixture of concerns about social and environmental impacts.

The case studies

The following three case studies have been selected to illustrate different types of eco-tourists, and also to assess the implications of various forms of organisation, product and activity within the context of ecotourism in New Zealand. All are quite typical of New Zealand, and have demonstrated financial sustainability because they have been operating in one form or another for some time – a feat not without some note in the world of small tourism business in New Zealand. The three cases studies are Mercury Bay Seafaris, Trek Whirinaki and Canoe

Safaris. The products being offered are, respectively, a dolphin watch opportunity, a two-day trek in the company of Maori, and a five-day canoe trip through the Wanganui National Park. They are all located in the North Island of New Zealand.

Mercury Bay Seafaris

Mercury Bay Seafaris have operated from Whitianga in the Coromandel since 1993 and is owned by Elizabeth and Rod Rae. Originally they had offered dolphin watching trips in the Kaikoura region. However, a mixture of concern about the growth of marine mammal watching activities and increased commercialisation of the operations, allied to a realisation that they were under-capitalised to compete with many of the newer companies, forced them to examine new options. They eventually sold their business, and after a search of more than twelve months, (during which time they were unemployed and survived on their savings), they eventually started their current operation. They are typical of many such small tourism businesses in New Zealand which are characterised by a love of a lifestyle and a want to communicate a passion to people. In the case of the Raes it is a love of dolphins, a want to communicate their interest to help other people understand dolphins and other sea animals, which is combined with long hours of work, business uncertainties and, at the end of it, an income which is less than if more conventional occupational paths had been chosen.

While the main focus of their interest is the 'Dolphin Quest', they offer two other products. Two reasons exist for this. First, there is a genuine desire by Rod Rae not to over-visit dolphins. Second, by offering two other types of trips, a mix within the total product portfolio is achieved which permits an appeal to a wider market. The two additional products still have an ecotourism theme. They are:

- The Seven Island Seafari, a boat trip of approximately 3 hours which visits various scenic bays, isolated volcanic rock formations, and permits snorkelling in an area which has a variety of fish including eagle rays. Also, penguins, gannets and other sea birds are usually seen; and
- The Glass Bottom boat trip of approximately 1½ hours duration. This is suitable for families with younger children, is entirely passive, and provides viewing of the sea bottom through glass.

When visitors book their 'Dolphin Quest' trip they are provided with written information about dolphins. The first page sets outs Rod's attitudes towards the trip, and it states:

> I would like to take this opportunity to talk to you about why you wish to make this trip. Most importantly, you as individuals need to think about **your** motivation for going to this expense, for the possibility of swimming with dolphins. If you think of this trip as an opportunity to

touch a wild animal or ride on its back you will be disappointed, indeed this sort of behaviour or a thrill seeking attitude we actively discourage. What is most needed out there is an open receptive mind, and extraordinary sensitivity.

The page ends with the statement:

If you require guarantees from us on sighting Dolphins in their natural environment you are probably not suitably prepared for this Trip.

We swim with Dolphins at Their invitation, They do not make appointments!

These last two sentences are also placed on a permanent sign on the interior of the boat so that all passengers are aware of the philosophy. The 'Dolphin Quest' poster outside of the Whitianga Visitor Information Centre, through which bookings are made, and the brochure advertising the trip also reinforce a message that one swims 'at the invitation of the Dolphins'. For Rod, humans provide entertainment for dolphins, if they want it, not the other way round.

Because Rod knows the waters well, it is rare that clients do not get an opportunity to view dolphins, and in many cases they do actually get a chance to be in the water near if not actually with dolphins. Therefore people are often able to hear the clicks and whistles of dolphins being carried through the water. For many this is indeed a highlight. In January 1996 the author spent a week asking clients before they went on the trip, what they expected to see, and then met them when they disembarked. None were disappointed. One German family had delayed their departure from New Zealand because of the opportunity to swim with the dolphins, something they had not been able to do at Kaikoura, and which for them was a 'special' opportunity. As it happened, again they were not able to do this, but they had been in the water and had actually heard the dolphins' sounds. They commented: 'the best thing was being able to hear the clicks of the dolphins in the sea', and all family members were highly satisfied with the trip.

The educational components of Rod's trip are clear in the written material that he produces, and such material is important for a number of reasons. It helps to overcome what he perceives as the 'conveyor belt' mentality of some operators. Thus he does not offer 'money back guarantees' that some marine mammal watch operators offer. To do this, he argues, is to make wild-life a commodity, while it also places commercial necessity above the interests of the animals. Also, because of the noise of the engines in what is quite a small boat (it carries ten passengers only), the written material means that if people miss a part of his commentary, they are still not being denied information.

That the boat is small raises the issue of being able to make the trips an economic proposition. Rod is aware that his engines are not ideal for the purpose; but investment in quieter engines requires more money. Revenue opportunities are also restricted by the length of both the tourist season and times when dolphins can be observed. The actual dolphin watch period is constrained by some important factors. The maximum period in a year when it is possible to run 'Dolphin Quest' has been estimated as being 212 possible operational days. The season for common dolphins is October to April, i.e. seven months. Bottle nosed dolphins are in the area all the year round and thus, arguably, the product could be extended, but Rod is aware of periods when dolphins need to be undisturbed. Weather is also a factor. Even during the summer months days can be lost due to not just bad weather, but also days when a high swell could mean discomfort to passengers. The Raes maintain records of operations, and data show 114 operational days for 1993/4, and 90 such days for the 'Dolphin Quest' in 1994/5. Given that these trips are running for so few days in the year, it reveals the vulnerability of an ecotourism operator not only to the problems of seasonality associated more generally with the tourism industry, but also to weather. Additionally, although there may be a given number of operational days, that does not mean that the trip actually runs on each of those days. The boat has to have a minimum of five passengers to cover running costs and to make a contribution to fixed costs, and if there are fewer than five passengers the Raes will cancel the trip for that day. They then offer passengers an alternative day, or an alternative trip, and through this mechanism each actual trip covers its costs.

They are able to adopt this policy because of the way in which they have financed their business. It is, in short, financed out of past savings and the sale of past assets. Their current boat was purchased from the funds realised by the sale of their business in Kaikoura, while the expenses of wetsuits for clients, cleaning facilities (the boat has to be taken out of water each day and cleaned to preserve its glass bottom), and other expenses were met from the proceeds of the sale of their house. That sale also financed them during the year in which they were seeking a good location for their business.

The financial vulnerability of small operators is further evidenced by the difficulty they faced when wetsuits to a value of NZ$4,000 were stolen. These had to be replaced, and this loss in the second year of their operation effectively turned trading profits into a loss. The Raes are also vulnerable to loss because high insurance premiums have meant that they are not insured as fully as they would like to be. Needless to say, the wetsuits now hang each night in their living room. Rod and Elizabeth also found that the reality of their operation bore little resemblance to their business plan, partly because the number of days of bad weather in their first year of operation was greater than anticipated. Thus, while hoping to move into profit in the second year of the business, the losses being carried forward from the first year of operation, combined with the theft of wetsuits, resulted in a further loss. Fortunately, in the third year of the business, while the season started late, high levels of demand continued well into the autumn. However, in late 1996/early 1997 the business was adversely affected at its busiest time by the arrival of two cyclones which lashed the Coromandel Peninsula. Favourable word-of-mouth reports, coverage by a major newspaper, recommendations by local motel operators and good weather combined to produce small profits in the third year of operation, but the cyclones put the business back in 1997.

Comments in the visitors' book indicate high levels of customer satisfaction – indeed many comments can only be described as 'enthusiastic' with some clients recording that the trip was indeed an experience of a life-time. Rod and Elizabeth have established a product that helps to create a better portfolio of tourist experiences for visitors to Whitianga. They have remained true to their ethics of what they view an ecotourism product to be. Currently they have a local monopoly position as they are the only Department of Conservation licensed operation in the Whitianga area. But they work long hours, and like many tourist operations have to earn enough during the main season to sustain them through the whole year. Other ways to greater economic security such as a larger boat that could carry more passengers remain beyond their current financial abilities, but at present they are beginning to feel more hopeful about their prospects than they have been for many years. However, their economic viability is dependent upon a local monopoly based on possession of a Department of Conservation licence. In 1997, the regional Departmental office had applications from seven other would-be dolphin watch operators.

Trek Whirinaki

Trek Whirinaki is also the result of an individual's perseverance and dreams. Chris Birt is an experienced journalist, and one furthermore who has written about and commented upon tourism developments in New Zealand for more than a decade. For five years prior to the development of Trek Whirinaki he cosseted and nurtured his own dream of being able to take tourists through Whirinaki Forest Park in the Central North Island. The Whirinaki Forest Park is immediately to the west of the Te Urewera National Park, and shares the same geographical features. Covering an area of 60,000 hectares it represents a prime example of podocarp rainforest, with *matai*, *miro*, *rimu*, *totara* and *kahikatea* trees – the last named being New Zealand's tallest native tree, reaching 65 metres. Below the canopy formed by all of these trees (which range from 40 to 65 metres in height) can be found a wide range of smaller native trees and ferns, including *tawa*, *kamahi*, northern *rata* and *rewarewa*, while the conditions support numerous mosses and lichens. The area now occupied by the forest parks prompted bitter dispute in the 1970s as conservationists sought to protect the unique botanical features of the area, and in 1984 the logging of forest fringes was halted. Local sawmilling then had to rely on exotic pine plantings (that is, non-native trees), but even that stopped when a total ban on logging was enforced in 1987.

The area is not simply rich in botanical life, for it is also the home of Ngati Whare, a *hapu* or sub-tribe of the *iwi*, Tuhoe, sometimes known as the Children of the Mist. The *tangata whenua* (original people of the land) came from the coast six hundred years ago, and old *Pa*, or Maori fortifications, are evident throughout

Whirinaki Forest. The combination of forest and Maori habitation has given rise to an interweaving of economic and social conditions unique to this area, but nonetheless representative of the situation in which other indigenous peoples have found themselves. First, there exists the folklore which Maori have established, which includes a detailed knowledge of the forest and its plant life extending to the medicinal uses of the plants. Second, the area became historically important for Maori as an area within which traditional *marae* life continued, with all that implies for concepts of self-identity. It was, for example, an area of religious re-birth based on Maori concepts in the early twentieth century. Third, an economic dependency emerged from the 1920s when logging in the area commenced, for the timber industry employed almost all of the adult male Maori population in areas like Te Whaiti, the major settlement of the area. Indeed, communities like that of nearby Minginui were solely dependent on the logging and sawmills for their livelihood. Consequently the abolition of logging in 1987 resulted in unemployment rates of over 90 per cent among Maori, and a heavy dependency upon benefit payments from various government agencies. Wholesale migration was not an option for Ngati Whare because of the historical and emotional ties of Maori to the land as described by Skerrett-White (1996) in the expression: '*I Ka Tonu Taku Ahi i runga i Toku Whenua*' (My fire has always been kept alight upon my people's land).

Chris Birt felt, therefore, that the development of an ecotourism product in the area would achieve a multitude of purposes. It would represent an economic opportunity for Ngati Whare and it would help people to better appreciate the importance of areas like Whirinaki, and the need to conserve such forests. The conservationist movement, it can be argued, needs not only a 'warm fuzzy' feeling to help sustain it, but also an informed opinion as to the ecological role of such forest areas, and their ability to act as reservoirs of bio-diversity which might represent as yet unused assets for humans. In the current situation in which New Zealand finds itself as it debates the respective role of Maori and European traditions, such a tour programme might help further understanding of the Maori perspective. A fourth reason, within the context of New Zealand tourism, was an increasing concern about the over-crowding of South Island guided treks, and hence the possibility existed that a new North Island trek might help relieve not only current pressures, but future demand as the number of overseas tourists increased.

The development of Trek Whirinaki, as already indicated, took more than five years before it was launched in 1993. There were three reasons for this. The first was the need for capital which meant that Chris had to continue working at his profession of journalism and publishing. Of more importance was his wish to ensure the product was of a high quality before it was presented to the marketplace. To attain the standard he wanted meant that it was essential that local Maori agreed with and were part of the concept as it involved their traditional lands. Additionally the tracks and camp sites had to be in areas that could sustain the impacts of visitors. The involvement of Maori is today evident in the experience that visitors have. Maori act as track guides providing an insight into local legends and myths, and giving information about the functions of plants and the history of the locality. The trek ends with a *marae* (meeting house) visit which is accompanied by the traditional Maori protocol of a *powhiri* (greeting), *hongi* (the rubbing of noses) and a *hangi* meal cooked in an earth oven in the traditional way. The purposes and symbolism of such protocol is fully explained to clients, and many find it an emotional experience of great intensity with which to finish their trek. Such current involvement was not easily won – elders (*kaumata*) had concerns about the possible impacts of tourists, the nature of their involvement with a *Pakeha* (European) operator, and about the types of knowledge which could be given to outsiders.

The actual trek programme provides two- and three-day tours, although the actual walking distance covered is quite short (15 kilometres). There are many stops for receiving information, and it is a walk easily handled by anyone of moderate fitness. Chris has also sought to ensure that the camp stops are far from spartan, while at the same time complying with his own wishes of minimal environmental impact and the need to meet legislative requirements. For example, the Local Government Act of 1974 states that no washing may be done in rivers or streams, while the 1956 Health Act (section 60) states it is an offence to pollute the water 'in such a manner as to make the water dangerous to health, or offensive,' while section 39 of the same act requires 'suitable appliances for the disposal of refuse water in a sanitary manner, and (c) sufficient sanitary conveniences'. These acts were reinforced by New Zealand's primary legislation concerning impacts upon the environment, namely the Resource Management Act of 1991 which specifically identifies the importance of water resources for special policies. Chris therefore had to face the fact that the provision of camp sites with running water (for example, showers are provided) needed to be carefully sited, planned and constructed in accordance with Department of Conservation standards. Fortunately the technologies required are simple and comparatively low cost. For example the disposal of plain washing water can be easily coped with by using a three-tiered plastic filter box about 40 cm square and 1 metre high. Water is poured in at the top and passes through a mesh sludge filter; it then drips through 'trickle' holes and various elevation and aeration shelves where oxygen is absorbed and minerals removed by centrifugation before finally passing through a purification block with a biodegradable acidity regulator. Finally, slow seepage into the soil is permitted so as not to cause any disturbance of soil patterns. In the case of water used for showers and drinking, filters are used to remove contamination by cysts carrying *giardia lamblia* which is found in animal faeces and can cause severe stomach upsets. Of necessity this overview of camp construction is brief, but nonetheless it indicates the considerations that every ecotour operator involved in providing camp accommodation in a forest location must face. The use of permanent camp sites is itself one way of reducing adverse impacts caused by using several sites, for the permanent site can be located in areas which are the most resistant to human impact. Additionally, human impacts are localised and not diffused. It should be noted in passing that New Zealand has a major problem with opossum, of which there are thought to be 80 million, and which threaten the flora of the forests, as well as being carriers of tuberculosis and a source of *giardia* cysts – hence concerns over water filters.

Trek Whirinaki has opted for a fly in–walk out programme for its treks. There are a number of reasons for this. First, it permits clients to have an overview of the vast mountainous terrain and to appreciate the density of the canopy. It also delivers clients quickly to the start of the trek, thereby reducing the time and effort required.

Since operations began in 1993, the number of clients to date has been modest, but overseas marketing is now beginning to generate increasing interest. Chris Birt at first introduced the trek to a number of key 'decision takers' in the New Zealand tourist industry through familiarisation tours. Such people included key marketing personnel from companies like Air New Zealand, the Mount Cook Group and Japan Travel Bureau. The linkage with these operators has proven useful, for Trek Whirinaki is now featured in brochures throughout the world. In doing this Chris has recognised a number of marketing realities. First, the need to charge realistic prices that reflect the quality of the product, and the commission structures of the industry. Arguably many small New Zealand tourist operators under-charge because they are used to a domestic market that they perceive is reluctant to pay realistic prices for quality products. This means that they have difficulties when marketing to the international market because they have not permitted sufficient margin for the payment of commissions to inbound tour operators or overseas wholesalers who are necessary links in the marketing chain. Because of his knowledge of the industry derived from many years of tourism-related journalism, Chris has avoided this problem with the result that he can deliver a commissionable product of a standard that permits an overseas wholesaler to be confident that its clients will be satisfied with their experience. Trek Whirinaki sustains the image of quality while further helping to increase the options these agents offer their customers. Chris has therefore solved a major marketing problem that many small ecotourism operators face.

Canoe Safaris

Canoe Safaris was started in 1979 and currently offers escorted canoe trips on the River Whanganui through the Whanganui National Park of five days duration, and a four-day Upper Mohaka rafting expedition on the Mohaka river in the same region. The Whanganui National Park is also a podocarp forest of *rata*, *rewarewa*, *rimu*, *tawa* and *akowhai* trees with tree ferns and plants clinging to steep river banks and gorges. Because the river offered navigable access to the interior and the Central Plateau of Tongariro, it is also an area rich in Maori and early colonial history. The *Te Atihaunui a Paparangi* people settled the valley from about 1100 AD, while the town of Wanganui at the mouth of the river became one of the major European settlements quite early in New Zealand's European history.

In 1891 a regular river boat service began carrying mail, freight and passengers to the settlers between Taumarunui and Pipiriki, and in the early twentieth century this river trip was compared, as a tourist experience, with the great river journeys of Europe. Many contemporary photographs show Edwardian ladies and gentlemen alighting at different points of the river, and the remains of these landing stages are pointed out to clients as they canoe down the river.

The Whanganui trip offers two main experiences. The first is an opportunity to canoe through a comparatively unspoilt area of natural vegetation; to view waterfalls, geological features and flora. The second is to learn about the area's early history, both Maori and *Pakeha* (Europeans). It is almost impossible not to do this. While canoeing it is easy to see the pole marks on the rock sides which the early settlers and Maori used to pole their boats upstream. The water course was changed in many places about eighty years ago as weirs were blasted to make possible the use of steamers on the river, and in doing so the Maori fishing traps were rendered useless. The tribulations and disputes of the past, and the present, are also alive on the River Whanganui. Clients have an opportunity to visit the 'Bridge to Nowhere', about 40-minute's walk from the river at Upper Mangapurua. This is one of the last remains of an ill-fated attempt to settle the land granted to returning soldiers from the First World War – an attempt which lasted over two decades. If they wish, clients can also land at Tieke Marae. Tieke was the site of an old *Pa*, or Maori fortification, and subsequently it became part of the Conservation Estate. However, in 1991 it was occupied by local *iwi* (tribe) in pursuit of a land claim, and is still being occupied. Visitors can land, but they have to comply with Maori protocol, including participation in a *powhiri* (formal welcome), a *hongi* (rubbing of noses) and the leaving of *koha* (a donation). The *iwi* are pleased to explain the reasons behind these protocols, and to share their food and hospitality. Today the *Marae* serves a practical purpose because urban Maori drug addicts and others are sent there to recover within a Maori support system.

Accommodation on the trip is based on camping at Department of Conservation sites. These sites provide running water and long-drop toilets, and flat areas for pitching tents. Cooking facilities and benches are also available at each of the sites en route. The total trip takes about five days, requiring four nights at such sites. It is possible to canoe the stretch of river used by Canoe Safaris in three days, and many independent canoeists do this, but Simon Dixon (the owner) stresses to his clientele that this is a holiday, not an expedition. In addition, the extra time makes it possible for groups to land on shore, visiting sites like the 'Bridge to Nowhere' as well as Te Tieke *marae*. In addition it allows time for clients to unload their canoes and retake the river rapids more than once if they want an extra 'adventure'.

For Simon, success has come through a continuous effort to be 'the best', for other operators also offer trips on the river, and, as stated, it is possible to canoe the river independently. Indeed some operators simply hire canoes and offer a pick-up service from where their client disembarks. Hence competition exists, and basic products can be offered with minimal barriers to entry. One barrier to competition is the need to obtain a DoC licence, but as just described, the canoe hire and pick-up service bypasses this obstacle. Additionally, unlike other areas, in the Whanganui National Park the DoC will grant a licence to any operator able to meet the conditions of showing proficiency. No quota system is operated. Simon has therefore to adopt a policy of 'being the best', but 'being best' does not come cheaply. Canoes have to be replaced frequently as Simon believes in offering the best design for the conditions the river poses; and thus older but still serviceable canoes may be replaced if better designed canoes become available. Additionally, tents have to be replaced, at the end of every second season at least. Consequently in each of the last two years replacement and upgrading of equipment has required investments of $20,000 in canoes and tents.

Canoe Safaris maintain an office, reception area and maintenance depot in Ohakune. This is to a high standard and clients start and end their trip at this base when tea, coffee and light refreshments are provided. It is next to a good 3-star quality motel with fourteen units and two chalets which also has a restaurant, so the complex meets clients' needs at the beginning and end of the trip. Staffing is a key factor and river guides are selected for their technical skills and personality, and are provided with reading materials and other information to learn the history of the region. Staff–client ratios are about one member of staff to five or six clients. The company also takes great care over catering and clients enjoy at each camp site in the evening three-course meals of a high standard, and in great quantity. Everything is designed to sustain the price charged, which while higher than that of some competitors, is not too much more and is readily justified by the extra quality. Some of that quality is achieved by attention to the margins. For example, the provision of good and varied menus does not cost significantly more than providing less exciting food. Care taken over the selection of good equipment and guides means that the guides tend to return for the following season because they have confidence in each other and the company.

The company has established a good reputation, and although brochure distribution is important many clients come on the trips through word-of-mouth recommendation. Recently Canoe Safaris diversified into new products and rafting trips involving overnight camping have been started on another river in the area. A potentially important development for the company is that for the 1996/7 summer season it is being featured in an Air New Zealand brochure which is aimed at the overseas market.

How economically viable is Canoe Safaris? The number of passengers taken on canoe and raft trips in 1995/6 was approximately 300 in total. In short, the business, while well established, and sustainable in the sense that it has continued since 1979, does not provide sufficient income for Simon himself to live and to invest in the business. This means that each winter he has to find employment elsewhere, and due to his commitment to Canoe Safaris it means that such employment is always of a temporary nature during the winter months.

Discussion

The first observation that can be made is that, in terms of the market segments to which these companies promote their products, none could be easily categorised using the segmentation devised by the DoC. The Dolphin Watch operation appealed to a broad spectrum of people. Perhaps the easiest match between product and categorisation was that of Canoe Safaris, where it might be said the product appealed to 'back-country comfort seekers' – people who wished to have access to what was normally not accessible, but with the assurance of skilled guides and the basics provided by DoC camp sites. Trek Whirinaki on the other hand seemed to appeal to the same market segment, albeit perhaps those with more limited time. It is of interest to note that it seems that in the initial stages of the product the clients were primarily motivated by the opportunity to trek in the rainforest, and the contact with Maoridom has come as an unexpected and, at times, intensely emtional experience. As the quality of this experience has become better known, the cultural component of the product is coming more to the fore, thus representing a component not used by the DoC in their classification.

What do these case studies indicate about ecotourism in New Zealand? The first comment to make is that they are generally indicative of the state of the business. Warren and Taylor comment that:

> The industry Eco-tourism operators are widely dispersed . . . Businesses are typically small, run by owner-operators. Some employ a small number of people – but employees are often family members and employed only on a part-time or seasonal basis . . . Some operators are passionately committed to passing on knowledge, educating tourists through the depth of information they provide . . . a significant proportion of operators have entered the industry under-capitalised and without fully anticipating all the costs of running a business.
>
> (1995: 5)

These case studies confirm this observation. Growth for ecotourism businesses of necessity must be constrained by the nature of the impact that increasing numbers of visitors have upon the environment and the quality of the experience being offered. Hence each of the companies involved has sought to achieve growth by establishing new products in new areas. Mercury Bay Seafaris achieve higher boat utilisation by offering three products and taking clients to different areas so as to restrict the number of daily visits to where dolphins are commonly found; Trek Whirinaki has plans whereby it can expand by using other footpaths and *marae* different to those being currently used, while Canoe Safaris are able to utilise another river. All three are fortunate that there are sites available for this to occur, and given the current levels of New Zealand tourism, and New Zealand's population of 3.6 million, such sites are sufficient to absorb higher levels of growth. A second common theme is that all operate under licensing arrangements with the Department of Conservation. While the Department does not intrude in the running of the businesses, the licensing does limit use of the natural areas, and hence imposes some, albeit few, constraints upon competition, which permits the companies to obtain some return upon their investment and effort. Given that all three examples are very dependent upon a seasonal business, this potential limiting of competition in their immediate area is important for financial viability. Yet, in one sense, all are in a highly competitive environment, for as Table 15.3 shows, there are a number of ecotourism operations in New Zealand which all compete for domestic and international visitors.

As for many businesses, growth is a natural outcome of success, and the question has to be asked as to why this is the case in ecotourism where, it might be thought, there would be good reasons for believing 'small is beautiful'. After all, small levels of operations mean reduced environmental impact and, arguably, enhanced visitor experiences. However, small levels of operation are insufficient to produce required levels of income, even for the least commercially inspired; while any potential growth which is not exploited means that an opportunity is created for competing companies. For example, if Rod Rae did not offer sightseeing trips, another local fisherman would, and this could undermine Rod's economic viability. If Canoe Safaris did not exploit the other rivers, other companies would. If Chris Birt did not negotiate with other *marae*, Maori from those *marae* would seek other partners for trekking holidays. Growth may not be sought in ecotourism, but failure to achieve growth threatens survival as a business. Additionally, there is always the thought on the part of the entrepreneur that growth opens the possibility of employing staff who can relieve the burden of administration, thereby permitting the entrepreneur to return to what it is he or she most enjoys. In conversations with the author, New Zealand eco-tourist operators have stated that they entered their business because of a love of their environment, and their businesses become a means of maintaining a life they want to lead while enabling them to share their enthusiasms with other people. Growth threatens this by absorbing more of their time in administration, but yet further growth holds the promise of appointing managers so permitting them to return to the lifestyle they wanted. Tensions thus exist within small ecotourism operations. Financial viability and competitive need requires growth, yet growth is arguably the antithesis of what ecotourism is supposed to represent.

The question arises as to whether financial viability is possible in a non-growth situation, and perhaps the only way this can occur is through external imposition of controls that inhibit further operations from existing or new companies. This would have the effect of establishing local monopolies who could charge accordingly. But practically, within New Zealand, how is this to be achieved? Further, should access to nature and an educative experience be confined to only those able to pay large sums of money? The Department of Conservation has a role to play in terms of issuing licences for the estate it controls, but as noted, the Department operates under financial constraints. Although the 1996 budget sought to address some of these by promising an additional $100 million over three years, part of this money was oriented towards improving visitor facilities; in short, the DoC was to encourage visitors. From a conservationist viewpoint this does have the merit of permitting some zoning activity to be carried out in indirect ways. However, not all ecotourism activities take place on Conservation Estate, and controls are limited. The Resource Management Act of 1991 is not beyond criticism in this respect, and in any case permits development where social enhancement can be shown to occur; that is, development can be permitted, albeit perhaps subject to some controls, if jobs are created, even if there is some environmental loss. Recent cases in the Coromandel have also shown that conservation groups face difficulties when seeking to enforce their rights under the legislation as they do not possess the financial resources available to business interests (see Ryan and Cheyne, 1996). Finally, there is a wider political tension as the structure of small ecotourism businesses seems to imply that financial viability can only be sustained through policies that inhibit the working of market forces. However, the recent history of New Zealand since 1984 has been one of a rejection of protectionism, a retreat from a central governmental role in a wide range of commercial and social issues, and a greater reliance on the market as an allocator of resources. If, therefore, ecotourism is to be seen as a special form of tourism to be exempted from current market policies by claims of environmental sustainability it needs to ensure that such claims can indeed be substantiated. Already it has been argued, particularly by Wheeller (1993), that ecotourism presents no solution to the problems of mass tourism, and indeed compounds the problems by introducing tourism to previously unspoilt areas. Additionally it has been said that small-scale tourism is not without disadvantages for certain groups of workers such as females (see Smith, 1995).

The three case studies illustrate that ecotourism which is educative and environmentally sensitive is being purchased at a cost not often considered – namely that small size means small financial reward to the entrepreneurs who show tenacity and initiative in maintaining their companies. In a sense they are typical of many small companies where the motivation of making money is second to the want to sustain a desired lifestyle and interest. Nationally it means that, as New Zealand attracts more tourists as evidenced in the first section of the chapter, any increase in ecotourism can only be sustained by introducing products in new areas. Generally, existing operators do not have the capital to diversify into new locations, so it is new entrepreneurs who are attracted to the industry. It raises a real issue as to whether ecotourism in New Zealand is condemned to a duplication of small companies increasingly dependent upon overseas markets and the marketing strategies of intermediaries in the channel of distribution over which they have little or no control. However, one way forward for such companies is the creation of coalitions of operators offering products in partnership with each other, and there are some signs that, encouraged by District and Regional Tourism Organisations, this is beginning to happen. This perhaps might be the next stage of a future story. Certainly there are no easy answers, and for ecotourism operators, the lack of such answers means that they are dependent upon their own resources, which, as demonstrated by the above case studies, in New Zealand, are often limited.

Acknowledgements

The author would wish to acknowledge the help of Elizabeth and Rod Rae, Simon Dixon, Chris Birt, the Smith Family of Auckland and other clients of the companies concerned. Additionally, the comments of Eric Laws and Gianna Moscardo have also been of help.

References

Bassett, B. (1996) *Tourism Research Bibliography 1994–1996*, Wellington, New Zealand, Tourism Policy Group, Ministry of Commerce, Te Manatû Tauhokohoko.

Blamey, R.K. (1995) *The Nature of Ecotourism*, Occasional Paper no. 21, Canberra, Bureau of Tourism Research.

Blamey, R.K. and Braithwaite, V.A. (1995) 'A Social Values Segmentation of the Potential Ecotourism Market', paper presented to International Geographical Union Symposium on the Geography of Sustainable Tourism, University of Canberra, 2–4 Sept.

Booth, K. (1996) 'The Intrusion of Noise by Scenic Flights at Milford and Mount Cook National Park', in G.W. Kearsley (ed.) *Towards a More Sustainable Tourism, Proceedings of Tourism Down Under II – A Tourism Research Conference*, Otago University, 3–6 December.

Bottrill, C.G. (1992) 'Ecotourism in British Columbia: theory and practice', unpublished MA thesis, Department of Geography, University of Canterbury, Christchurch, New Zealand.

Braun-Elwert, G. (1995) 'The Natural Quiet', in M. Hobbs (ed.) *New Zealand Outside Annual and Directory, 1996 edn*, Christchurch, Southern Alps Publications Limited, pp. 10–11.

Department of Conservation (1993) *Greenprint Overview – The State of Conservation in New Zealand*, Vol. 1. Wellington, Department of Conservation/Conservation Te Papa Atawhai.

Department of Conservation (1995) *Visitor Strategy*, Wellington, Department of Conservation/Conservation Te Papa Atawhai.

Eagles, P. (1992) 'The Motivations of Canadian Ecotourists', in B. Weiler (ed.) *Ecotourism: Incorporating the Global Classroom*, Canberra, Bureau of Tourism Research, pp. 12–19.

Econsult Pty Ltd (1995) *National Ecotourism Strategy Business Development Report* (for Commonwealth Department of Tourism), Melbourne, Econsult.

Forest and Bird (1995) *Conservation News*, August, Royal Forest and Bird Protection Society of New Zealand, Wellington.

Higham, J. and Kearsley, G.W. (1994) 'Wilderness Perception and its Implications for the Management of the Impacts of International Tourism on Natural Areas in New Zealand', in J. Cheyne-Buchanan and C. Ryan (eds) *Proceedings of Tourism Down Under Conference*, Massey University, pp. 505–30.

Hobbs, M. (ed.) (1995) *New Zealand Outside Annual and Directory*, 1996 edn, Christchurch, Southern Alps Publications Ltd.

Hobbs, M. (ed.) (1996) *New Zealand Outside Annual and Directory*, 1997 edn, Christchurch, Southern Alps Publications Ltd.

Juni, B., Cossens, J. and Barton, R. (1996) 'Ecotourism – An Examination of the Motivations of Ecotourism Visitors to New Zealand', in G.W. Kearsley (ed.) *Proceedings Towards a More Sustainable Tourism – Tourism Down Under II – A Tourism Research Conference*, University of Otago.

Kearsley, J. and O'Neill, D. (1994) 'Crowding, Satisfaction and Displacement: The Consequences of Growing Use of Southern New Zealand's Conservation Estate', in J. Cheyne-Buchanan and C. Ryan (eds) *Proceedings of Tourism Down Under Conference*, Massey University, pp. 171–84.

Marshall, D. (1996) Address to New Zealand Tourism Research Association meeting, at New Zealand Tourism Board, 27 March.

Newspoll (1994) *Ecotourism Study* (for Social Change Media Pty Ltd and Commonwealth Department of Tourism), Melbourne, Newspoll Pty Ltd.

New Zealand Tourism Board (1994) *International Visitor Survey, 1993*, Wellington, NZTB.

New Zealand Tourism Board (1996) *International Visitor Survey, 1995*, Wellington, NZTB.

Pearce, D. (1993) 'Comparative Studies in Tourism Research', in D. Pearce and R.W. Butler (eds) *Tourism Research; Critiques and Challenges*, London, Routledge.

Pearce, P. (1982) *The Social Psychology of Tourist Behaviour*: International Series in Experimental Psychology, Vol. 3, Oxford, Pergamon Press.

Ryan, C. (1995) 'Perceptions of the Rangitikei by Residents of Palmerston North', unpublished report, Department of Management Systems, Massey University.

Ryan, C. and Cheyne, J. (1996) 'The Resource Management Act and the Case of Taupo Bungy', in M. Oppermann (ed.) *Proceedings of Tourism in the Pacific Conference*, Wairikei Polytechnic, Rotorua.

Skerret-White, C. (1996) *Daughters of the Land, Nga Uri Wahine a Hineahuone*, Exhibition and accompanying publication, Bathhouse Museum, Rotorua.

Smith, V.L. (1994) 'Privatization in the Third World: Small-Scale Tourism Enterprises', in W. Theobald (ed.) *Global Tourism: The Next Decade*, Oxford, Butterworth Heinemann, pp. 163–173.

US Travel Data Center (1992) *Discover America: Tourism and the Environment: A Guide to Challenge and Opportunities for Travel Industry Business*, Travel Industry Association of America, Washington, DC.

Warren, J.A. and Taylor, C.N. (1994) *Developing Ecotourism in New Zealand*, Christchurch, New Zealand, Institute for Social Research and Development.

Wheeller, B. (1993) 'Sustaining the Ego', *Journal of Sustainable Tourism* 1(2): 121–9.

Yiannakis, A. and Gibson, H. (1992) 'Roles Tourists Play', *Annals of Tourism Research* 19(2): 287–303.

16

MANAGING TOURISM IN THE WET TROPICS WORLD HERITAGE AREA

Interpretation and the experience of visitors
on Skyrail

Gianna Moscardo and Barbara Woods

Introduction

Interpretation and the management of tourism

Interpretation can be defined as the process of communicating or explaining to people the significance of the place they have come to see, so that they enjoy their visit more, understand their heritage and environment better, and develop a more caring attitude towards conservation. This definition is derived from that set out by the Society for the Interpretation of Britain's Heritage (1996). Interpretation shares some of the features of environmental education with the important exception that it is primarily focused on people at leisure rather than in formal educational settings. This exception results in a stronger emphasis being given to the importance of enhancing visitor experiences in descriptions of interpretation (Knudson *et al.*, 1995). These key characteristics of communication, enhancing experience and creating conservation support amongst visitors suggest that interpretation could be an important strategy in the management of tourism.

Tourism has attracted much criticism, mostly concerned with the negative consequences of tourism for the places and people who act as hosts. Jafari (1990), in his history of research into tourism, notes the predominance of a 'cautionary platform' which refers to a large and vocal group of writers and researchers who have focused attention on the negative social, cultural, economic and environmental impacts of tourism. This criticism of tourism has come not only from within the world of academic researchers and cultural commentators, but also from communities, various interest groups within communities and from the world of government and policy makers. As a result of these criticisms there has been a substantial move towards applying the general principles of sustainable development specifically to tourism. In tourism research Jafari has noted a change towards a 'knowledge-based platform' where systematic and objective assessments of the sustainability of certain practices have become more common.

Definitions of sustainable tourism emphasise three important features.

- *Quality* Sustainable tourism provides a *quality experience* for visitors, while improving the *quality of life* of the host community and protecting the *quality of the environment* (Inskeep, 1991).
- *Continuity* Sustainable tourism ensures the *continuity of tourism* as an economic activity through the provision of satisfying experiences for visitors, *continuity of the natural resources* upon which tourism is based and the *continuity of the culture and lifestyle* of the host community (Wall, 1993).
- *Balance* Sustainable tourism *balances* the needs of tourists, operators, host communities and the environment (Nitsch and van Straaten, 1995; Bramwell and Lane, 1993).

According to Lane (1991: 2) ecologically sustainable tourism is tourism that provides visitors to a destination with 'an in-depth understanding and knowledge of the area, its landscapes and peoples'. Such an understanding should result in tourists who are concerned about and protective of the destination area. The similarities between these definitions of sustainable tourism and the definitions of interpretation previously described are clear.

At its simplest, interpretation can be seen as any activity which seeks to give tourists information about the place they are visiting and it can contribute to the management of tourism in several ways.

Interpretation can:

- educate visitors about the nature of the destination and inform them of the consequences of their behaviour thus encouraging them to act in more appropriate ways;
- develop visitor support for environmental conservation and management activities;
- relieve pressure on sites by encouraging visitors to go to less crowded or sensitive places and by providing them with alternative experiences;
- enhance the quality of visitor experiences adding value to tourism products.

Tourism and the Wet Tropics World Heritage Area

The Wet Tropics World Heritage Area (WTWHA) is an area of approximately 900,000 hectares in a region covering much of the coastal hinterland of north-eastern Australia between Townsville and Cooktown. Figure 16.1 provides a map of this region and the key locations which will be referred to in this text. The area was inscribed on to the World Heritage list in 1988 and conserves the largest area of tropical rainforest on the Australian continent. These tropical rainforests

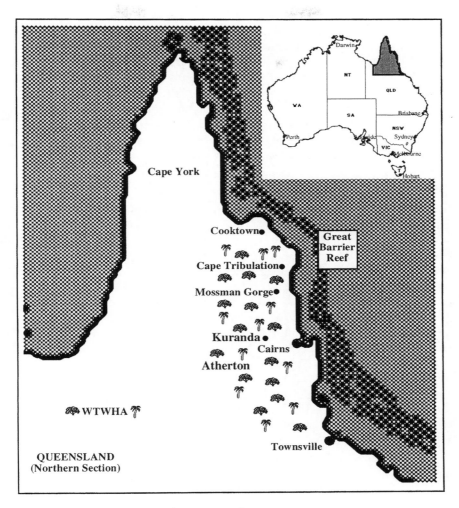

Figure 16.1 Map of Far North Queensland region

are one of the two major tourist attractions for this region, the other being the Great Barrier Reef World Heritage Area. It has been estimated that 1.4 million international and Australian tourists visit the WTWHA annually (Tillack, 1996), with a doubling of these numbers projected for the year 2001 (Driml and Common, 1996).

The *Draft Wet Tropics Plan* (Wet Tropics Management Authority [WTMA], 1995) highlights two major and related problems associated with tourism use of the area; crowding and congestion at a limited number of sites, and conflicts between tourist and local resident use. In a report on tourism prepared for the

Wet Tropics Management Authority (1992) this concentration of tourism use was also noted. The most popular places for tourism use are Cape Tribulation and sites along the road to Cape Tribulation, Mossman Gorge, Kuranda and the lakes of the Atherton Tablelands region. All these locations are in the northern section of the WTWHA reflecting the importance of Cairns as an accommodation base for tourism in this region (Driml and Common, 1996). There is growing concern amongst both managers and tour operators that the numbers of visitors to these sites may exceed the biophysical capacities of these sites and it is common to suggest that the region needs strategies for diverting visitors away from these high pressure sites (Driml and Common, 1996; WTMA, 1992).

The Skyrail Rainforest Cableway

In 1995 the Skyrail Rainforest Cableway was opened for business. This cableway extends 7.5 kilometres from Smithfield (a northern suburb of Cairns) to Kuranda (a shopping village in the rainforest) (see Figure 16.2 for a map of the route). Visitors travel in six-person cabins or gondolas just above the rainforest canopy and can stop at two stations along the route. One station, Red Peak, offers a rainforest boardwalk, while the other, Barron Falls, provides visitors with several lookouts over the Barron Gorge and Falls and a rainforest interpretive centre. The centre uses videos, interactive computer displays, touch tables and standard exhibits to provide visitors with rainforest information. The operation provides visitors with a unique perspective on the tropical rainforests of the region.

Much publicity has been given to the strategies used to manage the environmental impacts associated with both the construction and operation of Skyrail and a summary of these is given in Table 16.1. Of core interest to the present chapter is the extensive interpretation offered to visitors. In addition to the interpretive centre previously listed, a series of signs has been placed at both the stations and along the Red Peak boardwalk and trained interpretive staff are available to take guided walks and to provide impromptu talks and answers to visitor questions. Skyrail has the goals of providing a positive ecotourism experience which effectively communicates information about the rainforest to visitors, and relieving pressure on other WTWHA sites by providing an easily accessible opportunity for large numbers of visitors to experience rainforest. Skyrail thus offers the potential to reduce the crowding and congestion that has been noted as a major problem for some sites in the WTWHA.

The questions that were asked

This chapter reports on an analysis of the role of interpretation in the management of tourism using Skyrail as a case study. The core questions that this case study sought to answer were:

1. Does Skyrail's interpretation contribute to the management of visitors to the WTWHA? Do visitors to Skyrail learn about the rainforest? Does it relieve

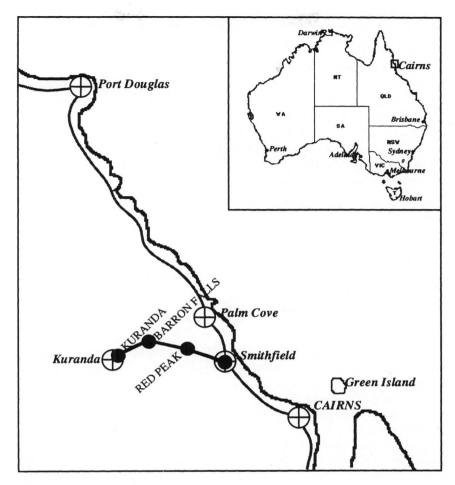

Figure 16.2 Map of Skyrail

pressure on other sites? What features of the interpretive program are particularly effective?

2. How much does the interpretive program contribute to the Skyrail visitor experience? Can interpretation enhance the quality of tourist experiences?

The method

An overview

Two methods were used to answer the questions set out in the previous section – surveying and observing Skyrail visitors. The survey section of the study was the

311

Table 16.1 Skyrail's environmental management strategies

Construction

1. All four stations were constructed in existing clearings.

2. No roads were required for construction (or operation).

3. Towers were placed by helicopter into 10 m × 10 m cleared areas. Tower placement was guided by the principle of avoiding rare or endangered species. Only the tops of the towers are visible.

4. The sites cleared for the towers were revegetated.

5. All construction personnel entered the sites on foot and and all boots and equipment was sterilised to ensure that weeds were not introduced.

6. Clivus Multrum toilets were installed along with rainwater collection systems

Operations

1. Gondolas are silent and pollution-free.

2. Visitors have direct access to the rainforest only at the stations which are on land previously cleared.

3. Solar powered communication systems are used.

largest component of the research and involved approaching Skyrail visitors at various points in their trip and asking them to complete a questionnaire. The target population for the research was English-speaking visitors, reflecting the time and other resources available for the research. The focus of the research was on comparisons between visitors surveyed before they began their Skyrail experience and after they had completed their Skyrail visit. The point of these comparisons was to estimate the impact of Skyrail on rainforest knowledge and planned rainforest activities. The post-visit survey also provided the researchers with an opportunity to gather information on visitor perceptions and evaluations of the Skyrail experience. The main types of information collected with the surveys are described in Table 16.2.

The observation study was specifically aimed at examining the relative effectiveness of the various interpretive components of Skyrail at capturing and holding visitor attention. At each of the stations a number of interpretive components were identified. In the interpretive centre, for example, eight separate exhibit areas were described and seven signs along the Red Peak boardwalk were used as separate interpretive components. An observer was stationed at each interpretive component for set periods of time and recorded the number of visitors who stopped at each point, as well as the number of visitors who passed by. This ratio of visitors stopping to visitors passing by is referred to as the attracting power of the interpretive component. The observers also recorded the behaviours and some characteristics of a sample of the visitors who stopped at each point. Visitor

Table 16.2 Variables in the surveys

Both pre-visit and post-visit surveys

1. Socio-demographics including age, place of residence and type of travel party.

2. Rainforest experience including visits to North Queensland and other rainforests, and prior Skyrail visits.

3. Skyrail visit behaviour including type of ticket and point of departure.

4. Rating of level of rainforest knowledge and interest in finding out more about the rainforest.

5. Ratings of the importance of major benefits from rainforest visits.

6. Intended activity participation in the region.

7. A series of twelve true/false statements about the rainforest.

8. Open-ended questions eliciting most notable features of the rainforest and questions to ask a rainforest scientist.

Post-visit survey only

1. Ratings of satisfaction with the overall experience and various components.

2. Skyrail trip behaviour including stations stopped at and interpretive components experienced.

3. Evaluative ratings of the interpretive components including how enjoyable, interesting and easy to follow they were.

characteristics included an estimate of age and who the visitor was with at the observed point. Visitor behaviours included length of stay and levels of interaction with the interpretation.

The samples

A total of 1,431 Skyrail visitors were approached and asked to participate in the research. Of those approached 9 per cent (125) could not speak English and 369 visitors refused to participate, giving a response rate of 72 per cent. Of the 937 completed questionnaires, 394 were completed by visitors who had finished their Skyrail visit and 304 were completed before visitors had begun their Skyrail visit. The remaining 239 surveys were completed by visitors who were only part way through their Skyrail visit. This latter group was included to allow for more detailed analyses of learning associated with the various interpretive components. For the purposes of the present study, however, the analyses reported were conducted with the pre- and post-visit samples only. Table 16.3 provides a profile and comparison of these two samples.

For each of the variables statistical tests were computed (T-test for age and Goodman and Kruskal Tau's for all the other variables) to check for any major

Table 16.3 Pre-visit and post-visit survey sample profiles

Variable	Pre-visit sample	Post-visit sample	Total sample
Usual residence			
Local area	16%	20%	19%
Other Australia	51%	45%	48%
International	32%	34%	33%
Travel party			
Tour group	25%	19%	20%
Independent	75%	81%	80%
Alone	9%	11%	10%
With children	11%	13%	12%
With friends	15%	13%	14%
With partner	40%	36%	38%
Been to Skyrail before	5%	16%	11%
Mean age	39 years	45 years	40 years

Table 16.4 Profile of observation sample

Variable	n	% of Sample
Gender		
Male	460	48
Female	498	52
Estimated age		
Under 21	24	2
21–30	184	19
31–40	207	22
41–50	211	22
51–60	229	24
Over 60	107	29
Observed party		
Alone	86	9
Couple	333	34
With children	130	14
Small group of adults	146	15
Tour/large group	268	28

differences between the samples. In most cases the samples were not significantly different. The exceptions were age and previous visits to Skyrail, with the post-visit sample older and more likely to have been to Skyrail before.

The observation studies involved a total sample of 4,360 Skyrail visitors. Most of these, however, were only recorded as they passed the interpretive components under study. More detailed observations were made of 936 visitors and a basic profile of these visitors is given in Table 16.4.

The effectiveness of the interpretation

In order to assess the effectiveness of the interpretation offered to Skyrail visitors it was necessary to first assess the exposure of Skyrail visitors to the main interpretive components. Table 16.5 provides both the percentages of the post-visit sample who experienced the various interpretive components and their mean satisfaction ratings for these components. The most commonly experienced features were the Red Peak rainforest boardwalk and the Rainforest Interpretive Centre and this latter feature was the most positively evaluated. Although fewer visitors interacted with the rangers, these interactions were highly rated by the visitors.

Table 16.5 Experience and evaluation of interpretive components

Interpretive component	% of post-visit sample who experienced it	Mean satisfaction rating (10-point scale)	% giving a score of 8 or higher
Red Peak Boardwalk	80	8.1	65
Rainforest Interpretive Centre	62	8.6	77
Skyrail Trip brochure	39	8.0	61
Talks with Rangers	39	8.4	70
Ranger-led tour	27	8.2	65

More detailed examinations of visitor use of the interpretive components revealed some different patterns of use associated with different visitors. In particular, significant differences were found between tour group members and independent visitors. Tour group members were significantly less likely to go on the Red Peak Boardwalk (67 per cent as compared to 83 per cent of independent visitors) and to visit the Rainforest Interpretive Centre (43 per cent as compared to 64 per cent of independent visitors). This was a substantial difference and it was therefore decided to treat these two groups separately in the other analyses.

The next stage of the research investigated the impact of the Skyrail experience on visitors' rainforest knowledge. This was assessed in four ways; a rating of self reported rainforest knowledge, a rating of interest in finding out more about the rainforest, the answers to a series of twelve true/false statements about the rainforest and a total correct score derived from the twelve statements. A series of t-tests and cross-tabulation analyses were conducted comparing the responses of independent visitors in the pre-visit and post-visit samples. The results of these tests are summarised in Table 16.6. There were significant improvements in the total scores, the ratings of self-reported knowledge and wanting to learn more and four of the twelve true/false statements for the post-visit respondents. No significant differences were found between pre- and post-visit tour group members for any of these measures of learning.

In addition to impacts on learning and rainforest knowledge, the researchers were also interested in the impacts that the Skyrail experience might have on planned rainforest activities. Table 16.7 provides a comparison of the pre-visit and post-visit samples in terms of their intentions to engage in various rainforest activities available in the region. Skyrail appears to have decreased visitor

Table 16.6 Summary of results of pre-visit/post-visit comparisons on rainforest knowledge measures

Knowledge measure	Pre-visit sample	Post-visit sample
Self-reported Knowledge (0, none – 10, a lot)	4.6	5.3
Interest in learning more (0, none – 10, very)	5.1	5.8
Total correct score for 12 true/false rainforest statements	7.2	8.1
Specific true/false statements		
Feeding wildlife helps threatened species survive	67%	81%
Cane toads are a natural part of the rainforest	65%	76%
A number of plants and animals living in the Wet Tropics are found nowhere else in the world	76%	90%
There are more than 40,000 insect species in the North Queensland rainforest	48%	58%

Notes
Figures for the first three measures are mean scores, figures for the specific statements are percentages of the sample giving correct scores. The correct answer for the first two statements is false and for the second two statements is true.

Table 16.7 Intentions to engage in rainforest activities for pre-visit and post-visit samples

Activity	% of pre-visit sample	% of post-visit sample
Short rainforest walks*	75	25
Wildlife viewing*	65	35
Rainforest day trips*	64	36
Canoeing	56	44
National Park sightseeing	54	46
Birdwatching	54	46
Overnight rainforest walks	51	49
Rainforest day walks	46	54
Bushwalking	48	52

* Significant differences at the 0.05 level indicated by Goodman and Kruskal's Tau's.

intentions for most other activities and this was especially the case for short rainforest walks, wildlife viewing and rainforest day trips. If these changed intentions translate into changes in actual behaviours then it is clear that the Skyrail experience will lessen pressure on other heavily used sites which usually provide short rainforest walks and are commonly used by day trip operations.

The other source of data relevant to evaluating the effectiveness of the Skyrail interpretation was the observations of visitor behaviour. In Table 16.8 the various interpretive components have been organised into three categories according to whether or not they were above average in terms of holding power and in terms of the time spent by visitors interacting with them. Attracting and holding visitor attention is a necessary, although not sufficient, condition for effectively communicating with them. Therefore those interpretive components which have both high attracting power and hold visitor attention for above average time are the most likely to be effective communication tools. In the case of Skyrail the most popular interpretive components were the Barron Falls Lookout (which provides visitors with spectacular views of both the Barron Falls and the Barron Gorge), three interactive computer displays, a touchtable and a video.

The value of interpretation for the visitor experience

The second set of questions that the case study was concerned with were those relating to the contribution of the Skyrail interpretation to the visitor experience. A series of satisfaction scores were used in the survey study including a rating of overall satisfaction with the Skyrail experience and satisfaction ratings for five major components of the experience. A summary of the mean scores for each of these scales is given in Table 16.9. As can be seen, visitors were very satisfied with their Skyrail visit. This is consistent with both the high percentage of visitors who would definitely recommend Skyrail to others (92 per cent) and the high levels of intention to return and repeat the Skyrail experience. Table 16.10

Table 16.8 Interpretive features organised according to ability to attract and hold visitor attention

Interpretive feature	% of visitors who stopped (in seconds)	Average time spent at feature by visitors
Above average attracting power and time spent		
Barron Falls Lookout	100	124
Interactive computer (sounds of the rainforest)	91	123
Rainforest video	77	116
Interactive computer (rainforest quiz)	73	187
Touch table and computer	85	82
Mural and seeds display	72	64
Above average attracting power, less than average time spent		
Barron Gorge Lookout 1	99	47
Barron Gorge Lookout 2	98	51
Diversity display in interpretive centre	80	36
Entry display in interpretive centre	75	35
'Rainforest Giants' sign	70	39
Less than average attracting power, and time spent		
'A hole in the canopy' sign	60	34
'Fight for life' sign	57	28
'Life in the gloom' sign	52	26
'Hitching a ride' sign	52	42
'Reaching towards light' sign	43	24
Touch table – cassowary	43	57
Touch table – seeds	26	47
Exit display in interpretive centre	24	29
Final boardwalk sign	13	10

Table 16.9 Satisfaction ratings for the total Skyrail experience and components

Rating of satisfaction with	Mean score (10 = very satisfied)	% giving a score of 8 or higher	95% confidence intervals
Total Skyrail visit	8.7	79	8.5–8.9
Friendliness of staff	8.9	84	8.7–9.1
Efficiency of staff	8.7	80	8.5–8.9
Service	8.6	79	8.4–8.8
Rainforest information	7.7	61	7.5–8.0
Value for money	7.7	61	7.5–8.0

Table 16.10 Intention to return to Skyrail for local residents and domestic and
              Australian tourists

---

1. Local visitors
    42% intend to return within 12 months
    11% intend to return after 12 months
    32% will possibly return

2. Domestic tourists
    58% intend to return if they return to the region
    10% intend to return within 12 months
    7% intend to return after 12 months
    14% don't know

3. International tourists
    62% intend to return if they return to the region
    16% don't know

---

provides the information on intentions to return for the local residents versus domestic and international tourists. What is noteworthy in this table is the large percentage of local residents who intend to go on Skyrail again within 12 months. Such intentions offer the hope of a strong, stable, and continuing local market for the attraction.

A multiple regression analysis was chosen as the statistical technique to further investigate Skyrail visitor satisfaction. This technique considers the relative contribution of various factors to a single outcome. In the present case it allowed the researchers to investigate the relative contributions of various components of the Skyrail experience to visitors' overall satisfaction. Specifically, the analysis looked at:

*   satisfaction with the service and staff (combined into a single variable because of high correlations between the scales),
*   interest in finding out more about the rainforest (an indicator of learning),
*   satisfaction with the provision of rainforest information,
*   age,
*   rating of rainforest knowledge,
*   the number of interpretive components experienced.

Table 16.11 provides the relevant statistical information on the results of this analysis. Of most importance is that the first four factors were all significantly and positively related to overall satisfaction. This would suggest that visitor satisfaction with Skyrail is most heavily influenced by satisfaction with staff, especially the interpretive rangers, by the level of visitors' interest in learning about the rainforest, visitor satisfaction with rainforest information and age. In summary, three of the four most important and significant factors were related to the interpretive experiences offered. This was supported by other analyses. A series of

Table 16.11 Results of multiple regression analysis to investigate overall satisfaction

Variables	Beta
Satisfaction with service/staff*	.454
Wanting to know more*	.171
Satisfaction with rainforest information available*	.135
Age*	.117
Amount of rainforest information	.097
Amount of interpretation experienced	.026

Adjusted R square was 0.43 and those marked with * were significantly related at the 0.01 level.

T-tests were computed comparing visitors who did, with those who did not, experience the major interpretive components. These tests indicated that people who did experience the Red Peak Boardwalk, the Rainforest Interpretive Centre and those who spoke with a ranger were significantly more satisfied with their Skyrail experience.

The lessons learnt

A summary of the results presented in the previous section provides a good starting point for setting out the lessons to be learnt for tourism management from this case study of Skyrail. The results indicated that Skyrail interpretation did make a positive contribution to the management of visitors to the WTWHA. In the first instance the interpretive program was successful at increasing independent travellers' rainforest knowledge which is consistent with one of the features of sustainable tourism listed in the introduction. The data do not, however, support a similar conclusion for tour group visitors. This does not necessarily reflect a failure of the interpretation with this group. Rather it appears that these groups have less time available for their Skyrail experience and consequently are significantly limited in their exposure to the interpretation available. This offers a challenge to the Skyrail management in their dealings with tour companies. In addition to increasing visitors' rainforest knowledge, the results also supported the argument that Skyrail offers a positive rainforest experience that can replace the desire for other rainforest activities. In particular, visitors who had experienced Skyrail were significantly less likely to say that they intended to go on short rainforest walks, rainforest day trips and wildlife viewing. These needs would appear to be met by the Skyrail experience thus lessening pressure on other sites which have much more limited capacity to cope with intense tourism use.

Both the observation and survey data highlighted the success of interpretive components which offered an opportunity for visitors to interact with the information available and to make choices which allowed them to personalise their experience. Interactions with the interpretive rangers, for example, was especially highly rated by visitors. Such interactions allow visitors to ask questions about information that is most interesting or personally relevant. This is consistent with research results in other contexts (Moscardo, 1996). Several studies of tourist–guide interactions have found that the guide's knowledge of the area they present was very important for successful interactions with visitors (Almagor, 1985; Roggenbuck and Williams, 1991; Geva and Goldman, 1991). The success of the interactive computer displays also supports the literature on effective interpretive techniques. It is notable that the computer programs used in the Skyrail interpretation program offered variety of sensory experience and opportunities for visitors to participate in the interpretation, as well as giving visitors control over the information that they received. Control, personal relevance and participation and interaction are some of the core principles for effective interpretation (Moscardo, 1996).

The second part of the analysis focused on the contribution of interpretation to the Skyrail visitor experience. In this case the data clearly supported the conclusion that the interpretation program, especially the guides, the Rainforest Interpretive Centre and the Red Peak Boardwalk, made a significant contribution to visitor's overall satisfaction with their Skyrail visit. A major lesson to be learnt from this case study is that effective interpretation can enhance the quality of visitor experiences. Interpretation can be used as a method for adding value to tourism products. Thus interpretation and tourism can be seen as being mutually beneficial activities and by working together they can support the development of more sustainable tourism.

Another important lesson to be learnt from this case study is the significance of conducting research into the social components of sustainability. Although interpretation and education are often listed in descriptions and discussions of sustainable tourism (see Gunn, 1994; Visser and Njuguna, 1992; United Nations Environment Programme, 1992), it could be argued that such inclusions are usually superficial. This reflects a general tendency in discussions of sustainable tourism to focus on the biophysical rather than the social and cultural aspects of tourism (Pearce, 1993; Pigram, 1990). Krippendorf has eloquently described the tendency for discussions of managing impacts to concentrate on physical solutions rather than management strategies aimed at improving the behaviour of tourists. As he puts it:

> The damage tourism causes to the people, economy and environment of the host area, especially in the long term, remains hidden from the tourist. He has been left out of all discussion on the subject, even though he is one of the main protagonists . . . They are therefore carefree and ignorant rather than devious.
>
> (1987: 43)

The present study demonstrates that tourism operations can be effective tourist educators and, further, that appropriate interpretation programs can lessen the negative impacts of tourism. According to Butler (1991) education offers a better strategy for managing tourism than many of the physical management options often suggested. He states that 'there is little doubt that in the long term it is probably the only solution which is likely to be broadly successful' (1991: 207). In the case of making more extensive use of interpretation in tourism management it could be argued that the sooner we start the sooner we will get there.

References

Special sources

Society for the Interpretation of Britain's Heritage (1996) 'Editorial notes', *Interpretation* 1, 2: 35.
Tillack, T. (1996) 'Mapping visitor movement patterns around the Wet Tropics World Heritage Area', unpublished Honours thesis, Department of Tourism, James Cook University.
Wet Tropics Management Authority (1992) *Tourism strategy*, Cairns: Wet Tropics Management Authority.
Wet Tropics Management Authority (1995) *Draft Wet Tropics plan*, Cairns: Wet Tropics Management Authority.

General references

Almagor, U. (1985) 'A tourist's 'Vision Quest' in an African game reserve', *Annals of Tourism Research* 12, 1: 31–48.
Bramwell, B. and Lane, B. (1993) 'Sustainable tourism: an evolving global approach', *Journal of Sustainable Tourism* 1, 1: 1–5.
Butler, R.W. (1991) 'Tourism, environment, and sustainable development', *Environmental Conservation* 18, 1: 201–9.
Driml, S. and Common, M. (1996) 'Ecological economics criteria for sustainable tourism: application to the Great Barrier Reef and Wet Tropics World Heritage Areas, Australia', *Journal of Sustainable Tourism* 4, 1: 3–16.
Geva, A. and Goldman, A. (1991) 'Satisfaction measurement in guided tours', *Annals of Tourism Research* 18, 2: 177–85.
Gunn, C.A. (1994) *Tourism planning*, New York: Taylor and Francis.
Inskeep, E. (1991) *Tourism planning*, New York: Van Nostrand Reinhold.
Jafari, J. (1990) 'Research and scholarship', *Journal of Tourism Studies* 1, 1: 33–41.
Knudson, D.M., Cable, T.T. and Beck, L. (1995) *Interpretation of cultural and natural resources*, State College, PA: Venture Publishing.
Krippendorf, J. (1987) *The holiday makers*, London: Heinemann.
Lane, B. (1991) 'Sustainable tourism', *Interpretation Journal* 49: 1–4.
Moscardo, G. (1996) 'Mindful visitors: heritage and tourism', *Annals of Tourism Research* 23, 2: 376–97.
Nitsch, B. and van Straaten, J. (1995) 'Rural tourism development: using a sustainable tourism development approach', in H. Coccossis and P. Nijkamp (eds) *Sustainable tourism development*, Aldershot, UK: Avebury.
Pearce, P.L. (1993) 'From culture shock to culture exchange', paper presented at the PATA/WTO Global Action to Global Change Conference, Bali.
Pigram, J.J. (1990) 'Sustainable tourism', *Journal of Tourism Studies* 1, 2: 2–9.

Roggenbuck, J.W. and Williams, D.R. (1991) 'Commercial tour guides' effectiveness as nature educators', paper presented at the World Congress on Leisure and Tourism, Sydney.

United Nations Environment Programme (1992) 'Sustainable tourism development', *UNEP Industry and Environment* 15: 1–5.

Visser, N. and Njuguna, S. (1992) 'Environmental impacts of tourism on the Kenya coast', *UNEP Industry and Environment* 15: 42–52.

Wall, G. (1993) 'Towards a tourism typology', in J.G. Nelson, R. Butler and G. Wall (eds) *Tourism and sustainable development*, Waterloo: University of Waterloo.

17

COMPETITION OR CO-OPERATION?

Small and medium sized tourism enterprises at the destination

Dimitrios Buhalis and Chris Cooper

This chapter provides an analysis of the problems and issues facing small and medium sized enterprises at the tourism destination, and a case study of Greek island tourism. Tourism destinations are dominated by such small enterprises which, whilst they allow a rapid injection of cash into local economies and provide a feeling of welcome and character to the visitor, are also typified by a lack of strategic vision and management expertise. Given that the nature of developments in the tourism sector on both the supply and demand side support larger enterprises, individual smaller businesses are faced with real problems of survival, although the sector is seen to have continuing potential in most destinations. In particular, their lack of a strong lobbying voice within the matrix of stakeholders at the destination means that they tend to lose out to the stronger voices of large enterprises and political groups in the planning and management of destinations (Heath and Wall, 1992). This chapter outlines the way forward for this very important sector of the industry and outlines the role of supporting agencies for small businesses at the destination.

Small and medium sized tourism enterprises at the destination

Small and medium sized enterprises (SMEs) dominate most markets world-wide. In Europe their significance is more obvious as they provide products and services for a wide range of cultural and national backgrounds. In addition, they provide employment opportunities and have contributed to stable economic progress by utilising innovative production procedures to deliver customised products and services. Consequently, SMEs are becoming the focus of attention by governmental and supranational organisations as well as attracting academic research (Giaoutzi *et al.*, 1988). Table 17.1 shows the distribution of enterprises and the

Table 17.1 Small and medium sized enterprises' contribution to the European Union economy

Type of firm	Employees	Firms in the EU	% of firms	Number of jobs	% of the EU employment	Total turnover (%)
Micro	0–9	14.5 m	93.05	26m	30	22
Small	10–99	1 m	6.42	22m	25	48
Medium	100–499	70,000	0.45	14m	16	
SMEs	**0–499**	**15,570,000**	**99.9**	**62m**	**70**	**70**
Large	500+	12,000	0.08	26m	29	30
Total		15,582,000	100	88m	100	100

Source: Adapted from EIM (1993), EC (1993) and Page (1993).

turnover in Europe. For classification purposes, SMEs are classified as all the enterprises with fewer than 500 employees, less than 75 million ECU net fixed assets, less than 38 million ECU net turnover and less than one third of the company held by a larger firm (EC, 1991: vi).

The destination can be regarded as the *raison d' être* for tourism, providing an amalgam of tourism products such as facilities, attractions and activities which respond to the needs and wants of the tourist. Destinations also attract and generate the tourism trip in the first place and consequently it is the tourism destination and its individual elements which comprise the most important elements of the tourism product. Most tourism destinations are based upon small and medium sized tourism enterprises (SMTEs) which provide a very wide range of tourism products and services. Consequently, the fortunes of both destinations and SMTEs are interrelated as the prosperity of one heavily depends upon the management and competitiveness of the other.

Clearly, the contribution of the small and medium sized tourism enterprises in the tourism industry is also very significant. Numerically they dominate the tourism sector throughout the world and they also provide a point of direct contact of the visitors with the destination and residents. In addition, they facilitate a rapid infusion of tourism spending into the local economy and therefore have a significant beneficial effect upon the host population and in turn, stimulate the multiplier effects of tourism activity at the destination. Moreover, SMTEs formulate value-added networks of products and service delivery which enhance tourist satisfaction: since a large number of SMTEs are involved in the delivery of tourism products and services, the destination is literally an amalgam of SMTEs which address consumer needs. It should also be noted that a significant proportion of these companies are small, even by the above definition. For instance, many guest houses, restaurants and souvenir shops are owned and staffed by the members of a single family.

In France for example, the size of tourism enterprises is clearly reflected in research on the micro-economic analysis of the tourism sector, undertaken by the

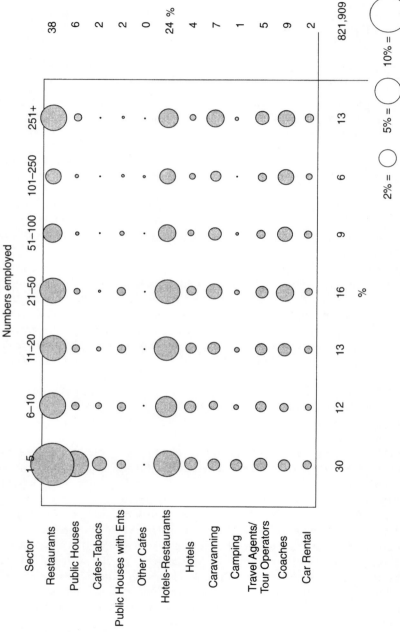

Figure 17.1 French tourism sectors – jobs
Source: pH Group (1993).

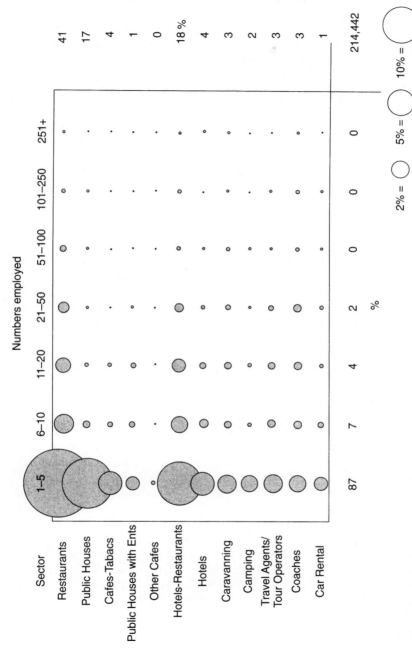

Figure 17.2 French tourism sectors – firms
Source: pH Group (1993).

pH Group on behalf of the European Union. Enterprises employing up to ten people represent 94 per cent of French tourism firms and 48 per cent of total employment in tourism. Only 19 per cent of employment is generated within France from enterprises with more than 100 employees as demonstrated in Figures 17.1 and 17.2 (pH Group, 1993). Statistics for the accommodation sector confirm this observation. In every single European country small independent and flexible hotels dominate the market. In total it is estimated that more than 90 per cent of hotels are SMTEs and are family managed (Shaw and Williams, 1990: 74; Go and Welch, 1991: 14; Sheldon, 1993: 636). In Switzerland for example, more than 90 per cent of hotels have fewer than 50 rooms and only 2 per cent more than 100 rooms, while in the UK the average hotel has 25 rooms. The average hotel and restaurant establishment in Greece employs 2.5 persons while the average number of beds/establishment on the Greek Islands is 63 (Buhalis, 1994).

Competitiveness of small and medium sized tourism enterprises

Porter suggests that competitive strategy 'is the search for a favourable competitive position in an industry' which in turn is a function of the attractiveness of the industry and the relative competitive position within that particular industry. He also argues that 'competitive strategy aims to establish a profitable and sustainable position against the forces that determine industry competition' (1985: 1). Competitiveness is, therefore, defined as the effort undertaken by organisations to maintain long-term profitability, above the average of the particular industry within which they operate or above alternative investment opportunities in other industries.

As most SMTEs are family owned and managed it is rare to identify rational entrepreneurs who regard their business as another investment opportunity (Brown, 1987). They tend to be emotionally involved with the enterprise and therefore they are reluctant to abandon it in difficult times. This means that tourism entrepreneurs rarely take into consideration the opportunity to diversify their portfolio with new enterprises, unless it complements their normal activities. An alternative entrepreneurial rationale is the opportunity that tourism-related enterprises offer to aesthetic locations such as Cornwall once sufficient capital has been accrued from the owner's former career. Thus, many such companies are managed by people who have little or no training in the tourism industry, and who may also lack local commitment. Furthermore, one of the motivations to open such a business is the potential to take long off-season breaks, thereby reducing the destination's ability to lengthen the season (Brown, 1987; Shaw and Williams, 1990). It is therefore important that SMTEs concentrate on their competitive position within their own industry and within a framework of competitive analysis. In the next section, a multi-level framework of competitive analysis is proposed for tourism enterprises, exemplified by a case study of SMTEs in the Greek Aegean islands.

Multi-level competitive analysis for small and medium sized tourism enterprises

Being able to understand the magnitude of the competition within the tourism industry is critical for SMTEs. However, research on the Greek islands has demonstrated that SMTEs are often myopic with regard to the competition (Buhalis, 1991). They tend to concentrate on local tourism product service providers, and fail to understand the global competitive environment where they compete against all the alternative leisure options for the consumer, as well as within distribution channels which attempt to reduce their profitability. Figure 17.3 demonstrates a 'competition pyramid' which places the interests of SMTEs in the centre. Five different levels of competition can be observed from the point of view of SMTEs.

Level 1 – competition from similar products and service providers at a destination

An example here could be a hotel in a destination competing with the rest of the accommodation sector of the destination; while a restaurant competes with all other catering facilities. Most entrepreneurs have focused their competitive efforts against their neighbouring SMTEs and therefore have failed to appreciate that they also compete against tourism products and services offered in alternative destinations or even with alternative leisure activities and spending opportunities at the tourists' place of origin. We argue that the emphasis of competition within destinations should be reduced and local enterprises should concentrate their efforts to attract consumers, profit margins and tourists' spending from the other levels of competition.

Level 2 – competition from similar or undifferentiated destinations

These destinations have established an image with the consumer which is easily substituted by alternative destinations. The most obvious example is the sun and sea product which can be found throughout the Mediterranean coast and islands as well as in Florida and other destinations. It seems that for some consumer segments, sun, sea and a reasonable accommodation establishment are all that are required, regardless of the location of the destination and the socio-cultural surroundings. Therefore, SMTEs compete with firms and alternative tourism destinations nationally and internationally.

Level 3 – competition from differentiated destinations

These destinations provide (or project the image of delivering) unique tourism products, based on their natural and or socio-cultural resources. Tourism

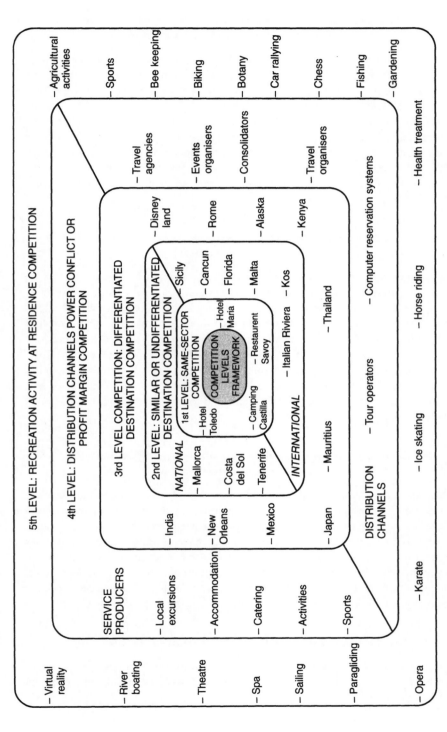

Figure 17.3 Framework for tourism competition multi-levels analysis

products in these areas are regarded as unique in the sense that they are not easily substitutable. Apart from areas with a distinct character, theme parks and cruise ships can be regarded as unique destinations in their own right, since they can attract consumers regardless of their geographical location.

Level 4 – competition in the distribution channel

This competition reflects conflict in the distribution channel and refers to the effort of each member of the channel to increase its profit margins against the profit margins of the other members. Since tourism market segments often have maximum price limits which they can afford and are willing to pay, the competition between members of the channel is fierce and powerful channel members can dominate and determine the distribution of the total profit margin. In addition, SMTEs compete with other tourism products/services providers at the destination for the tourist expenditure of the consumers. As a tourist might have a fixed holiday budget, tourism enterprises of all types compete in order to sell some of their products and services. For example, a tourism enterprise organising water sports at the destination competes indirectly for the holiday budget of the consumer with a firm providing night-time entertainment.

Level 5 – competition with alternative leisure activities

Such activities might have an element of training and education, or might be purely recreational. The development of rural tourism, sophisticated leisure activities and theme parks at the tourists' place of origin might effectively reduce the need for consumers to travel to particular destinations in order to enjoy tourism products. In addition, recreation services offered by new technology and its applications, such as virtual reality and computer games, might encourage people to stay at home and discover the destination from the convenience and safety of their armchair.

To summarise, this conceptual framework demonstrates that SMTEs compete on several levels. Although it seems natural to compete against enterprises of the same sector within destinations, SMTEs should not forget that the entire spectrum of local firms comprising the destination is ultimately the reason for travelling. Thus the amalgam of products and services provided by all SMTEs at the destination should maximise the satisfaction of consumer's needs and wants. In this sense, SMTEs need to co-operate at the destination level, and instead of competing between themselves, they could increase their competitiveness as a destination (or as the total tourism product) against substitute tourism products or factors which reduce their profitability.

Case study: small and medium sized tourism enterprises in the Greek Aegean islands

Based on research undertaken on the Greek Aegean islands, it is possible to analyse the competitive position of SMTEs through a strengths and weaknesses, opportunities and threats (SWOT) analysis. This approach clearly draws out the interrelationships and dependencies between SMTEs and the destination, and can be applied to SMTEs in most destinations and particularly Mediterranean holiday areas (Buhalis, 1992). This case study emphasises the structural, rather than functional, weaknesses, which are common in most SMTEs, and provides a way forward with the development of support systems for SMTEs.

Tourism in the Greek Aegean islands

The Aegean Archipelago is the largest island complex in Europe, accommodating 500,000 Greek people living on 95 inhabited islands. Apart from agriculture and an extensive fishing industry, the region's economy depends heavily on tourism, as its insular character deprives it of competitive advantages in other activities. Over one third of the 28,000 Aegean firms are tourism enterprises, all of which are SMEs, while almost two thirds of the Aegean labour force are employed in the service sectors, and most of them in tourism. Tourism contributes between 50 per cent to 90 per cent of the Gross Regional Product of the Aegean islands, depending on the level of development. The Aegean islands' wealth depends almost exclusively upon tourism activity, and therefore underlines the fact that supporting the Aegean destination and SMTEs is critical for the prosperity of the local population.

Tourism supply on the Aegean islands

The Aegean islands have been a traditional destination since the 1960s, offering a range of facilities. The traditional buildings and beautiful landscapes on the one hand, and the unspoiled sea on the other, make it a desirable destination. The major attraction is the classical 'sun, sand and sea' in combination not only with the cultural, natural and historical resources, but also the events which take place. In 1994, a total of 220,000 beds were provided in almost 9,000 registered enterprises. Nevertheless, due to the '*parahoteleria*' phenomenon, these figures are substantially underestimated. Thousands of catering, sports and entertainment enterprises of every type operate every summer. Essentially all tourism providers in the region are SMTEs, as the average hotel size is 66 beds per establishment, while only a few enterprises employ more than 100 people. The Aegean islands have a very complicated and well-developed transportation system which incorporates at least one major port on each island and 16 airports, most of which are capable of receiving charter flights directly from Europe. There is a great number of connections between the mainland and the islands, as well as between the islands themselves. The Aegean islands are also connected, by ferry services to the port of Piraeus in Athens, while a great number of island-to-island connections facilitate inter-island transportation. Accessibility and transportation connections between the Aegean islands improved recently with the introduction of modern, fast and convenient vessels, such as hydrofoils and catamarans.

The Aegean tourism products are distributed to the international market predominantly through European tour operators, through organised packages. In addition, there are more than 300 incoming travel agencies on the Aegean islands which act as tour operators' handling agents, and provide a variety of tourism services directly to consumers. In addition, the Greek National Tourist Office (GNTO) operates six information offices, while several local authorities provide tourist information as well as distribute promotional material.

Tourism demand for the Aegean islands

In 1993, around 48.5 million bednights were recorded for all types of Greek accommodation establishments, of which 75 per cent were international tourists. The Aegean islands have enjoyed a steady growth in tourism reaching 12 million bednights in 1991 and accommodating one quarter of all tourists arriving in Greece. The vast majority of the bednights spent in the area were by Europeans, especially British, Germans, Swedish, Finnish, Dutch and Austrians. The British and German tourists account for about 50 per cent of total bednights. Most tourists arrive on charter flights, and thus the destination is heavily dependent on tour operators for both their promotion to their target markets and their accessibility. The average length of stay in the country was 14 days, while the average expenditure per capita in 1993 reached a level of $354. Domestic and visiting friends relatives (VFR) tourism is also a significant contributor, although it is very seasonal and concentrated on July and August.

Strategic issues and weaknesses in Aegean island tourism

Despite its popularity and growth over the last forty years, the Aegean tourism industry has reached a stage where its competitiveness has become questionable. Several structural weaknesses have emerged, illustrating the need for strategic action in the planning and marketing processes of the region. The major weaknesses and threats for the Greek tourism product as well as the challenges can be summarised as the following:

- the image of Greece as cheap, simple, unsophisticated, undifferentiated, sun and sea destination;
- gradual deterioration of the tourism product and lack of re-investment in improvements;
- increase in tourism arrivals but decrease in tourism expenditure per capita;
- inadequacy of the Greek planning process;

- dependence upon major tour operators for promotion and distribution of the tourism product;
- a plethora of anarchically developed and behaving SMTEs, aiming for short-term profitability;
- inadequacy of infrastructure to serve the ever expanding demand;
- lack of co-ordination at the destination and disrespect for tourists' needs;
- lack of professionalism and training in both state and private tourism establishments;
- individualistic behaviour by SMTEs and unwillingness to co-operate on a destination basis;
- unsuccessful and inconsistent programmes of government intervention;
- a relatively unregulated environment, with near complete lack of control;
- development of tourism as a single regional development option;
- failure of the private sector to invest in long-term projects;
- deterioration of natural, social and cultural resources;
- SMTEs' inability to resist global concentration in the tourism industry;
- failure of both the private and public sectors to learn from internationally gained experience in tourism development and marketing;
- lack of tourism research to identify the impacts of tourism;
- negligence with regard to new tourism demand challenges.

As a result of these factors, and despite their unparalleled environmental and heritage resources the Aegean islands fail to attract the desired 'high-quality, high-expenditure' tourists, as they are increasingly unable to satisfy their requirements. Deterioration of the tourism product and image leads to a lower willingness to pay by consumers, which consequently, leads to a further fall in quality, as the industry attempts to attract customers with lower prices. Concentration of bargaining power in the hands of European tour operators, in combination with the inability of the Greek tourism industry to promote itself effectively, inevitably minimises the profit margins of SMTEs and their ability to yield decent returns on their investment. As a result, tourist expenditure per capita deteriorates gradually, while the volume of tourists increases.

A SWOT analysis of Aegean island tourism SMEs

The findings of the SWOT analysis for Aegean SMTEs are summarised in Table 17.2. As far as *strengths* are concerned, entrepreneurial activity dominates in the Aegean islands. For SMTEs, both the proprietor and their family are normally directly involved with every aspect of the business, reacting efficiently and promptly to any problem. Family involvement in running the enterprise provides considerable benefits, especially in having a very flexible workforce which might tolerate unsociable working hours whilst family members feel committed to the long-term prosperity of the enterprise. SMTEs can also capitalise in terms of personal relations with consumers, suppliers and the entire tourism industry in

Table 17.2 Greek tourism and SMTEs' strengths/weaknesses/opportunities/threats analysis

Strengths	*Weaknesses*
• Flexibility	• Management
• Tailor-made product delivery	• Marketing
• Entrepreneurial activity	• Information technologies illiteracy
• Family involvement	• Dependence upon tour operators
• Natural and cultural resources	• Supporting markets
• Strong local character	• Lack of economies of scale
• Personalised relationships	• Human resources management
• Labour loyalty and low turnover	• Education and training
	• Transportation and accessibility
	• Financial management and resources
	• Seasonality
	• Lack of standardisation
	• Lack of quality assurances
Opportunities	*Threats*
• European Union support	• Environmental degradation
• European redistribution of labour	• Concentration and globalisation
• Increase in tourism demand	• Oversupply
• Trends in tourism demand	• Lack of visibility in computer
• Low cost of living in periphery	reservation systems
• Information technology	• Infrastructure
	• Wars and terrorism
	• Political intervention

Source: Adapted from Buhalis (1991: 60c) and Cooper and Buhalis (1992: 108).

general, as their small size enables them to provide a personal touch to all products provided. In addition, natural and cultural resources are also significant assets for SMTEs, as the enterprises tend to be at the heart of the local community and thus benefit from local resources.

A number of *opportunities* arise from the external environment of the SMTEs. The European Union has taken several actions to support small and medium sized enterprises, while support in the infrastructure development of peripheral regions has significantly contributed to prosperity of SMTEs (EC, 1991, 1993; Thomas, 1995). This is particularly the case for Greece where the whole country is eligible for aid through the structural funds. In addition, SMTEs benefit from the continuous growth of global tourism demand as more people require holiday services. Emergent information and telecommunication technologies can provide strategic tools for the distribution of SMTEs' products and thus reduce their isolation. Moreover, new technologies offer opportunities for developing innovative tourism products such as teleworking for tourists who would like to spend time working during their stay at the destination. Finally, Greek SMTEs tend to achieve a cost advantage since the time of family members is not costed at realistic rates and they often operate in inexpensive peripheral regions which have lower inflation rates in comparison with metropolitan areas. Thus, they acquire significant cost-of-living advantages.

In terms of external *threats*, lack of know-how and funds effectively debilitates the attempts of SMTEs to deal with environmental problems (Carter and Turnock, 1992). Consequently, they often suffer severe damage and feel unable to take measures to remedy environmental pollution caused by inappropriate waste management and excessive usage of water, energy and other natural resources. Moreover, oversupply of tourism services in Greece and the lack of well-defined carrying capacity limits in several destinations has placed SMTEs in a disadvantaged position as they cannot achieve sufficient income.

The issue of infrastructure is closely related to the oversupply. Unplanned Greek destinations have limited infrastructure provisions which do not match increases in tourism plant. Consequently, there are pressures on the existing inadequate facilities at a destination level. Finally, like every other tourism enterprise, SMTEs suffer the impacts of wars and terrorism activities. Political instability nearby in Turkey, the former Yugoslavia, Albania and the Middle East has had an unfavourable effect upon SMTEs. However, they have more limited means of dealing with such situations and they are more vulnerable in accepting the implications. Political intervention, perhaps through the public sector and the legislative framework, as well as frequent changes of political leadership in Greece have often damaged the prosperity of SMTEs: as SMTEs do not have effective lobbying power, they cannot influence the political decisions which determine their welfare.

In terms of *weaknesses*, although entrepreneurs are normally an asset for SMTEs, it seems that a number of managerial problems often arise. Lack of operational management know-how creates inconsistency in the creation and delivery of the Aegean tourism product. This has direct implications for consumer satisfaction and the projected image of the SMTEs. More importantly, there is often a lack of strategic vision from SMTEs and it seems that the enterprise is an extension of the proprietor's domestic environment. Entrepreneurs recruit family members and relatives as personnel and suppliers, even though more appropriate/qualified alternatives can be found in the marketplace. Thus, the management of SMTEs clearly reflects the proprietors' family life and decision-making processes in a profit-making activity.

Marketing is another significant weakness for most SMTEs as they are often unaware of the techniques available. Thus, they follow a product-oriented rather than a consumer-oriented approach. Consequently, unco-ordinated, isolated and trouble-shooting marketing activities are occasionally undertaken by the Greek private and public sectors. Lack of marketing research limits SMTEs' knowledge of their consumers' needs and wants, and prevents them from identifying methods for improving services in order to meet consumers' expectations. Inability to execute and finance advertising campaigns and other promotional techniques provides minimal visibility for SMTEs in their markets. As a consequence of their management and marketing weaknesses, SMTEs suffer from dependency on the tourism distribution channels to promote and sell their product. In particular, tour operators have enormous power within the distribution channel and therefore control most of the marketing campaigns and the pricing of SMTEs. The marketing campaign and visibility of SMTEs in origin markets is often determined by the coverage, space, photographs and descriptions in tour operator's brochures. In addition, tour operators can influence the media with regard to editorial matter and they can often amend advertising campaigns organised by national or regional tourism organisations. It has frequently been observed that they can also determine and impose pricing of tourism products especially in regions with an oversupply of tourism enterprises. The accommodation sector frequently has to offer 40–70 per cent discount on the official/rack rates in order to attract visitors. In addition, tour operators can control the accessibility of destinations as they own most of the charter airlines which provide direct and inexpensive flights to destinations. However, whilst these small enterprises become much more visible to tourists on arrival at the destination, few are able to establish a strong niche appeal even at this stage of the tourist's decision process.

For those SMTEs located in peripheral areas there is usually little economic development in the supplying sectors and therefore destinations have to import essential material from elsewhere (Wanhill, 1997). Thus SMTEs often face delivery and purchasing problems and the transport cost for materials is problematic. In addition, lack of economies of scale essentially means that SMTEs have to pay higher prices for products than larger companies.

Lack of specialised personnel and inadequate training procedures mean that human resource management is a major weakness for Aegean SMTEs. In most SMTEs, personnel have to cover a wide range of positions, therefore loose job descriptions are often provided and multi-skilled personnel are required. The high labour turnover due to the seasonality of the tourism industry reduces the availability of qualified and experienced personnel and makes the delivery of the tourism product by SMTEs not only variable but also perceived as unprofessional.

Transportation and accessibility to SMTEs is a weakness in the Greek islands where 300 inhabited islands are connected by a varied and complex network of both ferry and air transport. The control of distribution channels over direct and inexpensive air transportation has already been mentioned. SMTEs often feel frustrated because they are unable to attract consumers simply because they cannot provide convenient, reliable and affordable transportation. The forming of charter air carriers by destination areas such as Turkey and Spain, as well as the emerging air transport deregulation policies in the European Union, might reduce this problem in the late 1990s.

SMTEs face significant financial constraints as on the one hand they are required to invest in fixed assets at the beginning of their operation and yet, on the other hand, they are not favoured by financial institutions since they normally have few assets. Consequently, they are forced to accept unfavourable financial deals. This is exacerbated by the seasonality problem. SMTEs have to produce adequate income within a limited period every year – often in a season of only twenty weeks. Lack of diversified investment in other economic activities forces proprietors to work intensively in the peak months and rest in the off-peak. This may mean that the time taken to pay back the investment is unfavourable.

It is therefore recognised that SMTEs are not simply smaller versions of larger tourism organisations but have distinct problems as well as managerial/owner cultures of their own. SMTEs have special needs and requirements which affect their competitiveness and prosperity (Lee-Ross and Ingold, 1994; Wanhill, 1997; Cooper and Buhalis, 1992; Thomas, 1995; Morrison,1994; Callan, 1989; Quinn *et al.*, 1992).

The need for a co-operation scheme for SMTEs: supporting agents for small and medium tourism enterprises

Throughout this case study it has been demonstrated that Aegean SMTEs face structural problems mainly due to their size, their inability to achieve economies of scale, their peripheral location and their position in the distribution channel. The weaknesses analysed above demonstrate clearly that the competitiveness of SMTEs is deteriorating as consumers become increasingly sophisticated and knowledgeable. On the other hand, vertical integration in tourism is threatening the existence and prosperity of the SMTEs.

Poon (1988) suggests that: 'there will be no place for the small stand-alone participants, but the world can become the oyster of the small, innovative, flexible and networked enterprise'. Networking will allow SMTEs to:

• pool their resources in order to increase their competitiveness;
• draw up strategic management and marketing plans;
• reduce operating costs; and
• increase their know-how.

Such networks may be developed gradually from co-operative schemes between similar tourism enterprises (horizontal integration) or by more sophisticated collaboration schemes where a group of complementary tourism enterprises draw together in order to formulate a tourism product or to improve a business function (vertical integration).

Hence, the ultimate aim should be the diagonal integration of the entire range of tourism enterprises in order to:

• maximise the benefits of tourism for the host population and local SMTEs by creating a total system of wealth creation;
• maximise the sustainability of local resources; and
• optimise benefits for the users of the destination area – namely residents and tourists.

The benefits of diagonal integration can be described as *synergies* referring to the:

> benefits which accrue to the management operation and organisation of interrelated activities, where each activity is capable of generating benefits that mutually reinforce each other. Each activity adds value to the other thereby making the whole output greater than the sum of the discrete parts.
>
> (Poon, 1993)

In the tourism industry, synergies can be utilised in order to increase total consumer satisfaction through integration of independently provided services. '*System gains*', that is, economies caused by creating engineering linkages among design, production, marketing organisation and management, may be made (Poon, 1993). SMTEs can benefit by integrating know-how and available resources in order to improve their business functions and in particular their marketing campaigns, their financial functions and their training. They may also benefit from '*economies of scope*', that is, lower costs associated with the joint provision of more than one product or service rather than producing each separately. With economies of scope, the individual production of two goods by one enterprise is less costly than the combined cost of two firms producing either good one or two. Co-operation between SMTEs for the joint production of a whole range of tourist products and services, such as sports training provision for tourists, or laundry and freezing facilities would have a number of benefits for both the SMTEs and the destination (Poon, 1989: 96).

Collaboration between Aegean SMTEs could be implemented and facilitated through a 'supporting agency for SMTEs' which should have a wide range of activities and responsibilities as demonstrated in Figure 17.4. Both the public and private sectors should co-operate closely in the development of such a supporting agency. The undeniable role of the public sector in planning infrastructure provision, incentives policy, personnel training, destination marketing, pricing co-ordination and industry control can be achieved through collective action and consultation with the private sector and specific interest groups. Moreover, the public sector would perhaps need to facilitate the availability of adequate funds and funding bodies for the purposes of the supporting agency. The agency should address the major weaknesses of Aegean SMTEs and identify ways to increase the competitiveness of the sector and achieve the strategic objectives of both the enterprises and the destination (Buhalis, 1991).

The role of a supporting agency for Aegean SMTEs

The supporting agency can facilitate *management* of the Aegean SMTE sector by providing consulting services in strategic planning, organisation and operational issues, financial management, accounting, marketing and technological issues. The consulting function could vary from providing expertise and genuine advice

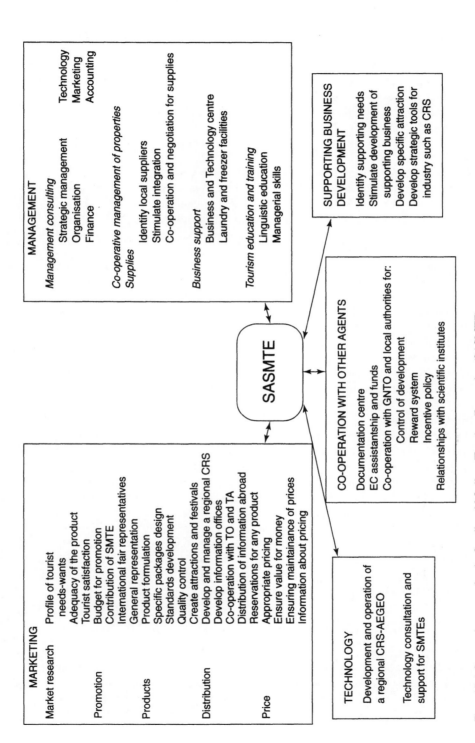

MARKETING

Market research	Profile of tourist
	needs-wants
	Adequacy of the product
	Tourist satisfaction
Promotion	Budget for promotion
	Contribution of SMTE
	International fair representatives
	General representation
Products	Product formulation
	Specific packages design
	Standards development
	Quality control
	Create attractions and festivals
Distribution	Develop and manage a regional CRS
	Develop information offices
	Co-operation with TO and TA
	Distribution of information abroad
	Reservations for any product
Price	Appropriate pricing
	Ensure value for money
	Ensuring maintainance of prices
	Information about pricing

MANAGEMENT

Management consulting
Strategic management   Technology
Organisation   Marketing
Finance   Accounting

Co-operative management of properties
Supplies   Identify local suppliers
             Stimulate integration
             Co-operation and negotiation for supplies

Business support   Business and Technology centre
                       Laundry and freezer facilities

Tourism education and training
Linguistic education
Managerial skills

SASMTE

TECHNOLOGY

Development and operation of
a regional CRS-AEGEO

Technology consultation and
support for SMTEs

CO-OPERATION WITH OTHER AGENTS

Documentation centre
EC assistantship and funds
Co-operation with GNTO and local authorities for:
    Control of development
    Reward system
    Incentive policy
    Relationships with scientific institutes

SUPPORTING BUSINESS
DEVELOPMENT

Identify supporting needs
Stimulate development of
supporting business
Develop specific attraction
Develop strategic tools for
industry such as CRS

Figure 17.4 Supporting agent for the Small and Medium Tourism Enterprises (SASMTE)

to undertaking a business function on behalf of the SMTEs. In addition, such agencies, might undertake the management of properties, supporting divisions or activities on a commission, rental or entrepreneurial basis.

Facilitation and rationalisation of the supplies required by SMTEs can also be provided. Co-ordination of the regional agricultural and manufacturing output should be undertaken in order to provide adequate supplies to the local tourism-generated demand and thus increase local tourism multiplier effects. Purchasing in bulk for the entire population of SMTEs on particular islands would increase the bargaining power of the SMTEs over their suppliers. The supporting agency can also assist the SMTEs by organising a business and technology support centre where expensive equipment could be available for both SMTEs and tourists who might need to communicate with them. In addition, the agency can provide key services and equipment for SMTEs, such as laundry and freezer facilities which can provide significant economies of scope.

Finally, the supporting agency can play a significant role in the training of executives and personnel by organising vocational training in both strategic and operational management skills. The emphasis should be on total quality management and a consumer orientation, alongside contemporary developments such as global distribution systems and environmental conservation techniques.

On the marketing side, a supporting agency can draw on expertise in order to undertake a co-operative marketing function for Aegean SMTEs and/or support with know-how the marketing function of each enterprise. Marketing research should drive the strategy of the SMTEs and the destination. Profiling target markets through segmentation will enable SMTEs and the islands to have a clear idea of consumers' needs and wants, the adequacy of the existing tourism product as well as the tourism satisfaction level.

Drawing resources together can help SMTEs to overcome their size problem and launch a comprehensive promotional campaign. The contribution of the GNTO can increase the budget of the campaign. An integrated promotional activity could include advertisements in the mass media, representation at international tourism fairs, public relations functions, direct marketing and other ways to motivate the target market.

The supporting agency could also assist in product formulation by creating customised tourist packages for specific segments. Combining and co-ordinating input by several tourism enterprises would enable the provision of total tourism products which would effect total tourist satisfaction. Standards should be designed and maintained through total quality management while formulation of local attractions, festivals and events would stimulate tourism demand and consequently reduce seasonality.

The distribution of the Aegean tourism product is becoming extremely sophisticated and SMTEs which fail to introduce an appropriate distribution mix jeopardise their competitiveness. Apart from selecting and establishing partnerships with reliable, stable and 'fair' intermediaries, a wide range of infor-mation and reservation facilities should be utilised at the place of tourist origin as well as on the islands. Comprehensive, reliable and accurate information should be available for each tourism product produced at the destination while the ability to resource any tourism product from the place of origin should be in place. 'Regional integrated computerised reservation information management systems' distributed independently and through global distribution systems could contribute significantly here (Buhalis, 1993, 1994).

Finally, a supporting agency can assist Aegean SMTEs to achieve 'fair' prices to yield an appropriate return on their investment and effort. While ensuring that the consumer receives value for money, SMTEs have to maintain adequate income in order to offer decent salaries to their employees, proper return on investment for their proprietors or shareholders and make a contribution to the sustainability of the natural, cultural and social resources consumed by tourists. Collaboration between SMTEs, aided by the supporting agency, and joint negotiation with tour operators would improve the bargaining power of SMTEs and reduce the ability of tour operators to determine their pricing policies. Finally, flexible pricing based on consumer segments and the season, could be formulated and published directly to the public.

A supporting agency can play a major role in incorporating technology in the tourism industry. As most SMTEs are unable to understand the potential of new technology and unable to invest in equipment, supporting agencies have a role in demonstrating the potential benefits as well as to provide technology consultation and support. Negotiation with information technology companies would reduce the price of equipment and could also result in the development of inexpensive purpose-built software for the needs of SMTEs. In addition a 'regional integrated computerised reservation information management system' can be launched in order to promote the entire spectrum of tourism products and also facilitate the distribution and pricing function of the SMTE (Buhalis, 1993, 1994).

Moreover, a supporting agency can also collaborate with other development agencies at a national or international level. First, they can provide a documentation and reference centre with relevant information for the major tourist markets, tourist statistics for the region, techniques available for SMTEs and technical reports which can facilitate the transfer of knowledge.

A supporting agency can lead fund-raising attempts through the European Commission and other potential sources for Aegean SMTE financial or technical support. Co-operation with the GNTO and island authorities could establish a system of development control in respect of carrying capacity limits, and the implementation of an incentives and rewards system based on total consumer satisfaction, employee training and development, environmental conservation and contribution to total quality tourist products. Lastly, collaboration with scientific institutions such as universities and research centres can provide specialised expertise and transfer scientific knowledge for the benefit of SMTEs.

Finally, a supporting agency should also promote the diagonal integration of the island destinations by identifying and promoting supporting businesses for the tourism industry. Essential agricultural and manufacturing inputs in the tourism product formulation should be incorporated and private or public enterprises which can provide these products on competitive terms should be identified. A supporting agency might develop a number of these businesses when economies of scale or scope (or perhaps other competitive advantages) can be achieved. In addition, a number of specific attractions such as theme parks, galleries, conference and cultural centres and golf courses can either be developed or facilitated by supporting agencies. Finally, the development of specific tools for the industry such as beach cleaning mechanisms or computer reservation systems would be an invaluable strategic tool for SMTEs in the Aegean.

A supporting agency: organisation, framework and barriers

Although a wide range of benefits can be easily identified from the development of a supporting agency, it is also quite apparent that Aegean SMTEs are reluctant to receive external advice or to participate in collaboration schemes, as they feel that they lose some of their independence and flexibility. Unfortunately though, current developments in both tourism demand and supply are eroding the competitiveness of SMTEs and jeopardising their position. Go and Welch (1991: 122) express this threat clearly for SMTEs in the hospitality sector by stating:

> romantic, distinctive and charming as the private country hotel may be in almost any country in Europe, all that will count for little if there are no customers. All hotels must ensure that they are put before their market in the most effective way.

Thus, a supporting agency should be based on the requirements of Aegean entrepreneurs and will hopefully reflect the importance of its existence for both their business and destinations. As a consequence, a 'bottom-up' approach should be followed. The voluntary participation of the entire population of tourism enterprises in the Aegean should be the goal, whilst strictly defined membership regulations should be respected and followed. SMTEs which fail to follow the main policies and regulations should be penalised by removal from the supporting agency.

The administration of the supporting agency should be undertaken by a committee composed of tourism enterprises, entrepreneurs, GNTO officials, island authorities and chambers of commerce. Furthermore, a number of scientists specialising in particular subjects should contribute as consultants. As the supporting agency should be perceived primarily as concerned with the tourism business, entrepreneurs have to be in the majority on the administrative committee while qualified personnel with sound professional experience should spearhead the management of the supporting agency.

Conclusion

Although SMTEs dominate the tourism industry in most European countries, their position is jeopardised by recent tourism demand and supply developments. These developments conspire to threaten the competitiveness and prosperity of small businesses in tourism. This chapter has outlined a Greek island case study drawing upon a multi-level competitive framework which enables SMTEs to assess their competitiveness not only against neighbouring tourism product suppliers but also against alternative destinations, members of the distribution channel, competitors for the tourists' disposable expenditure, and also recreational activities at the place of tourists' origin. In addition, the case study advocates the development of a 'supporting agency for SMTEs' as a means of addressing this issue. Collaboration between the private and public sectors will be required for the successful development of such an agency. The major responsibilities and advantages of the agency are the strengthening of the management and marketing functions of SMTEs and facilitation of the introduction of information technology and its utilisation. In addition, supporting agencies can co-operate with other development agents to attract assistance for SMTEs and also to establish and reward good operational practices throughout the industry. Finally, such an agency might encourage or develop supporting business in order to facilitate the operation of SMTEs and to deliver total tourism products to consumers. In other words, co-operation and networking become a surrogate for 'size' in the operation of small businesses in tourism by providing enterprises with the marketing 'reach' and management capabilities of much larger organisations.

We believe that the combination of this case study with an underpinning conceptual framework provides lessons and insights for other destinations worldwide. For example, as the tourism market matures with ever more demanding, discerning and knowledgeable consumers, destinations will constantly adapt to new market opportunities. Small businesses must be aware of these opportunities and respond to the needs of the consumer. Yet the very environment within which small businesses operate is problematic (Cooper, 1995). Often the political context of the destination is against tourism and public sector support for small businesses may be weak; indeed the small business community at the destination is often poorly organised politically. In addition, the market intelligence available at destination level is often poor and businesses find it difficult to elicit the needs of their consumers and to break into new markets. The establishment of powerful networks amongst small businesses is a credible way forward at the destination level and will allow the creation of wealth on the supply side and delivery of total tourist satisfaction on the demand side. Failure to appreciate the need for collaboration will lead to isolation with severe consequences for the prosperity of SMTEs.

References

Brown, B. (1987) 'Recent Tourism Research in South East Dorset', in G. Shaw and A. Williams (eds) *Tourism and Development: Overviews and Case Studies of the UK and the South West Region*, Working paper 4, Department of Geography, University of Exeter.

Buhalis, D. (1991) 'Strategic Marketing and Management for the Small and Medium Tourism Enterprises in the Periphery of the European Community. A Chapter for the Aegean Islands in Greece', MSc Dissertation, University of Surrey, UK.

Buhalis, D. (1992) 'SWOT Analysis for the Small and Medium Tourism Enterprises: The Chapter of the Aegean Islands, Greece', paper presented at the Conference of Hospitality Management Education: Hospitality and Tourism Industry Research Conference, Birmingham Polytechnic, April.

Buhalis, D. (1993) 'Regional Integrated Computer Information Reservation Management Systems as a Strategic Tool for the Small and Medium Tourism Enterprises', *Tourism Management* 14 (5): 366–78.

Buhalis, D. (1994) 'Information and Telecommunications Technologies as a Strategic Tool for Small and Medium Tourism Enterprises in the Contemporary Business Environment', in A. Seaton, R. Wood, P. Dieke and C. Jenkins (eds) *Tourism – The State of the Art: The Strathclyde Symposium*, Wiley and Sons, Chichester.

Callan, R. (1989) 'Small Country Hotels and Hotel Award Schemes as a Measurement of Service Quality', *The Service Industry Journal* 9(2): 223–46.

Carter, F.W. and Turnock, D. (eds) (1992) *Environmental Problems in Eastern Europe*, Routledge, London.

Cooper, C. (1995) 'Barriers to the Delivery of Service Quality in Declining Destinations', in R. Teare and C. Armistead (eds) *Services Management*, Cassell, London, pp.70–74.

Cooper, C. and Buhalis, D. (1992) 'Strategic Management and Marketing of Small and Medium Sized Tourism Enterprises in the Greek Aegean Islands', in R. Teare, D. Adams and S. Messenger (eds) *Managing Projects in Hospitality Organisations*, Cassell, London, pp.101–25.

EC (1991) *Impact of Completion of the Internal Market on the Tourism Sector*, Technicon Consultants, European Community, DG XXIII, November, Brussels.

EC (1993) *Communication from the Commission: The European Observatory for SMEs – Comments by the Commission on the First Annual Report (1993)*, European Community (COM993) 527 final, 5 November, Brussels.

EIM (European Institute of Management) (1993) *The European Observatory for SMEs, First Annual Report, Report Submitted to the European Community*, DGXXIII, Small Business Research and Consultancy, Netherlands.

Giaoutzi, M., Nijkamp, P. and Storey, D. (1988) 'Small is Beautiful – The Regional Importance of Small-Scale Activities, in M. Giaoutzi, P. Nijkamp, and D. Storey (eds) *Small and Medium Sized Enterprises and Regional Development*, Routledge, London, pp.1–18.

Gilbert, D. (1990) 'Strategic Marketing Planning for National Tourism', *Tourism Review* 1: 18–27.

Go, F. and Welch, P. (1991) *Competitive Strategies for the International Hotel Industry*, The Economist Intelligence Unit, Special Report No. 1180, London.

Heath, E. and Wall, G. (1992) *Marketing Tourism Destinations*, J. Wiley, Chichester.

Lee-Ross, D. and Ingold, T. (1994) 'Increasing Productivity in Small Hotels: Are Academic Proposals Realistic?', *International Journal Hospitality Management*, 13(3): 201–7.

Lowe, A. (1988) 'Small Hotel Survival – An Inductive Approach', *International Journal of Hospitality Management*, 7(3): 197–223.

Morrison, A. (1994) 'Marketing Strategic Alliances: The Small Hotel Firm', *International Journal of Contemporary Hospitality Management*, Vol.6(3), pp.25–30.

Page, A. (1993) 'Global Small Business – A Focus for Economic Recovery', paper presented at the conference Overcoming Isolation – Telematics and Regional Development, 30 April–2 May, University of the Aegean, Business Administration Dept., Chios Island, Greece.

The pH Group (1993) *Microeconomic Analysis of the Tourism Sector*, Report for the European Union, DGXXIII, Brussels.

Poon, A. (1988) 'Flexible Specialisation and Small Size – The Case of Caribbean Tourism', SPRU, DRC Discussion Paper No. 57, University of Sussex, Brighton.

Poon, A. (1989) 'Competitive Strategies for New Tourism', in C. Cooper (ed.) *Progress in Tourism Recreation and Hospitality Management*, Vol. 1, Belhaven Press, London, pp. 91–102.

Poon, A. (1993) *Tourism, Technology and Competitive Strategies*, CAB International, Oxford.

Porter, M. (1985) *Competitive Advantage*, Free Press, New York.

Quinn, V., Larmour, R. and Murray, A. (1992) 'The Small Firm in the Hospitality Industry', *International Journal of Contemporary Hospitality Management* 14(1): 11–14.

Shaw, G. and Williams, A.M. (1990) 'Tourism, Economic Development and the Role of Entrepreneurial Activity', in C. Cooper (ed.) *Progress in Tourism, Recreation and Hospitality Management*, Vol. 2, Belhaven Press, London, pp. 67–81.

Sheldon, P. (1993) 'Destination Information Systems', *Annals of Tourism Research* 20(4): 633–49.

Thomas, R. (1995) 'Public Policy and Small Hospitality Firms', *International Journal of Contemporary Hospitality Management* 7(2/3): 69–73.

Wanhill, S. (1997) 'Peripheral Area Tourism: A European Perspective', *Progress in Tourism and Hospitality Research* 3(1): 47–70.

WTO (1991) 'Seminar on Small and Medium Sized Enterprises', Report on WTO meeting in Milan, 16–17 December, World Tourism Organisation, Madrid.

18

FACILITATING TOURIST–HOST SOCIAL INTERACTION

An overview and assessment of the culture assimilator

Philip L. Pearce, Edward Kim and Syamsul Lussa

This chapter deals with a ubiquitous issue in tourism: the quality and success of tourist–host interaction. The range of views on this topic is striking. At one end of the spectrum we have consistent calls from politicians and tourism executives claiming that tourism can help overcome many real prejudices, foster new bonds of fraternity and become a real force for world peace (Canan and Hennessy, 1989). The optimism of some speechmakers has a parallel voice in pessimism: some media commentators and church groups see tourism as a series of short, unhappy encounters where people feel disgruntled, dissatisfied and, finally, disgusted with one another (Crick, 1989; Lovel and Feuerstein, 1992). In this chapter initial emphasis will be placed on exploring an understanding of tourist–host contact. The core case study presented here will focus on an evaluation of one strategy to make such encounters more effective.

Two dynamic issues in global tourism trends enhance the importance of improving our understanding and management of tourist–host encounters. As tourism grows, a wider variety of source countries are contributing visitors to the global market place. In particular, the rapidly growing source markets of the Asia Pacific region which includes new tourism generating nations such as China, Korea, Taiwan, Indonesia and Malaysia must be considered as posing new social interaction management challenges for the visited communities (World Tourism Organisation, 1996).

Further, survey evidence suggests that visitors from Westernised countries are becoming increasingly interested in cultural experiences including meeting hosts and appreciating their way of life (Hawkins, 1993). The combination of these two forces of change highlight the need to understand culture contact issues from both the perspective of the host communities and the new kinds of visitors coming to culturally different destinations. The objective of this chapter fits well with one of the global tourist research agenda items identified by Ritchie (1993) when he states that we need to 'identify the major sources of tourism friction between developed and developing countries with a view to establishing their seriousness, generalizability and possibility of resolution' (Ritchie, 1993: 213).

It is valuable to set the tourist–host encounter literature in the context of the social impacts of tourism and cultural tourism more generally. It is common practice among tourism researchers to divide the impacts of tourism into three categories: physical, economic and sociocultural (Archer and Cooper, 1994; Mathieson and Wall, 1982). While such a division can be useful in describing and analysing tourism impacts it should be remembered that the boundaries of each category are not always clear. Instead, the consequences of tourism are often interrelated, and examples from the sociocultural sphere may have implications for the economic and environmental components. In defining social and cultural impacts themselves some further clarification is also needed. Smith (1978) was one of the first authors to highlight cross-cultural and social interaction issues in tourism development. Many of the authors in Smith's volume noted that communities experienced both short-term social impacts as well as longer-term cultural changes which were augmented by, but not limited to, tourism effects. According to Fridgen, social impacts can be thought of as 'changes in the lives of people who live in destination communities' (1991: 92) while cultural impacts can be conceived as 'the changes in the arts, artefacts, customs, rituals and architecture of a people' (1991: 97). In further discussion Fridgen (1991) suggests that social impacts are associated more with direct contact between residents and tourists while cultural impacts are longer-term changes which result more from tourism development. This time-scale separation of social and cultural impacts is one valuable way to differentiate them.

Another perspective on culture and its relationship to tourism comes from an exploration of definitions of culture (Richards, 1996). Rather than choosing from the myriad definitions available to explain culture, three themes from this diverse literature can be identified. These themes include:

- culture as a value system relating to intellectual, spiritual and aesthetic development;
- culture as summarising the whole 'way of life' of a people;
- culture as the works or products of intellectual and artistic activity (Williams, 1983).

These three variants on the term 'culture' are all useful in assessing the tourism–culture relationship and ultimately, therefore, in understanding social impacts. The direct contact experiences between tourists and hosts are most likely to occur when tourists seek to gaze on the visited culture, an activity which van den Berghe (1992) has defined as perceiving the local people as tourees, that is visual objects to be inspected and appraised by the visitor (cf. Urry, 1990). Additionally, when tourists are interested in the first and third themes defining culture they will interact with a range of service personnel and these individual encounters will also form a part of the spectrum of the tourist–host interaction experience.

Two of the dominant themes for assessing tourism in the 1990s have been to regard the phenomenon as a system and to emphasise the sustainability of tourism (Gunn, 1993; Lane, 1991; Mill and Morrison, 1992). The systems approach is powerful in understanding tourist–resident interaction because it identifies both the impacts of the encounters on tourists themselves as well as on the host community. The four core phrases and elements usually listed as the key points of ecologically sustainable development are:

- maintaining the resource base;
- maintaining the productivity of the resource base;
- maintaining biodiversity and avoiding irreversible environmental change;
- ensuring equity in and between generations (see Ecologically Sustainable Development Working Group, 1991; Bruntland, 1987).

The sociocultural resource base is an implicit component which is marginalised in terminology such as 'resource', 'productivity', 'biodiversity' and 'environmental change'. In the original formulations of sustainability (Bruntland, 1987) socio-cultural resources are included in the total sustainability equation (Reynolds, 1992; Dearden and Harron, 1994; Wight, 1994). Rather too frequently, however, the sociocultural base is not discussed and analysed nor are any strategies developed when sustainability issues are being pursued by researchers.

The present chapter seeks to move the sociocultural component of ecologically sustainable tourism to centre stage. In particular, it will seek to outline ideas and strategies for facilitating positive host–guest interaction in tourism.

Culture shock

The origin of the term 'culture shock' is credited to Oberg (1960) who summarised a number of American expatriate experiences in Africa. Oberg suggested that travellers who are well intentioned towards their hosts may be frustrated when they are in a new cultural environment. They may become frustrated because they lose their familiar signs, cues and symbols in interacting with local communities. The consequence of these difficulties is that the individual reacts to the uncomfortable environment by developing a disapproving attitude to the local people. In turn, these attitudes can result in behavioural conflicts. For many, the core of the problem of cultural shock is language but other behaviours such as how to greet people, how to maintain friendships and how to act in novel situations have been identified as part of the background to the generation of culture shock problems. The initial information on culture shock was broadly conceived for all cross-national interactions. For tourists the situation may be more subtle because contact is briefer and more specific.

Furnham and Bochner (1986) argued that even when tourists have a clear goal, such as relaxation or viewing specific sites, they still experience some degree of stress. Culture shock for the pleasure traveller, they argued, may be a lesser malaise than for the diplomat or businessperson, but this is not always the case and many reactions of tourists are not markedly different from those of foreigners. As suggested earlier, not only do tourists experience culture shock, but also the encounter may be stressful for the host population because they are confronted with new behaviours and novel values.

Pearce (1982) noted that tourists appear to have maximum social and psychological impact on their hosts when the host communities have had limited exposure to other cultures. The substantial literature on the social impacts of tourism is reviewed in Pearce *et al.* (1996) and some of the problems identified in this multidisciplinary and extensive body of knowledge may be traced to unsatisfactory interpersonal encounters which, in turn, are due in part to the core culture shock issues addressed in this chapter.

In all the discussions of culture shock a number of consistent consequences of the problem emerge. Individuals suffering from cultural shock are diagnosed as being confused, anxious and puzzled by the way others behave. Not only does this state lower the satisfaction of the visitors (or hosts) but it is seen as a precursor of intolerance and occasionally aggression (Smith, 1978). The need for cross-cultural training to reduce the impacts of culture shock has been recognised for a considerable period of time. In the United States there has been concern with the effectiveness of military forces and diplomatic personnel and systematic attempts to introduce such individuals to the culture of the visited community resulted in a program of research in the 1970s (Triandis, 1972). In the last two decades the rise of business interests in the Asia Pacific has prompted a number of commentaries and popular publications on better business relations in specific cross-cultural contexts (see for example James, 1995 and Naisbitt, 1997). Tourists, too, are targets for advice on how to behave through guidebooks and codes of conduct although there is some recent analysis suggesting that the guidebooks are ethnocentric and prejudice-confirming (Bhattacharyva, 1997).

Analysis of cultural contact

Before proceeding to a direct discussion of the tactics used in recent decades to reduce culture shock, it is important to provide a conceptual foundation for these strategies. Two specific conceptual frameworks will be discussed. The co-ordinated management of meaning (CMM) is a theory of human communication which identifies cross-cultural interaction at six different levels of sophistication and complexity (Cronen and Shuter, 1983). According to Cronen and Shuter the fundamental goal of communicating is to achieve co-ordination where co-ordination is the perception that the pattern of interaction makes sense to the participants. They do not suggest that each conversation or interaction is perceived identically by the participants but rather that people are satisfied if they feel there is a degree of understanding in the exchange of messages. The six levels of CMM theory are useful in terms of identifying the diversity and range of cross-cultural contact difficulties which may be identified in tourist–host encounters.

At the first level, CMM theory suggests that there can be a set of problems associated with what they have defined as context, that is the verbal and non-verbal behaviour available to the participants. The context level of communication deals with how clearly people understand one another's speech, gestures, posture and associated signals which form the building blocks of communication.

A second level of potential difficulty is defined as the speech acts, or the way meaning is imputed to forms of address. Implicit in this level of difficulty are such concerns as the status of the participants and the levels of formality that are involved in the ways in which people address one another.

A third level of concern is defined as an episodes level of analysis. Here the sequences of behaviour are seen as providing some potential problems. For example, some cultures have highly ritualised forms of greeting which occur frequently throughout a day while other communities may have an initial greeting at the start of a day's proceedings and then fail to acknowledge people formally on subsequent encounters. Episodes embrace a range of topics of interest to tourists and include such matters as arrangements for eating, sight seeing, photography, tipping, souvenir purchasing and gift giving.

The relationship level in CMM theory places attention on the nature of social bonds between people and the reciprocal rights and expectations which attend these relationships. This level of analysis is of particular interest when visitors and local residents form friendships and attachments since more opportunities for misunderstanding arise as the intimacy of the contact develops.

The life scripting level in CMM theory refers to the way people perceive themselves in action, an important difference being individuals from cultures who see themselves as very independent versus individuals from more communal cultures whose identity is chiefly governed by the attitudes and expectations of their close kin and reference groups. Further, the individual's relationship to the physical environment may be markedly different at this level, with some views seeing the environment as integral to individual and group identity while other cultures view the environment as a resource to sustain human well-being.

The final level of analysis provided by CMM theory is concerned with cultural patterns, the essential ways of knowing and acting that define the larger community and this may relate to such broad-scale items as what is perceived as honesty, justice or equity within a society. The kinds of issues which arise in the cultural pattern arena include such acts as payment for services, which Westerners might regard as bribery and Eastern cultures might see as a justified and required process to foster business and personal relationships.

The significance of the CMM approach to tourist–host encounters is to recognise that not all of the difficulties which tourists experience are going to be amenable to one kind of remedial activity. Language and non-verbal behaviour difficulties, for example, may require one form of education and treatment while the more complex issues in terms of life scripting and cultural patterns may require substantial education about the nature of societies and their religious, interpersonal and business practices.

A complementary conceptual scheme to that provided by CMM theory is the analysis of social situations (Argyle *et al.*, 1981). While the CMM analysis helps understand the diversity of encounters and their levels of complexity, the social situations approach focuses on identifying particular influences which are operating during any given encounter. A logical chain can be distinguished in social situational analysis amongst eight components which are the core elements of any social interaction. These elements are represented in Table 18.1.

A brief explanation of these eight elements is provided in the following section.

Table 18.1 Definition of the eight features of social situations

Features	Brief definition
Goals	Goals may be seen as the purposes or ends which direct social behaviour.
Rules	Rules are the shared beliefs which regulate behaviour. The existence of many unstated rules is most clearly shown in the opprobrium attached to rule-breaking behaviours.
Roles	Roles are the duties or obligations which attend the social positions people occupy.
Repertoire of element	The sum of behaviours which are appropriate to that situation.
Sequences of behaviour	The ordering of the repertoire of behaviours. Sequences may be very fixed or very fluid.
Concepts and cognitive structure	Concepts and cognitive structure may be thought of as the shared definitions and understandings needed to operate in the social situations.
Environmental settings	Environmental setting consists of the props, spaces, barriers, modifiers (viz. the physical units and their arrangement) which influence the situation.
Language and communication	The interest in language and speech in this context focuses on how things are said, the code of speech, vocabulary, and social variation inherent in language.

Source: Argyle *et al.* (1981)

Goals

Most social behaviour is goal-directed, and cannot be understood until the goals are known. Situational goals are related to individual forms of motivation. People enter situations because they anticipate being able to attain certain goals. Situations provide occasions for attaining goals. An individual may be motivated to pursue more than one goal, and these goals may help, interfere with or be independent of each other. Additionally, conflict between the goals of the participants may generate substantial communication difficulties.

Rules

Rules are shared beliefs which dictate which behaviour is permitted, not permitted or required. Rules are generated in social situations in order to regulate behaviour so that the goals can be attained. The existence of many unstated rules is most clearly shown in the negative social reactions attached to rule-breaking behaviours.

Roles

Every situation has a number of specified roles which provide the individual with a fairly clear model for interaction. Roles can be defined as encompassing the duties or obligations or rights of the social position. Roles involve a great number of expectations about the actions, beliefs, feelings, attitudes and values of the person holding that role. Situations produce role-systems in order that situational goals can be achieved.

Repertoire of elements

The 'repertoire of elements' can be described as the sum of behaviours which are appropriate to a situation. Repertoires will vary with situations in a number of areas of social behaviour. There exist several different kinds of elements: verbal categories and speech content, non-verbal communication and bodily actions. Elements provide the steps needed to attain goals.

Sequences of behaviour

The elements of behaviour in a situation may occur in a distinctive sequence. There are a number of sequences that occur in cross-cultural situations, but mistakes in the sequences of behaviour in greeting, eating and farewell encounters can be highlighted.

Concepts and cognitive structure

Concepts and cognitive structure vary between situations, and are developed to enable performers to attain situational goals. Different situations require sets of concepts, such as those needed to play cricket or visit restaurants (cited in Argyle *et al.* 1981). Sometimes there is an elaborate conceptual build-up, which must be mastered in order to cope with the situation. The goals of a situation may be pursued by different methods, aided by different sets of concepts.

Environmental settings

The environmental setting consists of the physical features of a setting and their arrangements. Boundaries are enclosures within which social interaction takes place. All boundaries contain props that are necessary within that boundary. Each prop has a particular social function and there is often a special social meaning and symbolic significance attached to it. Modifiers are physical aspects of the environment that affect the emotional tone of the behaviour being enacted. Spaces refers to the distances between people and objects. Spatial behaviour can be investigated in relation to four basic phenomena: privacy, personal space, territoriality and crowding.

Language and communication

Language and communication can be used in a variety of ways in different situations. The various aspects of language and communication, such as vocabulary, codes of speech and voice tone, are partly situation-specific. There are certain features of the social variation inherent in language that are situation-specific while others are applicable to all or many situations.

Armed with these two conceptual approaches to social encounters it is possible to analyse any single instance of difficulty between tourists and hosts according to both the level of the encounter in CMM theory and the key social situation features which need attention in order to solve the problem for the interacting parties. Kim (1996) lists thirty-seven problems associated with tourist–host encounters when Korean travellers come to Australia.

The scenarios associated with these encounters in Table 18.2 can be categorised in terms of the CMM approach and the social situation variables of most relevance. Table 18.3 provides four examples of this joint analysis of problematic social encounters.

Indonesia – a case study in tourist host interaction

The previous sections of this chapter have concentrated on identifying the nature of culture shock and outlining some conceptual schemes to help analyse and describe cross-cultural interaction for tourists and hosts. The present section focuses on one remedial attempt to improve Australian travellers' ability to have successful social encounters with Indonesian hosts.

Table 18.2 Problems associated with encounters by Korean visitors in Australia

No.	Problems associated with encounters
1	Confusion when selecting TV or video using a TV choice button in a hotel
2	Being led to a table around a corner in a restaurant considered to be a poor treatment
3	Strong preference for 'Soju' (Korean alcohol) even on an overseas holiday
4	Eating dried squid or fish in a closed environment (in a bus, on an airplane)
5	Breaking into a line of waiting people
6	Different exchange currency rate is thought of as a rip off
7	Trying clothes on in public without shame
8	Touching a baby's penis or bottom
9	Blowing noses loudly in public
10	Different bedding system with no covers tucked in
11	Smiling or laughing when confused or embarrassed
12	Getting in elevators before others get off
13	Staring and pointing at people different from themselves
14	Not filling out items which are required in their customs declaration cards
15	Disturbing others in public such as restaurants, or hotel foyers by using loud voices
16	Being rude to service personnel in hotels
17	Bumping into others in a crowd
18	Grabbing at someone's clothes to get attention or tapping the person's arm
19	Smoking anywhere without considering those around them
20	Not saying 'Excuse me', or 'I'm sorry' in public crowded settings
21	Not holding the door for the person behind them
22	Removing seat belts and standing up in an airplane immediately upon landing
23	Having facial expressions that are unemotional and dull
24	Using the middle finger to point or for counting number two with index finger
25	Gargling noisily after a meal and burping
26	Demanding discounts on all merchandise at stores
27	Frustration with Aboriginal contacts
28	Not being given advance notice for changing schedules
29	Holding hands with the same sex while walking
30	Slurping loudly while eating soup
31	Carrying a large amount of cash for shopping, thus becoming a major target for thefts
32	Misusing man-made facilities (elevator/emergency exit)
33	'Stealing' a child's nose in fun and showing it to him by placing one's thumb between the index and middle finger (a childish game in Australia but seen as a crude symbol by Koreans)
34	Using only one hand to give things to and receive things from elders
35	Drinking 'Soju' (Korean alcohol) anywhere they happen to be
36	Not trying to learn English and expecting Australians to speak Korean
37	Littering in public instead of in the bins provided

Table 18.3 Four Korean–Australian encounters illustrating CMM theory and social situations analysis

1   Encounter: confusion when selecting TV or video using a TV choice button in a hotel.

Explanation with social situation analysis and CMM theory.
    Some Korean visitors use the pay-for-viewing video service in hotel rooms incorrectly. Such visitors with limited language skills select the pay-for film and entertainment options without realising they have done so. They are then upset at the additional charges on their hotel bill. The core social situation element here is the cognitive element of not understanding the system of payment and video selection while the CMM theory casts this encounter as an episode in which the sequence of behaviour is misunderstood.

2   Encounter: breaking into a line of waiting people.

Explanation with social situation analysis and CMM theory.
    Waiting in line is not a familiar social behaviour for Koreans. According to CMM theory this encounter can be understood at the level of social bonds and relationships. In public places Koreans do not have any need to show deference or respect for strangers but seek to fulfil their own needs. By way of contrast social obligations are very strong towards family and work colleagues. In social situation terms breaking into a waiting line is rule-breaking behaviour.

3   Encounter: using only one hand to give things to and receive things from elders.

Explanation with social situation analysis and CMM theory.
    In Korean society deference to elders is signalled by the non-verbal behaviour of giving and receiving items (e.g. drinks) with two hands. The CMM theory would categorise this encounter as being at the contest or non-verbal level of analysis while the social situation analysis would again identify the role and rules of the behaviour.

4   Encounter: different exchange currency rate is thought of as a rip off.

Explanation with social situation analysis and CMM theory.
    This encounter illustrates, in CMM terms, a cultural pattern issue. Korean society is unused to variability in currency exchange rates and as a more centrally controlled economic society there are strong forces dictating money exchange rates. In the context where different banks, hotels and exchange agents offer different rates, in social situations terms the goals of the hosts are being questioned and the freedom and flexibility allowed in a fully commercial capitalist system are interpreted as exploitative.

Callan *et al.* (1991) suggest that there are several basic approaches to cross-cultural training. The first method focuses on information or fact-oriented training, where trainees are given facts about the country such as climate, the economy and its cultural history, usually through lectures, films, readings, group discussion and textbooks. Traditional guide books for tourists fall into this information or fact-oriented training approach.

A second method focuses on attribution training where explanations of the behaviour of other people is involved. Attribution training includes learning about the causes, or the reasons which lie behind the behaviours of the visited culture. A third technique is known as cultural awareness training since it focusses on the trainee's own culture. Here the task involves making would-be travellers more aware of their own implicit beliefs and customs, the things they do automatically, so that when confronted with a new cultural experience they are better able to see that their own behaviour is culturally embedded.

The fourth technique involves various versions of role playing. Here trainees are introduced to the culture or some simulation of it. The role plays may focus on particularly difficult situations which the traveller is likely to encounter, possibly working with members of the contact culture or 'old hands' who have successfully lived and worked in the country to be visited.

The focus of this case study is on one attribution training method known as the Culture Assimilator. In this approach to cross-cultural interaction trainees read a series of carefully selected scenarios and, from a multiple choice format, they are asked to pick the best explanation for the behaviour relating to the critical incident. Usually there are 25 to 40 incidents which are seen to illustrate the most important features of cross-cultural interaction likely to be encountered. Trainees read through the information and are asked to not only suggest the correct answer to the situation but also to appreciate the explanation or attribution which lies behind the answer. By proceeding through all of the episodes it is hoped that the would be traveller internalises the cross-cultural issues in the Assimilator enabling them to perform more successfully in the new culture.

Lussa (1994) constructed a Culture Assimilator for Australians travelling to Indonesia. Constructing a Culture Assimilator is a complex task. Randolph et al. (1977) suggested that it is both expensive and time-consuming. At core, it involves collecting a diversity of incidents and writing succinct episodes based on these incidents. The episodes in the Culture Assimilator are effectively short stories capturing the themes and possibilities involved in the situation. Additionally, the Assimilator must have answers which are equally plausible as well as adequate explanations for the key answers. Finally, a process of validation needs to be added to the construction process. In the validation step both the host party and the visitor are required to trial the Culture Assimilator to ensure that the correct meanings are being given and that the episodes are easily understood.

In terms of layout and presentation Triandis (1976) suggested that each episode of a Culture Assimilator should normally occupy six pages: the first page contains the critical short story and is followed by a question; the second page conveys four possible answers to the episode; while the last four pages provide the explanations for each answer.

The Culture Assimilator (which is also referred to by some commentators as a culture sensitiser, perhaps a more appropriate name) has some of the following advantages. It can accommodate various social situations and a range of levels of cultural interaction as indicated in CMM theory. The materials can be presented in a booklet form involving a high level of flexibility for conducting the training schedule. The training can be conducted individually or in a group at any time, and in a variety of settings, and the approach can be combined with other approaches to culture training such as the use of guide books. Additionally, the approach has a set of research findings indicating its effectiveness in the diplomatic, military and international volunteer fields. These findings all concur that a well-constructed Culture Assimilator is both viewed positively by the trainees and generally evaluated as effective in the field situation (Furnham and Bochner, 1986).

In the study conducted by Lussa, twenty-one critical incidents were selected to identify key issues when Australian tourists interact with Indonesian people. The diversity of Indonesian culture and behaviour represented one of the problems in selecting and constructing the Culture Assimilator in this context. A decision was made to focus on common Indonesian cultural situations rather than those specific to one location such as Bali, Sulawesi or different parts of Java or Sumatra. Sixty-three Australian travellers to Indonesia read the Indonesian Culture Assimilator (ICA) booklet and a further 122 Australian travellers who did not read the booklet were treated as a control group. The control group members were shown to be comparable to the travellers on key demographic and motivational variables. Both of these groups were used to overview and assess the cross-cultural contact difficulties they had while they were in Indonesia. An example of the format of the Indonesian Culture Assimilator is given in Table 18.4.

In evaluating the Indonesian Culture Assimilator seven questions were constructed. The travellers who used the assimilator were asked to assess overall whether the ICA was helpful, whether the episodes were too complicated, whether the topics were relevant and whether or not the episodes were too long, the explanation too simple, the ICA was fun or entertaining to read and whether the ICA had taught the respondent a great deal. The results may be interpreted as a mixed evaluation of the Indonesian Culture Assimilator. The relevance, helpfulness and appeal of the ICA were perceived favourably but there was a concern that the episodes were too complicated, overly long and the explanation sometimes too simplistic.

Both the control group of subjects and those who had received the ICA were asked to report on their difficulties in interacting with Indonesian hosts. The control group in particular reported a diversity of cross-cultural contact problems. They reported language and communication barriers, problems with the street hawkers and tourist exploitation, problems with touching, theft and a broad range of tourist physical-facility-related problems. The results from the control group may be compared with the cross-cultural contact difficulties experienced by those who had the benefit of the Indonesia Culture Assimilator. This material is presented in Table 18.5.

In terms of the reported cross-cultural difficulties for the two groups, the data from the Culture Assimilator tourists is encouraging. Only a limited number of

Table 18.4 Example of format of the Indonesian Cultural Assimilator

13. A CROWDED SITUATION

On the way from Jakarta to Bandung, Christian travels by train. At the time, there are many passengers on the train. Some of them are standing because all seats are already occupied by the other passengers.

Christian is one of the lucky passengers on that train because he still can enjoy sitting down. However, because he has long feet, it is very difficult to move his legs among bags, parcels, etc. under the seat in front of him.

In order to make himself more comfortable, he puts his left leg on someone's bag. Unfortunately, on his right hand side there is no space to put his right leg because the parcel in front of him is as high as the seat. He tries to move a little bit to the right by using his right leg but he cannot because a young Indonesian, Tono, is standing there. He then asks Tono to help him move the parcel. But Tono does not understand what he says. Holding a drink in his right hand and a camera in his left, Christian tries to indicate by using his right leg that he wants the parcel moved.

Tono does not do anything. Moreover, he seems to pretend that he does not understand.

Question: Why does Tono not want to help him?

Answer:

1. *Because Tono sees that his bag is being damaged by Christian who deliberately put his left leg on it.*

2. *Tono is jealous because Christian, a foreigner, has a seat while he does not.*

3. *Tono does not really understand what Christian means.*

4. *Tono considers that Christian asking for help by pointing using his legs is impolite.*

(detailed explanation of the correct answer is on the next page)

Table 18.5 Cross-cultural contact problems Australian travellers to Indonesia: Open-ended responses

	With Culture Assimilator training $N = 63$	*Without Culture Assimilator training* $N = 122$
	Number of incidents	
1. Language communication barrier	4	22
2. Street hawkers	2	16
3. Being exploited	2	5
4. Touching	–	3
5. Theft	–	2
6. Cultural expectations (unknown)	–	2
7. Facilities	–	2
8. Getting lost	–	2
9. Beggars	–	2

Incident 13: Rationales for Alternative Explanations

(1) You choose 1. *Because Tono sees that his bag is being damaged by Christian who deliberately put his left leg on it.* The story does not indicate that this is the root of the problem, please respond again.

(2) You choose 2. *Tono is jealous because Christian, a foreigner, takes a seat while he does not.* Some old Indonesians who were involved directly in the fighting for the independence of the country are very resentful when seeing Westerners. The reason behind this is simple. They consider that any Westerner is Dutch. However, this group is now very small. Younger Indonesians like Tono appreciate their guests more. Please look more closely at the critical incident and choose again.

(3) You choose 3. *Tono does not really understand what Christian means.* The story suggests that Tono just pretends not to know what Christian means. Please make another selection.

(4) You choose 4. *Tono considers that the way Christian is asking for help by pointing using his legs is impolite.* The best and maybe the only accepted way to point to something is to use the right hand. The more polite way is to show your fist while your thumb directs the things you want to show. Using other parts of your body such as the left hand is not accepted. Moreover, if you use your feet, this is considered very rude. This is the best explanation.

Pointing to something by using a leg or foot is considered very rude.

problems are reported by those who had the benefit of reading the ICA. For those who did not read the ICA, more problems and a broader range of problems appear. It is important, however, to interpret these findings with considerable caution. First, the interaction difficulties are self-reported problems, not observer or objective accounts of tourist behaviour. There may be some tendency for those who have used the ICA to think of themselves as more knowledgeable and hence report fewer problems. Second, the number of tourists in the sample is small and possibly those who returned the questionnaire in its fully completed form may have been the most interested and receptive to the communication. Nevertheless, these results represent a beginning for the further development of this kind of approach for tourists.

The comments of the ICA users were also noteworthy. In addition to the previously reported 7-scale items concerned with assimilator effectiveness, a number of additional open-ended remarks were provided. Several people noted that they would have preferred a video rather than a text- or story-based format. Others remarked that the booklet was useful as a refresher while travelling, a comment that would not apply so well to a video version, but indicating that a summary of the rules or points would also be helpful.

Overall evaluation of the Indonesian Cultural Assimilator

Although the overall results of the analysis of the Indonesian Culture Assimilator indicated that respondents reacted quite positively to the concept, several important limitations must be noted. The study, despite a comprehensive attempt to identify core difficulties for tourists, missed two of the main problems, street hawkers and tourist exploitations, which were experienced by both the control and the ICA samples. These two problems were seen to be particularly important because their commonality across both groups indicates that they are additional issues to be added to any Indonesian Culture Assimilator. The failure of the collection procedure to identify a common social situation of concern has been supported by previous researchers who note that one of the difficulties of collecting critical incidents is to make sure that all of the areas of concern are covered. Distribution difficulties encountered in this pilot study should also be noted. The study was conducted with the assistance of Garuda airlines who suggested to passengers *en route* to Jakarta or Bali that they could use the ICA; 300 booklets were distributed and while only 20 per cent of the passengers completed the evaluation of the ICA, a number of others may have used sections of it without participating in the study.

One suggestion emerging from the study is that an in-flight video form of Culture Assimilators could be considered for contemporary travellers. In-flight videos could present more dynamic and behavioural aspects of cross-cultural education. An added advantage of the video material is that it provides a more stimulating and possibly succinct presentation than the text. Again, however, the co-operation of airlines and tourist businesses would have to be more direct and committed than was evident in the initial study.

Conclusion

The recognition of cross-cultural difficulties as an area of concern in con-temporary tourism is a valuable step in enhancing the sustainability of tourism as a world-wide activity. The culture contact experience of many visitors is akin to being physically lost, since the usual landmarks of how to deal with the other people are missing, direct pathways to achieving goals are blocked, and frustration, anger and blaming others can be significant outcomes. By recognising that cultural problems come at different layers, or levels, in the communication process it can be suggested that the need to empower visitors involves a range of tasks. A social situations appraisal to understanding culture shock emphasises that visitors lack certain kinds of skills and behavioural knowledge to deal with the situation thus suggesting a range of training and remedial approaches. Teaching visitors simple routines or exposing them to factual information may not be enough to generalise their learning. The Culture Assimilator, even in its rudi-mentary text-based version can tackle a range of cross-cultural contact problems varying in level and skills.

The trialing of the Culture Assimilator as presented in the present chapter represents an interesting opportunity for tourism professionals. The technique is not without problems and the evaluation of the approach will require larger samples, more diverse locations and a continued attempt to ensure that the most critical incidents are included in the instrument. Nevertheless, as a strategy for providing non-judgemental and more sophisticated understanding of the culture contact situation the instrument has some possibilities. In particular, the possible use of the Culture Assimilator approach as a combined effort with other techniques such as tourist behaviour codes, sets of visitor rules and sound advice from professional guides may ease the pain of culture shock and promote sustainable sociocultural tourism encounters.

The need for tourists to prepare themselves for sociocultural contact may be seen as analogous to visitors with an interest in exploring physically risky locations. Few adventure tourists would think of undertaking high-risk travel without prior training and appropriate equipment. It is equally appropriate for culturally committed visitors to prepare themselves mentally for similarly hazardous journeys, and even for those visitors who are less committed to seeking contact with local people there are enough minor difficulties to justify the effort of preparing themselves for the experience.

The sociocultural components of tourism are not incidental to the further development of the world's largest and most international phenomenon. The human elements of tourism are the felt, personal and direct quality-of-life components which have the potential to bind cultures together. If the goal is sustainable tourism then paying attention to the human resource development needs of tourists and of those with whom they come in contact is vital. An important new direction for culture shock training and Culture Assimilator development might involve developing such instruments for hosts as well as improved versions for tourists.

References

Archer, B. and Cooper, C. (1994) 'The positive and negative impacts of tourism', in W.F. Theobald (ed.) *Global Tourism*. Oxford: Butterworth-Heinemann.

Argyle, M., Furnham, A. and Graham, J.A. (1981) *Social Situations*. Cambridge: Cambridge University Press.

Bhattacharyva, D. (1997) 'Mediating India. An analysis of a guidebook', *Annals of Tourism Research* 24(2): 371–89.

Bruntland, G.H. (1987) *The Bruntland Report. Our Common Future*. World Commission on Environment and Development. Oxford: Oxford University Press.

Callan, J.V., Gallois, C., Noller, P. and Kashima, Y. (1992) *Social Psychology*, 2nd edn. Sydney: Harcourt Brace Jovanovich.

Canan, P. and Hennessy, M. (1989) 'The growth machine, tourism and the selling of culture', *Sociological Perspectives* 32: 227–43.

Crick, M. (1989) 'Representations of international tourism in the social sciences: sun, sex, sights, savings, and servility', *Annual Review of Anthropology* 18: 307–44.

Cronen, V.E. and Shuter, R. (1983) 'Forming intercultural bonds', in W.B. Gudykunst (ed.) *Intercultural Communication Theory: Current Perspectives*. Beverly Hills, CA: Sage.

Dearden, P. and Harron, S. (1994) 'Alternative tourism and adaptive change', *Annals of Tourism Research* 21(2): 81–102.

Ecologically Sustainable Development Working Group (1991) *Final Report – Tourism*. Canberra: Australian Government Publishing Service.

Fridgen, J.D. (1991) *Dimensions of Tourism*. East Lansing, MI: Educational Institute of the American Hotel and Motel Association.

Furnham, A. and Bochner, S. (1986) *Culture Shock*. London: Methuen.

Gunn, C. (1993) *Tourism Planning*, 3rd edn. New York: Taylor and Francis.

Hawkins, D.E. (1993) 'Global assessment of tourism policy', in D.G. Pearce and R.W. Butler (eds) *Tourism Research. Critiques and Challenges*. London: Routledge.

James, D.L. (1995) *The Executive Guide to Asia-Pacific Communications*. St Leonards, Australia: Allen and Unwin.

Kim, Y.J. (1996) 'Korean tourists' cross-cultural experiences in Australia'. *Conference Proceedings Travel and Tourism Research Association*, Las Vegas.

Lane, B. (1991) 'Sustainable tourism: a new concept for the interpreter', *Interpretation Journal* 49: 1–4.

Lovel, H. and Feuerstein, M. (1992) 'The recent growth of tourism and the new questions on community consequences', *Community Development Journal* 27(4): 335–52.

Lussa, S. (1994) 'The development of an Indonesian Culture Assimilator for Australian tourists', unpublished Master of Administration (Tourism) thesis, James Cook University.

Mathieson, A. and Wall, G. (1982) *Tourism: Economic, Physical and Social Impacts*. London: Longman.

Mill, R.C. and Morrison, A.M. (1992) *The Tourism System*. Englewood Cliffs, NJ: Prentice-Hall.

Naisbitt, J. (1997) *Megatrends Asia*. London: Nicolas Brealey.

Oberg, K. (1960) 'Culture shock: adjustment to neo-cultural environments', *Practical Anthropology* 17: 177–82.

Pearce, P.L. (1982) *The Social Psychology of Tourist Behaviour*. Oxford: Pergamon Press.

Pearce, P.L., Moscardo, G. and Ross, G.F. (1991) 'Tourism impact and community perception: An equity-social representational perspective', *Australian Psychologist* 26(3): 147–52.

Randolph, G., Landis, D. and Tzeng, O.C.S. (1977) 'The effects of time and practice upon Culture Assimilator training', *International Journal of Intercultural Relations* 1(4): 105–20.

Reynolds, P.C. (1992) 'Impacts of tourism on indigenous communities', in C.P. Cooper and A. Lockwood (eds) *Progress in Tourism, Recreation and Hospitality Management*, Vol. 4. London: Belhaven Press.

Richards, G. (ed.) (1996) *Cultural Tourism in Europe*. Wallingford, Oxon: CAB International.

Ritchie, J.B.R. (1993) 'Tourism research: Policy and managerial priorities for the 1990s and beyond', in D.G. Pearce and R.W. Butler (eds) *Tourism Research. Critiques and Challenges*. London: Routledge.

Smith, V.L. (ed.) (1978) *Hosts and Guests*. Oxford: Blackwell.

Triandis, H.C. (1972) *The Analysis of Subjective Culture*. New York: Wiley.

Triandis, H.C. (1976) 'The Culture Assimilator: An approach to cultural training', in H.C. Triandis (ed.) *Variations in Black and White Perceptions of the Social Environment*. Urbana: University of Illinois Press.

363

van den Berghe, P.L. (1992) 'Tourism: The ethnic division of labor', *Annals of Tourism Research* 19: 243–49.

Urry, J. (1990) *The Tourist Gaze: Leisure and Travel in Contemporary Societies*. London: Sage.

Wight, P.A. (1993) 'Sustainable ecotourism: Balancing economic, environmental and social goals within an ethical framework', *Journal of Tourism Studies* 4(2): 54–66.

Williams, R. (1983) Keywords. Fontana: London.

World Tourism Organisation (1996) *WTO's International Tourism Overview. A special report from the World Tourism Organisation*. Madrid: WTO.

19

A GLOBALISED THEME PARK MARKET?

The case of Disney in Europe

Greg Richards and Bill Richards

The effects of globalisation, or the increasing integration of national economies, are particularly evident in leisure and tourism. The global spread of homogeneous consumer products has, for example, been labelled McDonaldisation (Ritzer, 1993) and Disneyfication (de Roux, 1994). The growth of the theme park, and in particular the market leader in the global theme park industry, Walt Disney, has made the Disney parks one of the most talked about symbols of global cultural homogenisation. Boje (1995: 997) has even been prompted to pose the question 'Who is better known, Jesus Christ or Micky Mouse?' As Altman (1996: 43) comments: 'Walt Disney occupies a special place in the global culture of the second half of the twentieth century. With the advent of the mass media, Disney characters are household figures all over the globe'. The Disney phenomenon is particularly interesting from a management perspective, not only because Disney itself is often praised as an example of excellent management (Peters and Austin, 1985), but also because Disney parks are frequently cited as examples of postmodern styles of tourism consumption (Bryman, 1993).

Disney is a highly successful global leisure corporation, which has expanded its activities across a wide range of leisure markets, including theme parks, media, resorts, publishing and merchandising. An important factor in the success of Disney has been the integrated nature of its products, with synergies between film and television, between media and theme parks, and between theme parks and hotel and resort operations. All of these products relate to leisure markets, and as Gratton and Taylor (1988) suggest, it is knowledge of leisure markets which are the essential management competence in the leisure industry. The worldwide expansion of Disney in the field of media seem to suggest that Disney products have a universal appeal, which can transcend national market boundaries and cultural distinctions (Altman, 1996). Confidence of the Disney management in the global appeal of the Disney product was a key factor in the decision to enter first the Japanese and then the European theme park markets. While the media products have been successfully globalised, however, the opening of EuroDisney in Paris provided a major test for the global strategy of Disney and the universal appeal of its products. Would the Disney formula work in Europe? Could the Disney park, the epitome of American culture, be transplanted into the cultural heart of France?

This chapter examines the globalisation of Disney theme parks, and assesses the managerial problems involved in developing transnational tourism and leisure products. In particular, the chapter analyses the role of market knowledge in the success of such products.

What is a theme park?

Although the literature relating to theme parks has blossomed in recent years, there is still little agreement over what a theme park actually is. A wide range of definitions have been proposed, but no single definition has been established. Some authors take a fairly wide view of theme parks, arguing that theme parks are simply 'family entertainment complexes' or 'modern leisure parks' developed from the 'basic fairground approach' (Bruce *et al.*, 1986). Other authors have taken a far narrower approach, insisting that a theme park must have a clearly identifiable theme or themes (Mintel, 1990). If the narrow definition is used, however, many parks which are commonly referred to as 'theme parks', would not qualify as such. Disneyland, for example, in common with the other Disney parks, is based not on a single theme, but on a series of loosely related themes in different 'worlds' or areas of the park, and in many other parks the 'theme' is so generalised as to be almost invisible. The International Association of Amusement Parks and Attractions (IAAPA, in Tourism Research and Marketing, 1993: 50) defines a theme park as a facility 'with a definable and non-changing location which has as its primary draw, hard rides, and which also has a theme'. For IAAPA, therefore, it is the presence of rides which is central to the theme park concept, and the theming plays a supporting role. The problems of this definition become obvious when you consider a park such as Legoland, now open in Denmark and the UK, and soon to open in America. This park has a single theme, but lacks thrill rides, and would therefore not fit IAPPA's definition of a theme park. Yet the appeal of Legoland is very wide, and it can offer formidable competition in the family market to established 'theme parks', such as Thorpe Park in the UK.

Our approach (Tourism Research and Marketing, 1993) has been to define a number of features of theme parks which seem to be common to most attractions identified as theme parks, and which reflect the distinctive management parameters of a theme park operation.

The last point in Table 19.1 has recently been added to the definition (Tourism Research and Marketing, 1996) to distinguish existing theme parks from the new generation of indoor simulation-based entertainment centres, such as Sega World in London. Such developments indicate how dynamic the theme park market is, and how definitions also have to remain flexible to retain their usefulness.

Table 19.1 Features of a theme park

- Primarily an outdoor attraction.
- A visitor destination in its own right.
- Based on rides, which are operated as a single management unit.
- An all-inclusive entry charge or ride charge, which covers the use of all major facilities in the park (pay-one-price system).
- Constructed around the needs of visitors, rather than relying on pre-existing natural or built features.
- Focused on entertainment rather than education.
- Based on physical experiences rather than simulation.

The global theme park market

Theme parks evolved out of the amusement parks and funfairs which developed in North America and Europe at the end of the nineteenth century. In his analysis of the European development of theme parks, McEniff (1993) identifies theme parks as having roots in fixed fairgrounds, pleasure gardens and seafront developments. In the United States, amusement parks began to develop close to major conurbations during the nineteenth century, and these gradually developed into more spectacular and boisterous attractions. By the peak in the amusement park boom in the 1920s the three amusement parks in Coney Island near New York were attracting 20 million visitors a year (Weinstein, 1992).

The American amusement parks declined during the depression of the 1930s, and it was during this time that Walt Disney developed his idea for a new type of family entertainment complex. Disney's ideas crystallised in the early 1950s, when he hit on the idea of opening a park which would act as a backdrop for a TV series entitled 'Disneyland', and which would also enable visitors to meet their favourite Disney characters in the flesh. In searching for inspiration Disney visited many existing parks, including de Efteling in the Netherlands, where Disney allegedly borrowed some of the fairy tale elements which were to be so successful in Disney's parks. In 1955 Disney opened the first 'theme park' in Anaheim, California. Disneyland differed from its competitors in having a coherent theme, which for Disney was based on his film characters. Disneyland was an immediate success, and attracted 4 million visitors in its first year. The success of Disney sparked a rash of imitators in North America.

The 1960s and 1970s saw rapid growth in terms of theme park development and attendance, and by 1980 theme and amusement parks in North America were attracting 125 million visitors a year (Loverseed, 1994). The theme parks were successful because they met the need for family entertainment in a market where travel was expanding rapidly. Disney expanded its theme park operations with the opening of Disney World (1971), EPCOT (1982) and MGM Studios (1989), all based in the 6,000 acres of Florida swampland that Disney had bought for $50 an acre.

In Europe, the theme park 'boom' began much later, with the major growth occurring in the 1980s. Early centres of theme park development were the Benelux countries and Germany, where a number of amusement parks had opened in the 1960s and 1970s. These parks gradually developed into fully fledged theme parks. UK theme park development started in the late 1970s, while France only joined the rush in the 1980s.

Although some smaller amusement parks opened in Asia in the 1970s, major theme park development in Asia and Australia only began to take off in the 1980s. Asia is now the fastest growing area of theme park development, with growth rates in the early 1990s of up to 20 per cent a year.

The theme park market exhibits a similar pattern to many other global markets, with developments originating in the 'core' markets of North America and Europe, and gradually spreading to peripheral regions. The growth of the market in recent years has been characterised by increasing geographic spread and physical scale of facilities, increasing professionalisation and the growing application and sophistication of technology. The level of investment required to develop new theme parks has consequently risen. At Universal Studios in Hollywood, for example, MCA recently invested $75 million in the 'Back to Future' ride, and $110 million in recreating 'Jurassic Park'.

As the theme park market has matured, a growing proportion of the parks have been incorporated into major theme park companies and entertainment corporations, such as Disney, Warner Brothers, MCA, Paramount, Anhauser Busch, Six Flags, the Tussaud Group and Walibi. As competition has increased in their own domestic markets, these operators have looked to international markets to provide the necessary business growth. Disney were the first major corporation to expand abroad with the opening of Tokyo Disneyland in 1983. Although the park was not owned by Disney, it was an almost complete copy of Disneyland in California, with very few concessions to the Japanese market. As a PR executive for Tokyo Disneyland commented: 'we really tried to avoid creating a Japanese version of Disneyland. We wanted the Japanese visitors to feel they were taking a foreign vacation by coming here, and to us, Disneyland represents the best America has to offer' (Van Maanen and Laurent, 1992). Tokyo Disneyland, like the original, was an instant success, attracting 10 million visitors in its first year of operation, and it is now the most popular theme park in the world, with almost 16 million visitors in 1995.

The success of Tokyo Disneyland, and the growth potential identified in the relatively immature European market persuaded Disney that there would be a market for a Disney park in Europe. The idea of expanding the theme park operations abroad fitted well with the aggressively expansive company policy during the early 1980s. Over 200 European locations were researched, before a site at Marne la Vallée, near Paris, was chosen.

The plan

The contracts for the construction of EuroDisney at Marne la Vallée, 32 km east of Paris, were signed in 1987. The choice of Paris as a site for the Disney park had been clinched by financial incentives from the French government in the form of grants, tax breaks and infrastructure improvements, which amounted to a quarter of the projected development costs of Ffr 22 billion ($4.2 billion). Having granted tax breaks to Disney, the French government was actually forced to lower the rate of value added tax for all parks from 18 per cent to 7 per cent. In return for these advantages, Disney promised to make EuroDisney the only Disney park in Europe for at least five years, to make French the principal language in the park, and to award the bulk of building and equipment work to French and European companies.

The overall development was planned to cover almost 2,000 hectares, which would be developed in three stages. In the first stage the main elements would be the Magic Kingdom theme park, a resort complex with six hotels offering a total of 5,200 rooms, and assorted commercial and residential developments. The second phase would feature a park similar to the MGM Studios complex at Disney World, in Florida. Forecast to cost Ffr 18 billion ($3.38 billion), the second park would generate a further 8 million visitors a year when it opened in 1995. A third park, modelled on EPCOT, was scheduled to open in 2000, accompanied by a convention centre, a water park and 13,000 additional hotel rooms.

In order to finance the development, EuroDisney was launched on the stock market in 1989, with the aim of raising Ffr 5 billion ($950 million) million. In selling the park to investors, much play was made of Disney's enviable track record, and its successful transplant of the Disney formula to Japan. The prospectus forecast that the park would make pre-tax profits of Ffr 200 million ($38 million) in its first year of operation, rising to Ffr 1,500 million ($300 million) in 1996 when the second stage of the development would be complete. Shares in the French holding company, EuroDisney SCA were launched at 70Ffr ($13), and quickly rose in value to over $18 by mid-1990. A peak of $26 per share was reached in early 1992, just before the opening of the park (Tourism Research and Marketing, 1993).

Table 19.2 Visitor attendance at Disneyland Paris

	Visitors (millions)	
	Actual	*Forecast*
1992/3	10.5	11
1993/4	9.5	
1994/5	8.8	
1995/6	10.7	
1996/7	11.8 (forecast)	20

The forecast attendance for the first year of operation was 11 million visitors, about 50 per cent of whom were expected to come from France. These estimates were regarded as conservative by some observers, who expected the visitor numbers to be closer to Disney's 'optimistic' scenario of 15 million visitors a year (Table 19.2). The hotels and catering operations were expected to contribute 50 per cent of total turnover for the park, the same ratio as in the American parks. The hotels were expected to have an average 85 per cent room occupancy, also similar to that achieved in Disney's American resorts.

The reality

Fairly soon after the extravagant opening in April 1992, the projections which had been made for the operational and financial performance of the park began to be questioned. It rapidly became clear that not only would the park fail to meet even its pessimistic visitor targets, but that it would sustain serious financial losses. By 1993 EuroDisney's debts had mounted to over Ffr 20 billion ($3.8 billion), with rising interest charges and disappointing revenue figures contributing to a loss of over Ffr 300 million ($60 million) in the first year of operation, rising to over Ffr 5 billion ($1 billion) in the second year. By late 1993 the share price had reached a low of $6 (Tourism Research and Marketing, 1994).

A number of other assumptions in the original plan also proved false. Only 30 per cent of the visitors in the first year were French. There was also a steep fall in foreign visitors after the first year, as the novelty factor began to wear off. UK visitors dropped from 600,000 in 1992 to just over 300,000 in 1993. The hotels performed far worse than had been expected. In the first three months the average occupancy level was only 70 per cent, already well below the forecast 80 per cent, and during the winter months occupancy levels fell to disastrous levels. The average room occupancy for the first year of operation was only 55 per cent. Catering and merchandising sales were also disappointing, with average spending levels being way below those of Disney's American parks. Disney had expected that merchandise sales would mirror those in America, with an emphasis on high value, high quality items. European spending patterns proved to be far more downmarket.

Disney also misunderstood the importance of bus and coach traffic in the European tourism market. No facilities were provided for coach drivers, which is a major consideration at most other large European parks. Disney expected most public transport business to come via the TGV and the Channel Tunnel, not understanding that rail is relatively expensive for the theme park public.

Tried and trusted Disney management strategies also failed to translate well to the French situation. The French 'cast members' 'met the show metaphor with cynicism and resistance' according to Boje (1995). There was also friction over the American style of management and the strict Disney dress code. As a result labour turnover up to August 1993 was about 50 per cent, over twice the level achieved at Disney's American parks (Bryman, 1993).

The recovery

Shortly after the opening of the park, Disney faced significant problems in terms of falling visitor numbers, poor image and publicity and, most seriously, rapidly rising debts. All of these problems required swift and decisive action from the company.

The most significant changes were financial. In order to inject more equity into the company, a rights issue of almost 600 million EuroDisney shares at Ffr 10 ($2) each was made, which raised almost Ffr6 billion ($1.2 billion). Almost half the shares were taken up by the Walt Disney parent company, increasing their financial stake in EuroDisney. A moratorium on interest charges was arranged with the banks, reducing the interest burden. The Walt Disney parent company agreed to waive their royalty fees from 1993 to 1998, and to reduce the royalty fees by half from 1998 to 2003, and similar reductions were also made in the management fee. The parent company therefore injected significant new capital into EuroDisney, reducing the debt burden, and by reducing their fees, they also eased the operational costs of the park. Disney also found an injection of capital from Saudi Prince Al-Waleed, who agreed to take a significant stake in the company. Prince Al-Waleed has invested heavily in hotels and entertainment, tending to buy into failing businesses when the price is low. As with many of his other adventures, the Prince's Disney adventure also brought about the required confidence boost, followed by the desired boost to the share price.

In order to try and improve the financial position of the company, a number of cost-cutting measures were also instigated. Staff numbers were cut from 14,000 to 12,000 in the first winter of operation, and by 1993 winter staff levels had been further reduced to 10,000 (Tourism Research and Marketing, 1993). Hotel rates were reduced in 1993 by 30 per cent during the winter. Room occupancies improved to over 60 per cent in mid-1994, and rose to just under 80 per cent in 1995/96. Much of the improvement was generated by promoting the hotels as corporate conference venues, which, in spite of the entertainment connections of the venue, has been quite successful. In spite of the improvement in occupancies, and the closure of one of the hotels, the hotels are generating far less revenue than Disney had forecast even in its pessimistic forecasts.

A further problem experienced in Paris was the fact that the park opened with too few attractions. The Paris park had only 15 major attractions in 1992, compared with 45 in the California version. New attractions have therefore been developed, including Space Mountain, opened in 1995 at a cost of Ffr 400 million ($80 million). Based on a Jules Vernes story, this 'white knuckle ride' catapults visitors to the height of a five-storey building in less than 2 seconds, with an acceleration of 1.3 g.

In a turnaround from the initial catering policy, sit-down restaurants were transformed into fast food outlets. The increased flexibility of the catering outlets allowed Disney to deal with two problems. First, the habit of the European visitors of wanting to eat at the same time, instead of 'grazing' throughout the day like many Americans. Second, the tendency for food purchase decisions to be dictated by children, who were likely to influence their parents to visit the nearest food outlet, rather than making a decision based on the quality or type of food available. In a significant departure from Disney policy, alcohol was also introduced at some outlets, in order to cater for the French insistence on drinking with their meals.

A further departure from company policy was made with the introduction of differential admission prices at different times of year. The initial adult admission price of Ffr 225 ($43) was over twice the price of Alton Towers, the most expensive theme park in the UK (see Figure 19.1), and markedly higher than the Disney parks in America. By 1994, although the high season adult price was raised to Ffr 250, a shoulder price of Ffr 225 and a low season price of Ffr 175 brought the admission price significantly closer to the competition. The lower prices were accompanied by a change in the name of the park from EuroDisney to Disneyland Paris. It was claimed that the prefix 'Euro' was seen in a bad light at a time of political unrest over the unification of Europe. There was also a significant change in the management structure, with a predominantly American team being replaced by a largely French group headed by Philippe Bourguinon.

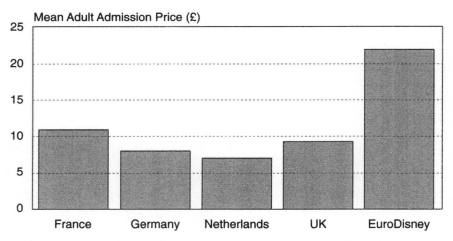

Figure 19.1 A comparison of EuroDisney admission prices with those of other European theme parks 1992
Source: Tourism Research and Marketing (1992).

These changes had a marked effect on the operation of the park. Visitor numbers recovered substantially in 1995 as it became clear that the park would stay open, and the product was made more attractive through the opening of new attractions and lowering of prices. Attendances increased from less than 9 million in 1994/5 to 10.7 million in 1995/6, just short of the park's original target of 11 million visitors. The improved performance of the park was rapidly reflected in better financial figures. Following substantial operating losses in the first two years of operation, the park began to make an operating profit in the 1994/95 financial year, and profits had risen to Ffr 200 million ($40 million) by the end of 1995/96.

The recriminations

What were the causes of this apparently disastrous performance? Lower than expected visitor numbers clearly contributed to the problems of EuroDisney, but given the fact that the first year attendance was only 5 per cent less than forecast, this is not sufficient to explain the scale of the problems experienced. A number of explanations have been offered for Disney's 'failure'. The most common explanations fall into four categories: locational, cultural, managerial and financial.

Locational

The simplest explanation advanced is the fact that Paris does not have a suitable climate for the operation of a year-round theme park. In the original share prospectus, only average annual temperatures were given, providing a somewhat rosy picture of the Parisian winter. In fact, the operation of Disneyland Paris has been highly seasonal, with 70 per cent of visitors coming in the summer months (Taylor and Stevens, 1996). Winter attendances have been below expectations, in spite of the fact that the under cover facilities were judged by consumers to be better than those of comparable parks in the UK (Tourism Research and Marketing, 1994). As Disney's CEO Michael Eisner himself remarked, when asked if Paris was the right location:

> We feel we found the perfect place. It is within 20 minutes of both De Gaulle Airport and the mass transit train. It is only when, on December 17 standing in the snow, that one thinks: maybe Spain would be nicer? Maybe Greece would be nicer?
>
> (quoted in Boje, 1995: 1019)

It has also been suggested that the hotels at Disneyland have suffered from being too close to Paris, although the recovery of hotel occupancies following the price cuts seems to refute this.

Cultural

Much of the criticism levelled at Disney has centred on the American culture of the company and its products. As Walt Disney himself said:

> Disneyland will be based upon and dedicated to the ideals, the dreams and hard facts that have created America. And it will be uniquely

equipped to dramatize these dreams and facts and send them forth as a source of courage and inspiration to all the world.

(quoted in Bryman, 1993)

There were many in France who did not appreciate having Disneyland as a source of inspiration, including Culture Minister Jack Lang, who denounced Eurodisney as a 'Cultural Chernobyl'. Altman (1996) argues that while Disney has spent much time and effort stressing first the 'European' and later the 'French' distinctiveness of the park, Disneyland Paris is actually a much more faithful reproduction of the original Disneyland than the Tokyo park. In a France which considers itself to be 'under American cultural siege', Disney became symbolic of American cultural imperialism, which added to the resistance to the park from the French consumer.

Disney have tried to deflect criticism of the company's image as a bridgehead of American imperialism by introducing French culture in a number of areas of the operation. The pronunciation of Micky Mouse was changed to 'Meekay', and the menus were altered to better meet French tastes. Disney also tried to deflect criticism about its alleged 'cultural imperialism' by promoting French cinema in the United States (van 't Zelfde, 1996). In spite of these apparent concessions to French culture, French visitors still make up only 40 per cent of all visitors, a much lower proportion than the original forecasts had suggested.

Cultural miscalculations also extended to a lack of understanding of the European consumer. As van 't Zelfde (1996: 94) explains:

> Disney made some mistakes as far as their estimation of the European public is concerned. Europeans turned out to be more frugal than American theme park visitors, and far less people were prepared to stay in the adjoining hotels than had been hoped. Europeans also caused problems for the park because they insisted at eating around midday, rather than 'grazing' all through their visit as many Americans do. The French visitors in particular were dismayed to find that there was no alcohol available in the park.

Managerial

Disney started out with what might be identified as an 'ethnocentric' (Go and Pine, 1995) attitude towards managing EuroDisney. The original management team was 90 per cent American. According to McGrath and MacMillan (1995), Disney management fell into the trap of not examining the assumptions they had made in planning the new park. As they point out:

> new ventures are undertaken with a high ratio of assumption to know-ledge. With ongoing businesses, one expects the ratio to be the exact opposite. Because assumptions about the unknown generally turn out to

be wrong, new ventures inevitably experience deviations – often huge ones – from their original planned targets.

(1995: 44)

Disney made a range of huge assumptions about the performance of EuroDisney, including attendance levels, occupancy levels at the hotels, sales of merchandise and pricing, most of which turned out to be wrong. Many of these faulty assumptions can be attributed to a lack of attention being paid to the cultural differences between European and American consumers, but different styles of management also caused problems for Disney.

The problems encountered by Disney in Paris stimulated some major managerial changes, including the appointment of Frenchman, Philippe Bourguignon, as Chairman. There was a shift from an 'ethnocentric' to a more 'regiocentric' style of management, as the largely American management team was replaced by a largely European one, with almost half the members coming from France and only a third from America. The basic argument was that the French managers were closer to the European consumer and the European employees of EuroDisney, and therefore understood the context within which the park had to function far better. As Bourguignon himself said soon after his appointment:

The blame for the calamitous first year is on its excessively pro-American stance. Although most of our visitors enjoy an authentically American atmosphere, it was just a little too much for them to have to celebrate Halloween, which is why we have decided to place the accent on events that will seem more familiar to Europeans.

(Quoted in Tourism Research and Marketing, 1993: 34)

In an attempt to underline its cultural change, 'Disney decided to admit their mistakes with a cathartic zeal. Admissions of arrogance, an excessively pro-American stance, [doing] too much too soon . . . have been carefully delivered with convincing sincerity' (Tourism Research and Marketing, 1995: 52). Such past mistakes could usefully be blamed on the old management, providing a useful contrast with the changes brought about by the new 'French' management.

Financial

In the view of Taylor and Stevens (1996), locational and cultural factors are inadequate to explain the performance of Disneyland Paris. They argue that the basic 'mistake' lies in the financial structure of the EuroDisney operation, and its relationship to the parent Walt Disney company. Walt Disney took a 100 per cent interest in the management company, but only a 49 per cent share of the operating company. In addition, Disney only paid Ffr 10 ($2) per share for their interest in EuroDisney, while the other shareholders were asked to pay Ffr 70

($13) per share. Walt Disney therefore gained a 49 per cent share in EuroDisney by providing only 11 per cent of the capital. In addition, Walt Disney was to be paid a management fee and royalties by EuroDisney, which would come directly out of operating expenses. In other words, the Walt Disney company not only had little to lose financially if things went wrong, but this deal also guaranteed the parent company income even if EuroDisney operated at a loss.

These conditions, which were extremely favourably for the Walt Disney Corporation, were extremely unfavourable for EuroDisney investors. The company was under-capitalised, and saddled with large royalty payments to the parent company and faced with lower than expected revenues. Not only did the financing structure work against EuroDisney, but the economic climate also turned against them. Poor economic conditions in the early 1990s meant that the profits which Disney had envisaged from land sales and commercial lettings around the park did not materialise.

Evaluating Disneyland Paris

Of the various explanations advanced for the failures of Disneyland Paris, which is correct? Although it is tempting to target locational, cultural, managerial or financial factors as the single cause of Disney's problems, the truth is likely to involve a mixture of these explanations. As McGrath and MacMillan (1995) point out, Disney displayed a general arrogance in assuming that the strategies employed in their American operations would translate effortlessly into a European context. Disney had arguably been lulled into a false sense of security by the success of Tokyo Disneyland, which, in spite of the very different cultural context, was able to function much like its American counterparts (Van Maanen, 1992). Secure in the knowledge that they could not fail, Disney proceeded to develop Disneyland Paris without a full understanding of the European market or the European theme park consumer.

This lack of market knowledge proved costly, as the Disney management were unable to anticipate the problems of seasonality and hotel occupancy caused by the Paris location, or the fierce cultural resistance to the Disney management style. Most crucially perhaps, they failed to appreciate that Europe is still composed of a series of individual national markets, which, in spite of some convergence in the wake of the Single Market, still exhibit very different consumption patterns. The vast differences between the forecast and actual visitor origin profile for the park underline this problem. The mis-reading of the European consumer by Disney tends to suggest that, as Gratton and Taylor (1988) suggest, it is market knowledge which is crucial in leisure management. Knowledge of leisure consumption, however, seems not to transfer faultlessly from one world region to another.

Although Taylor and Stevens (1996) contend that the financial structure adopted by Disney to develop Disneyland Paris was the major cause of its problems, if Disney had got its market assessment right, the financial problems might never have surfaced. What the financial structure did do, however, was to substantially lower the risks involved in the Disney investment. This arguably compounded the problem of limited market knowledge, since the lower risk level may also have reduced the care exercised in assessing this vital aspect of the development.

One also needs to be critical about the extent to which Disney failed in the case of Disneyland Paris. Although the company did not meet its financial targets, its achievements have still been considerable. The park added about 30 per cent to the volume of the European theme park market almost overnight (Tourism Research and Marketing, 1993). The park was completed on time, to a very tight schedule. It is now the biggest tourist attraction in Europe, and the only attraction that currently draws a 'European' visitor market, rather than a regional market with a few overseas visitors as an added extra.

Although the operational and organisational qualities so much admired by Peters and Austin (1985) are clearly still present, it is the marketing expertise which seems to have let Disney down particularly badly in Paris. Gratton and Taylor (1988) have pointed out that knowledge of markets is the key competence which leisure managers possess. The lessons from the Disney case may indicate that there are some aspects of leisure markets which are not yet transnational. Differences in the time/space relationships between North America and Europe play a significant role in consumer demand. In particular, the greater availability of holiday time for Europeans is arguably a cause of differing consumer behaviour as far as theme park visiting is concerned. For Americans, with 10–15 days holiday a year, a visit to a theme park is often regarded as a holiday, rather than a day trip. Europeans, with 5 or 6 weeks holiday, can afford to take a 3-week main holiday and a shorter second holiday, and still have days left to visit theme parks. Theme parks in Europe are therefore seen as day trips rather than holiday destinations. US theme parks therefore generate a much higher level of overnight stays, and consequently hotel revenue than their European counterparts (Richards, 1996).

Altman (1996: 55)summarises:

> in crossing borders one should watch the basic assumptions and be careful with business objectives. Markets are perhaps becoming global and business may be converging, but there are sufficient localized intervening variables to make things go wrong. Relying on financial, economic or marketing analysis will not do: one needs to pay attention to societal and cultural issues as well.

What this case study clearly illustrates is that managerial problems can have a range of different causes, and they can also be analysed in a range of different ways. In many cases, a purely managerial or economic analysis will be insufficient, since the social and cultural context within which the organisation operates will be largely ignored. Recent experience of theme park development seems to suggest that the chances of success for transnational development are greatest in those regions where the cultural distance between the host culture and the home culture of theme park developer is smallest. The experience of Disney in France indicates that a truly global theme park market may not yet exist, although the speed with which Disney has responded to its initial problems may mean that the birth of the global theme park company is not so far away.

References

Altman, Y. (1996) 'A theme park in a cultural straightjacket: the case of Disneyland Paris, France', *Managing Leisure* 1: 43–56.

Boje, D.M. (1995) 'Stories of the storytelling organization: a postmodern analysis of Disney as "Tamara-land"', *Academy of Management Journal* 38: 997–1035.

Bruce, C., Rhodes, J., Venner, S. and Warwick, N. (1986) 'Time out on a theme park', Polytechnic of North London, 21pp.

Bryman, A. (1993) *Disney and His Worlds*. Routledge, London.

Go, T. and Pine, R. (1995) *Globalization Strategy in the Hotel Industry*. Routledge, London.

Gratton, C. and Taylor, P. (1988) *Economics of Leisure Services Management*. Longman, Harlow.

Loverseed, H. (1994) 'Theme parks in America', *Travel and Tourism Analyst* June: 51–63.

McEniff, J. (1993) 'Theme parks in Europe', *Travel and Tourism Analyst* September: 52–73.

McGrath, R.G. and MacMillan, I.C. (1995) 'Discovery driven planning', *Harvard Business Review* July–August: 44–54.

Mintel (1990) 'Theme parks', *Leisure Intelligence* 1.

Peters, T. and Austin, N. (1985) *A Passion for Excellence*. Random House: New York.

Richards, G. (1996) 'Time for a holiday?' *Rechtshulp* 6/7: 2–10.

Ritzer, G. (1993) *The McDonaldization of Society*. Pine Forge, Thousand Oaks.

de Roux, E. (1994) 'Eviter la Disneylandisation', *Le Monde* 23 May.

Taylor, R. and Stevens, T. (1996) 'An American adventure in Europe: an analysis of the performance of Euro Disneyland (1992–1994)', *Managing Leisure* 1: 28–42.

Tourism Research and Marketing (various years) *Theme Parks: UK and International Markets*. Tourism Research and Marketing, London.

Van Maanen, J. (1992) 'Displacing Disney: some notes on the flow of culture', *Qualitative Sociology* 15: 5–35.

Van Maanen, J. and Laurent, A. (1992) 'The flow of culture: some notes on globalization and the multinational corporation', in S. Ghosal and D.E. Westney (eds) *Organization Theory and the Multinational Corporation*, St Martin's Press, New York.

Van 't Zelfde (1996) 'Enviromentality at Disneyland Paris', in B. Bramwell, I. Henry, G. Jackson, A. Goytia, G. Richards, G. and J. van der Straaten (eds) *Sustainable Tourism Management: Principles and Practice*. Tilburg University Press, Tilburg, 87–102.

Weinstein, R.M. (1992) 'Disneyland and Coney Island: reflections on the evolution of the modern amusement park', *Journal of Popular Culture* 26: 131–64.

20

STRATEGIES FOR OPTIMISING REVENUES FROM SHORT BREAKS

Lessons from the Scottish hotel markets

David A. Edgar

This chapter explores the strategic dimensions of enhancing revenue from commercial elements of the Scottish short break market from an industrial economics approach. The focus of the case study is on hotels and commercial activity, and the chapter seeks to highlight a methodological paradigm for consideration by destination managers which can be used to explore the potential of maximising short break yield from an understanding of the dynamics of the market structure and organisation strategies. The chapter begins with an examination of the structure of the short break market in Scotland and the marketing strategies adopted by hotel operators. The research methodology is then described and optimum marketing strategies are identified for various types of hotel operations. Finally, implications for destination management are discussed.

Short break market structures

Definitions of short breaks

Despite the fact that the short break phenomenon is not new, and not restricted to the UK (Loverseed 1992, Cockerell 1989, Potier and Cockerell 1992), there is still no standardised definition for short break, short holiday, short holiday break, short break holiday or bargain break. A short break is essentially characterised by the duration of stay (Lohmann 1991), and the form of accommodation used (BTA 1989).

Since the 1940s short holidays have been described as 'trips of up to three nights away from home, primarily for holiday purposes' (Beioley 1991). While this duration of stay is the most commonly accepted for defining short break holidays (BTA 1989, UKTS 1991, Law 1990, 1991, Beioley 1991, Bailey 1989, Davies 1990), a more current stream of thinking is that short breaks are actually characterised by a duration of stay of one to four or five nights (Schidhauser 1992, MEW Research 1994).

These new definitions of short break holiday while adding to the incomparability of data sources do not fully establish why definitions have changed. One possible explanation is that these changes reflect the changing nature of the short break market from an off peak market to a recognised market in its own right. As most secondary data is in the one to three night format, and considering the supply side nature of the research, it was deemed that for the present study the duration of stay adopted for definition purposes should be stays of one to three nights' duration.

The vast majority of short break holidays utilise friends and relatives (VFR) lodging for accommodation, however, the revenue value of this supply sector is minimal when compared to commercial accommodation. This difference in type of accommodation, i.e. commercial or non-commercial, forms the essential difference between the short holiday and the short break. Short holidays are commonly referred to as all holidays of duration one to three nights (BTA 1989, Beioley 1991, Bailey 1989) in all types of accommodation, while short breaks are referred to as holidays of one to three nights taken in commercial accommodation. Some definitions specify the type of accommodation as hotels (MSI 1991, Davies 1990), hotels and guest houses (Euromonitor 1987) or simply paid-for accommodation (Bailey 1989, Beioley 1991, MEW Research 1994).

The definition of a short break adopted for the purposes of this chapter is one that specifies accommodation, time period, and activity. The following definition is adopted :

> Hotel packages of one to three nights which for a single price together with accommodation includes one or more of the following: meals; transport; entertainment; or a programme of activities.

This definition has been adopted and used previously in the author's work. (Edgar 1992a, 1992b, 1993, 1995, Edgar and Littlejohn 1994, Edgar *et al.* 1994, Crichton and Edgar 1995)

Characteristics of the short break market

Short breaks are to a great extent a creature of the nature of hotel operation cost structures, environmental change, and the off peak seasonality drive for demand (Edgar *et al.* 1994). The short break market is widely recognised as a key growth market for hotels and has distinct market characteristics. Major city centre hotels traditionally targeted the business market, meaning high occupancy percentages between Monday and Thursday, and creating 'slack periods' Friday to Sunday. With the advent of recessionary periods, city centre operators (dominated by hotel groups) recognised the potential of short breaks as a means of gaining additional revenue. Therefore, by using an attractive pricing policy the short break market evolved (Davies 1990).

The short break concept has gradually evolved from its traditional off peak image to represent a market in its own right. Although the key characteristics of the markets for short breaks vary with the theme of the break, short breaks tend to have the following general properties (MSI 1991, BTA 1989, Euromonitor 1987, Beioley 1991, Bailey 1989):

- short breaks tend to be less seasonal
- the average length of stay is around two nights
- travel is by car
- typical customers are between 16 and 34 and of socio economic group ABC1
- breaks are not planned very far in advance of travel, but are booked in the month of travel, direct with the hotel
- the most common reason for taking short breaks was to 'attend a special event' with most important features sought being good hotel accommodation, peace and quiet, and lots to see and do

In more recent years a greater degree of market segmentation has arisen with more sophisticated marketing strategies (Vierich and Calver 1991) and selective use of distribution channels.

The growth of the short break market is shown in Figures 20.1 and 20.2. Major periods of growth of trips and nights (Figure 20.1) were between 1983 and 1985,

Figure 20.1 UK short break market 1980–90
Source: BTS/UKTS (1980–91).

381

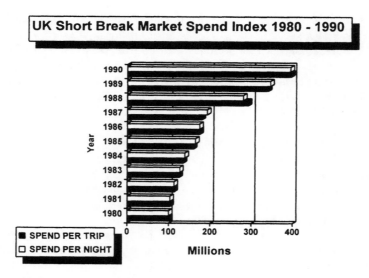

Figure 20.2 UK short break market spend index 1980–90
Source: BTS/UKTS (1980–91).

and 1986 and 1988. Further, when compared to other markets, short breaks have actually been growing in market share of all holidays throughout the ten-year period. Perhaps the greatest value of short breaks is seen in Figure 20.2, indicating a growth in value of short breaks.

Between 1980 and 1990 the growth in spending per night and per head has been around 300 per cent, 200 per cent of this achieved in the latter part of the 1980s. Hence, combining Figures 20.1 and 20.2, it becomes clear that while trips became shorter and more frequent they also became considerably more valuable per trip, thus increasing the market attractiveness from a supply perspective, especially with higher ancillary spending and greater 'holiday stability'.

From the development of the commercial element of short breaks, hotels have undertaken a key role. The next section focuses on the nature of hotel units and the key elements of market structure influencing hotel activities in short break markets and the resulting structural elements.

Implications of short breaks for hotel operations

The nature of the hotel unit

The cost structure of hotels means that, like airlines, marginal pricing can be an important strategy to boost occupancy and increase profitability. Because hotels, being property-based businesses, have high levels of fixed to variable costs, marginal changes in occupancy may produce significant differences in profitability. This explains significant levels of discounting at certain times where occupancy may be low. Increased profitability may be gained from contribution to fixed costs which would otherwise be lost, or by encouraging the lower rated customer to increase consumption of supplementary hotel services such as meals, or by a combination of both these factors.

Low occupancy may be related to the location of particular units, meaning that hotels in resort areas would need to fill certain months that were off peak, while city centre hotels may need to fill beds during weekends. This differing seasonality results in a wider spread of available seasons for short breaks, as well as providing a higher standard of accommodation, especially in city centre locations. In addition, it gives rise to the concept of opportunity cost, whereby developing in one market the operator must forsake another. These opportunity costs must be carefully considered, and the factors of production (land, labour and capital) correctly balanced for the selected market.

From a demand-oriented approach short breaks are a logical progression in that they allow a marketing emphasis, identifying major price and activity segment differences, and forming a means of further segmenting and branding hospitality and tourism markets. From both supply and demand approaches, the basic resource, a hotel, its services, location and tourism attractions conferred upon it, are the same; what may change are the design and timing of the breaks, and the marketing methods used.

Short breaks are therefore economically worthwhile as long as variable costs are met. In reality, many short breaks provide the opportunity to secure considerable additional supplementary revenue and therefore contribute significantly to profits. Senior and Morphew (1990) have argued that the survival of traditional operators will depend upon their ability to extend their horizons and move from short-term tactics, such as pricing and product augmentation, to long-term strategies. The provision of short breaks provides such a strategy. Hence, as customers become more sophisticated in manipulating the current pricing system of hotels, hotels will eventually be forced to modify their pricing structure (Hanks et al. 1992) and can react by combining the key elements of yield management through the provision of short breaks.

With the increased interest and developments in the market, strategies designed to capture and protect market share have begun to emerge and mature. The result is a market containing a wide range of segments, dominated by large hotel groups and needing continual market innovation. In economic terms, the market reflects high degrees of monopolistic competition tendencies.

Strategic dimensions of short break provision

Given the monopolistic competition nature of the short break market in Scotland, this section explores the nature of competition in the market before highlighting main strategies adopted by hotel groups and determining structure–strategy associations.

Competing in the short break market

Teare *et al.* (1994: 6) propose that:

> hospitality services (of which hotels are a core element) have both
> functional and expressive roles to fulfil. The consumer is primarily
> concerned with the desire to satisfy basic functional needs (e.g. hunger
> and thirst). These are accompanied by more expressive, psychological
> needs, driven by consumers' lifestyles and prior experience. They may
> also be motivated by aspirations to experience surroundings beyond
> current lifestyle expectations.

Short break suppliers are competing for market share as opposed to organic
growth, and consumer satisfaction is derived from different kinds of service
experience and interaction, unique to the occasion and situation. Competition in
the short break markets is therefore both dynamic and complex. Hence hotel
groups are having to compete directly with one another in the same locations (Dev
1990) – a fundamental change for an industry traditionally reliant on locational
specificity.

Growth and competition in the industry mean that market share is no longer
assured. Hotel groups have been forced to follow other sectors of the tourism
industry with the implementation of marketing strategies (Tarrant 1989, Meidan
and Lee 1982), often based on product differentiation, growth in new markets,
high value for money, or emerging brands (Olsen 1993). The concept of competitive advantage is at the heart of such marketing strategies. Competitive
advantage can be gained from a range of sources, including technology (Dollar
1993), differences in supply (Leamer 1993), or product differentiation (Hummels
and Levinsohn 1993). Companies have to concentrate on such activities and
competencies to build sustainable competitive advantages and therefore create
value greater than competitors (Bronder and Pritzl 1992, Overstreet 1993,
Braithwaite 1992, Dev and Klein 1993).

Strategies in the UK short break market

Strategies adopted by hotel groups in the short break market consist of two types,
defender and predator. Defender strategies are essentially adopted to protect
existing market share, through reducing industry attractiveness or creating raised
standards in the market. Predator strategies seek to gain market share through
expanding the market base and are often used in conjunction with defender
strategies. Table 20.1 provides a summary of the prime strategies adopted in
terms of defender and predator strategies.

Previous research by the authors has provided evidence that some strategies
result in better performance than others (Edgar and Littlejohn 1994, Edgar *et al.*
1994). Additionally, the research has exposed deliberate attempts by hotel

Table 20.1 Short break prime strategies

Category	Prime strategy	Description of strategy
Defender	Pricing	Emphasis on price reductions or alternatively focus on marginal pricing
	Differentiation	Emphasis on unique location, market level, or facilities available
Predator	Distribution	Used to reach target segments, often involving a third party and incurring considerable commission levels
	Segmentation/ packaging	Consisting of a bundle of goods and services, composed to simplify market complexity, segmenting the market and allowing wider varieties of break to be offered
	Promotion	Most evident in branding, brochure production, and advertising power. Used to create awareness and gain brand loyalty

companies to raise the complexity of the market through additional packaging and segmentation strategies (two strategies providing lucrative rewards in terms of short break revenue). The result is a short break market that is complex, dynamic and highly segmented, and strategies that are increasing in sophistication and innovative in nature. This raises the question – do certain structure–strategy combinations result in enhanced performance?

Research methodology

The case study described in this chapter forms a development of current research being undertaken by the author, investigating linkages between elements of the industrial economics paradigm, namely market structure, organisation strategy and resulting performance.

As secondary sources of data were relatively scarce and existing data was prescriptive and highly demand-oriented in nature, and as the research focuses upon the strategic dimensions of the market, a supply perspective to analysis was required. This supply perspective compensated for the lack of data through constructing a hotel groups database. The database highlighted by regions the location, short break offerings and performance differences of major UK hotel companies operating in Scotland (representing approximately 5 per cent of the Scottish hotel room stock and approximately 70 per cent of 'commercial' short break revenue generation).

To explore the strategic dimensions of the market, semi-structured telephone interviews were conducted with twenty key executives of the most active groups, representing 78 per cent of the sample room stock. The interviews were conducted to gather information on three key sets of variables – market structure characteristics, strategic choices and resulting performance.

Enhancing performance of hotels in short break markets

Structure–strategy–performance relationships

The concept of a strategic 'group' of firms, essentially defined as a set of firms competing within an industry on the basis of similar combinations of scope (Cool and Schendel 1987), customer patronage (Harrigan 1985) and commonality of strategies in setting key decision variables (Hunt 1972, Caves and Porter 1977), provides a framework for determining the relationship between structure, strategy and performance within and between industries (Dess and Davis 1984, Hergert 1983, Barney and Hoskisson 1990). Since Hunt (1972) introduced the term 'strategic groups', a growing body of literature has adopted this analytical concept to determine relationships between industry competitors (Porter 1980, Hatten 1974, Cool 1985, McGee and Thomas 1986, Meyer and Rowan 1977). Such studies have been further developed to demonstrate performance linkages (Newman 1978) and to suggest forms of competitive benchmarking through strategic modelling (Kumar *et al.* 1990) and frameworks for group formation (Fiegenbaum and Thomas 1990).

The potential of strategic group concepts as an analytical decision-making tool are obvious. However, there are many ambiguities surrounding its application (Hatten and Hatten 1987). Key issues relate to the definition and existence of 'strategic groups', the absence of longitudinal empirical analysis, and their relation to performance variances (Cool and Schendel 1987, 1988). Hence, although strategic group mapping can be a useful way of tracking industry dynamics (Harrigan 1985), with clusters of firms in strategic space (collections of structure–strategy associations) being identified and group membership defining the essential characteristics of a firm's strategy (Reger and Huff 1993), the explanatory power of the strategic group concept is fundamentally dependent on the strength of the scheme adopted to 'operationalise' strategy (Thomas and Venkatraman 1988).

As such, the framework used for this chapter is an adaptation of the 'Group Competitive Intensity Map' (McNamee and McHugh 1987) displaying strategic groups evolving from key structural and strategic variables. The strategic framework used, based upon work by McNamee and McHugh (1987: 124) assumes that:

> any distinctive characteristics (in this case performance) which members of any strategic group display are a function not just of the strategies that they follow but also the structures they possess, and consequently the competitive location of any firm, or strategic group into which it falls, should evolve from co-ordinates based on both strategy and structure.

The model framework used in the present analysis is shown in Figure 20.3, with the horizontal axis representing the identified prime marketing strategies and the

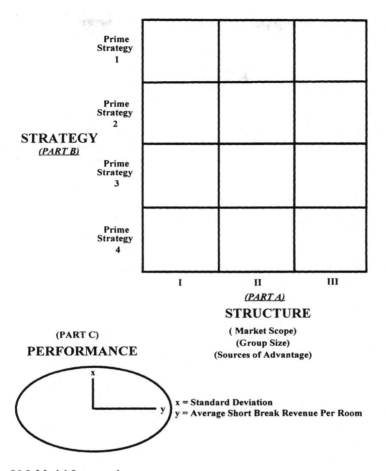

Figure 20.3 Model framework
Source: Edgar and Litteljohn (1994).

vertical axis identifying structural variables relating to the hotel groups. The resulting zones represent strategic space containing clusters of hotel groups. These clusters are then related to performance through the use of ellipses formed by plotting the total short break revenues of the cluster group, against the standard deviation of the figures for those groups. The framework is therefore composed of the key variables of structure, strategy and performance.

Structure

The structural determinants in the model (see Figure 20.3, part A) are taken as variables relating to market scope, group size and sources of competitive advantage. The scope of the market represents the number of short break segments targeted by hotel groups. Market scope was categorised on three levels, narrow, medium and wide scope (adapted from Porter 1985). Table 20.2 shows the number of the Scottish hotel groups in each of the levels and an indication of the number of market segments represented by each level.

Work by Porter (1980) and the Boston Consulting Group (1972) suggests that hotel group size is a key structural and performance determinant. For the purposes of this chapter, group size is determined by calculating the concentration ratios (Scherer 1980) of hotel group room stock and dividing the ratios into three concentration categories, shown in Table 20.3. Concentration ratios provide an indication of relative market presence and therefore are more representative of hotel group size than market share measures.

The final measure of structure relates strategy adopted and performance achieved to the number of sources of competitive advantage. Sources of competitive advantage are the means by which firms can compete against each other and are fundamental to strategic choices. In general terms, the more sources of advantage, the more strategic options are available – however, this does not imply that such sources of advantage are sustainable. Sources of competitive advantage

Table 20.2 Hotel groups short break market scope 1991–3

Scope	Market segments	No. groups
Narrow	1–3	6
Medium	4–6	8
Wide	7–9	2

Table 20.3 Size of hotel groups by concentration ratios 1991/2

Group size/room share	Ratio range	No. groups
High	10 + *	4
Medium	5–9.9	4
Low	0–4.9	8

* One group ratio was 32.5 on the first analysis, emphasising market dominance in room holdings, this group has been included in the 'High' category and has a similar asset structure to others in the group, making it a legitimate member of that cluster (Mascarenhas and Aaker 1989). The ratios were then re-calculated omitting this group and groups categorised accordingly.

Table 20.4 Number of sources of competitive advantage – hotel groups 1991–3

Category	No. sources	No. groups
Few	1–2	9
Average	3–4	2
Many	5–6	5

are widely recognised as a key structural variable in determining strategy and therefore resulting performance (Porter 1980) and should therefore be accounted for. Table 20.4 indicates the number of sources of competitive advantage and the number of groups obtaining these sources, split into three categories of few, average and many.

Strategy

As it is recognised that strategic groups may change with organisation levels (corporate; business; functional) giving rise to the concept of 'overlapping' strategic groups (Fombrun and Zajac 1987), and that such groups may be the result of combinations of strategies that are internal or external in orientation (Wright *et al.* 1991), the research sought to establish what the hotel groups identify as their prime short break strategy with other secondary strategies supporting the prime one. As such, the prime strategies consist of the organisations' variables that cannot easily be changed (Tang and Thomas 1992) or replicated (Mobility Barriers – Caves and Porter 1977), thus forming the horizontal axis of strategy (Figure 20.3 part B).

Performance

This measure is represented in the model by a ellipse constructed from a horizontal axis of 'average short break revenue per room' (ASBRPR) and a vertical axis of 'the standard deviation (spread of performance) of average short break revenue per room per annum 1991/2' within the cluster (Figure 20.3 part C). This approach displays the stability of the cluster relating to structural and prime strategy variables, and allows conclusions to be drawn as to the potential of movement in strategic space.

Models of performance

Using the prime strategies previously described in Table 20.1 and the structural measures of market scope, company size and number of sources of competitive advantage, a series of specific models of short break performance were described and these are shown in Figure 20.4. Each model can be analysed in turn and general conclusions drawn.

Market scope

From Figure 20.4(i) it can be seen that all the companies adopting a distribution strategy appear to also have narrow market scope and considerably lower performance (ASBRPR was £967).

The greatest stable revenue earning cluster appears to be composed of companies implementing segmentation/packaging strategies, with medium

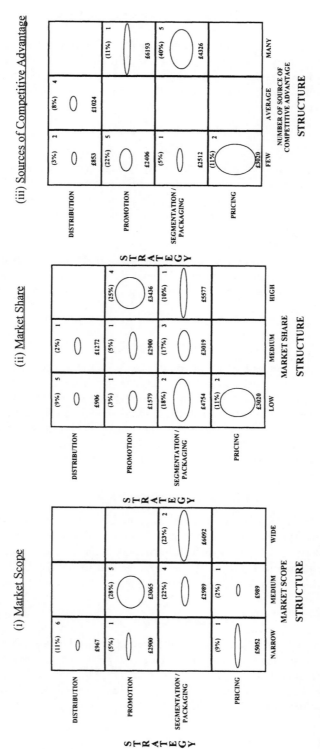

Figure 20.4 Models of short break performance

market scope (ASBRPR of £2,989), or wide market scope (ASBRPR of £6,092). The *wider the market scope the more revenue appears to be generated.* This is the only prime strategy adopted by companies with a wide scope, and none of the companies adopting this strategy have a narrow scope. This area warrants further investigation, particularly in the field of entry barriers (Porter 1980).

Most companies adopting a promotion strategy are of medium scope with an average short break revenue per room per annum of £3,065 but a very large spread of data indicating instability in the cluster performance. This would provide opportunities for members of this cluster to enhance performance by attempting to shift in strategic space to a strategic cluster offering higher financial rewards.

The remaining three companies are not members of specific clusters; however they do reveal the individual hotel group market scope–short break strategy–performance relationship, with pricing strategies in a narrow scope producing considerably good performance per room (presumably in volume terms, as a percentage of total revenue, or an upward niche pricing policy), and pricing in a medium scope indicating poorest performance, perhaps indicating a need to focus more on niches to improve performance.

Market share

While the clusters are considerably more widespread and generally smaller in nature, some patterns do emerge. Hotel groups with a *high market share* of room stock offered to short break markets, adopt promotional strategies essentially in the form of branding, with a performance level of £3,436 and a wide spread of data (high ellipse). Other hotel groups with high market share adopt packaging strategies and achieve considerably higher performance at £5,577. This indicates a potential option for companies in the promotional strategy space to perhaps move into a segmentation/packaging strategy space. In general, high market share companies only adopt segmentation/packaging or promotion prime strategies.

Of the *medium market share* clusters, the results indicate poorer performance by the hotel group adopting a distribution strategy and comparable performance between those adopting packaging and promotion strategies. None of the medium share clusters adopt pricing strategies. The *low market share* clusters indicate the full range of strategies employed, with segmentation/packaging showing £4,754 and low skew. This indicates stable results and high revenue potential. Pricing strategies produce for some hotel groups high performance based on small size. Promotion strategies are used less by low share hotel groups and result in poorer performance than other companies adopting promotion strategies with a higher market share.

Again, the poorer performance of clusters adopting distribution strategies is highlighted, indicating that most hotel groups using distribution as their prime strategy are of low share, and perform less well at £906 with fairly consistent results (low skew). Such hotel groups should perhaps consider moving in strategic space from a prime strategy of distribution to one of segmentation/packaging, supported by distribution. Segmentation/packaging and promotion strategies appear to be adopted across the range of hotel group market shares, with higher performance in the segmentation/packaging strategy clusters and more of an emphasis on market share in the promotion strategies with performance improving with market share.

Number of sources of competitive advantage

From Figure 20.4(iii), it is evident that hotel groups adopting a segmentation/ packaging strategy tend to have many sources of competitive advantage and achieve higher performance than the hotel groups that have fewer advantage sources. This may be due to the composition of the packages and areas of potential synergistic advantage. As such, this area warrants further, more detailed investigation, and forms a separate study. In contrast, hotel groups adopting a promotion strategy appear to have fewer sources of competitive advantage and achieve lower performance. In general, it would appear that the sources of competitive advantage have a considerable affect on the type of strategy adopted at the strategic choice stage and may reflect capability to implement the strategy.

Of the hotel groups adopting distribution strategies, performance is again relatively stable. Higher performance (although comparatively poorer than other strategies adopted) is achieved by clusters with an average as opposed to few sources of competitive advantage. The main source of competitive advantage being image or brand. Both pricing strategy hotel groups have few sources of competitive advantage and result in a wide spread of data.

Structure–short break strategy–short break revenue: summary

The result of this analysis has found considerable differences between hotel group market performance.

- Distribution strategies are adopted in narrow market scope only and perform better when combined with medium market share and average sources of competitive advantage. Overall however, this is the worst performing strategy.
- Promotion strategies perform best in medium market scope structures, with high market share and many sources of competitive advantage.
- Segmentation/packaging strategies appear to perform best in wide market scope, high market share market structures and with many sources of competitive advantage.
- Pricing strategies' performance is enhanced in niche markets (narrow market scope) and with low market share and few sources of competitive advantage.

These outcomes imply the potential for hotel companies to move between clusters by changing strategy and/or structure in attempts at improving short break performance. This process is called movements in strategic space.

Movements in strategic space

Teece (1984: 289) suggested that 'a firm needs to match its capabilities to its ever-changing environment if it is to attain its best performance'. Based on this grounding, it could be argued that if companies operating in the short break market could simplify the environment and determine areas potentially more lucrative, then they would attempt to revise their mix of resources (capabilities) in order to change strategy, thus moving in strategic space and enhancing performance. Following on from this, it could be argued that the methods put forward in this chapter (although not fully tested) go some way towards simplifying the environment and thus allow general 'implications' to be drawn with regard to company movements in strategic space – on both a theoretical and a practical level.

The following two examples are suggested based upon this premise.

1. Companies adopting a pricing strategy, operating with a medium scope, low market share and few sources of competitive advantage should, in an attempt to improve performance, move towards promotion strategies, with a longer-term view of increasing scope and adopting packaging strategies. Alternatively, as a short-term response they may consider narrowing their scope to enhance performance.
2. Distribution should be avoided as the prime strategy. Companies adopting these strategies should attempt to move strategic space first by increasing scope, followed by a move towards packaging strategies, or perhaps, if it is not possible to widen scope, they may consider maintaining a narrow scope and adopting a pricing strategy.

In general, packaging strategies would appear to be a strategy available to any size of company operating in, or able to operate in, a market with medium to wide scope. Promotional strategies would be the direction for companies with few sources of advantage, a medium scope but with a medium to high company market share of rooms. Distribution would be recommended as a support strategy, not prime, and pricing recommended for companies with low market share, narrow scope and few sources of competitive advantage.

Moving towards these strategies will be in incremental steps and will require some form of environmental scanning to monitor competitors' reaction, but should enhance performance in the short break market.

Implications for destination management of short breaks

This chapter has provided a model to examine the relationships between marketing strategies, operation characteristics and performance which was applied to hotel groups operating in short break markets in Scotland. The use of the model in this particular case study highlighted a range of potential areas of performance enhancement for short break marketing through the development of an understanding of the dynamics of strategy and strategic space. The contention here is that organisations charged with the management of destinations may also employ such approaches to understanding the dynamics of strategy in short break markets and thus capitalise on the lessons learned by hotel groups operating in Scotland.

Market structure dimensions for consideration are dependent upon specific destination characteristics but may include a range of dimensions including location, activity base, season or market segment. It is anticipated that the strategic dimensions of the market will be largely unchanged and still centre around the prime strategies identified. The author believes this work offers considerable contributions to the management of destinations and hopes that the approach and developed model framework will stimulate further discussion and thought, and thus develop more strategic thinking in terms of tourism destination management.

References

Bailey, J. (1989) 'Short holidays', *Insights* (magazine of the English Tourist Board/British Tourist Authority) B1–B7.

Barney, J. and Hoskisson, R. (1990) 'Strategic groups: untested assertions and research proposals', *Managerial and Decision Economics* 11: 187–98.

Beioley, S. (1991) 'Short holidays', *Insights*, (magazine of the English Tourist Board/British Tourist Authority) B29–B38.

Boschken, H.L. (1990) 'Strategy and structure: reconceiving the relationship', *Journal of Management* 16(1): 135–50.

Boston Consulting Group (1968, 1970, 1972) *Perspectives on Experience*, Boston, MA BCG Ltd.

Brathwaite, R. (1992) 'Value-chain assessment of the travel experience', *The Cornell Hotel and Restaurant Administration Quarterly*, October: 41–9.

Bronder, C. and Pritzl, R. (1992) 'Developing strategic alliances: a conceptual framework for successful co-operation', *European Management Journal* 10(4): 412–21.

BTA (1989) *The short break market*, London, British Tourist Authority/English Tourist Board, November.

British Tourism Survey (BTS) (1980–9), British Tourism Authority, London.

Caves, R. and Porter, M.E. (1977) 'From entry barriers to mobility barriers', *Quarterly Journal of Economics* 91 (May): 241–61.

Cockerell, N. (1989) 'The short break market in Europe', *EIU Travel and Tourism Analyst* 5: 41–55.

Cool, K. (1985) 'Strategic group formation and strategic group shifts: a longitudinal analysis of the US pharmaceutical industry, 1963–1982', Unpublished doctoral dissertation, Purdue University West Lafayette, Indiana.

Cool, K. and Schendel, D. (1987) 'Strategic group formation and performance: the case of the US pharmaceutical industry', *Management Science* 33 (9): 1102–24.

Cool, K. and Schendel, D. (1988) 'Performance differences among strategic group members', *Strategic Management Journal* 9: 207–23.

Crichton, E. and Edgar, D. (1995) 'Managing market complexity for competitive advantage: an I.T perspective', *International Journal of Contemporary Hospitality Management* 7(2/3): 12–19.

Davies, B. (1990) 'The economics of short breaks', *International Journal of Hospitality Management* 9(2): 103–6.

Dess, G. and Davis, P. (1984) 'Porter's (1980) generic strategies as determinants of strategic group membership and organisation performance', *Academy of Management Journal* 27: 467–8.

Dev, C.S. (1990) 'Marketing practices of hotel chains', *The Cornell Hotel and Restaurant Administration Quarterly* November: 54–63.

Dev, C.S. and Klein, S. (1993) 'Strategic alliances in the hotel industry', *The Cornell Hotel and Restaurant Administration Quarterly* 34(1): 42–5.

Dollar, D. (1993) 'Technological differences as a source of competitive advantage', *AEA Papers and Proceedings* 83(2): 520–5.

Edgar, D. (1992a) 'Commercial short holiday break markets in Scotland', unpublished report for the STB, Napier Polytechnic, March.

Edgar, D. (1992b) 'A model for analysing the commercial short holiday break market in Scotland', *CHME Research Conference Proceedings*, Birmingham Polytechnic, April.

Edgar, D. (1993) 'Commercial short holiday breaks: an analysis of market provision and supplier strategies in Scotland', *CHME Research Conference Proceedings* Manchester Metropolitan University, April.

Edgar, D. (1995) 'The strategic gap – a short break multi-site perspective', in N. Johns (ed.) *Productivity Strategies for Service*, Cassell, London.

Edgar, D. and Littlejohn, D.L. (1994) 'Strategic clusters and strategic space: the case of the UK short break market', *International Journal of Contemporary Hospitality Management* 6(5): 20–6.

Edgar, D.A, Littlejohn, D.L., Allardyce, M. and Wanhill, S.R.C. (1994) 'Commercial short holiday breaks: the relationship between market structure, competitive advantage and performance', in A. Seaton (ed.) *Tourism: State of the Art*, Wiley and Sons, Chichester, pp. 323–42.

Euromonitor (1987) 'Short break holidays, market research GB', *Euromonitor* Publ. Ltd 26 (April): 15–22.

Fiegenbaum, A. and Thomas, H. (1990) 'Strategic groups and performance: the US insurance industry, 1970–84', *Strategic Management Journal* 11: 197–215.

Fombrun, C. and Zajac, E. (1987) 'Structural and perceptual influences on intraindustry stratification', *Academy of Management Journal* 30: 33–50.

Hanks, R.D., Cross, R.G. and Noland, R.P. (1992) 'Discounting in the hotel industry: a new approach', *The Cornell Hotel and Restaurant Administration Quarterly* February: 15–23.

Harrigan, K. (1985) 'An application of clustering for strategic group relationships', *Strategic Management Journal* 6: 55–73.

Hatten, K. (1974) 'Strategic models in the brewing industry', PhD dissertation, Purdue University, West Lafayette, Indiana.

Hatten, K. and Hatten, M. (1987) 'Strategic groups, asymmetrical mobility barriers and contestability', *Strategic Management Journal* 8: 329–42.

Hergert, M. (1983) 'The incidence and implications of strategic grouping in US manufacturing industries', Unpublished Doctoral Dissertation, Harvard University.

Hummels, D. and Levinsohn, J. (1993) 'Product differentiation as a source of competitive advantage', *AEA Papers and Proceedings* 83(2): 445–9.

Hunt, M. (1972) 'Competition in the home appliance industry 1960–1970', unpublished PhD dissertation, Business Economics Committee, Harvard University.

Kumar, K., Thomas, H. and Fiegenbaum, A. (1990) 'Strategic groupings as competitive benchmarks for formulating competitive strategy', *Managerial and Decision Economics* 11: 99–109.

Law, J. (1990) 'Short breaks', *Travel Trade Gazette* 1931 (November): 27–30.

Law, J. (1991) 'UK short breaks', *Travel Trade Gazette* 1950 (March): 51–5.

Leamer, E.E. (1993) 'Factor supply differences as a source of competitive advantage', *AEA Papers and Proceedings* 83(2): 435–9.

Lohmann, M. (1991) 'Evolution of shortbreak holidays', *Tourist Review* 2: 14–22.

Loverseed, H. (1992) 'The North American short break market', *EIU Travel and Tourism Analyst* 4: 48–65.

Mascarenhas, B. and Aaker, D. (1989) 'Mobility barriers and strategic groups', *Strategic Management Journal* 10: 475–85.

McGee, J. and Thomas, H. (1986) 'Strategic groups: theory, research and taxonomy', *Strategic Management Journal* 7: 141–60.

McNamee, P. and McHugh, M. (1987) 'Mapping competitive groups in the clothing industry', in *Developing Strategies for Competitive Advantage*, Harvard Business School, Boston.

Meidan, A. and Lee, B. (1982) 'Marketing strategies for hotels', *International Journal of Hospitality Management* 1(3): 169–78.

MEW Research (1994) 'Short break destination choice', *Insights* (magazine of the English Tourist Board/British Tourist Authority) January A77–A94.

Meyer, J. and Rowan, B. (1977) 'Industrialised organisations: formal structures as myth and ceremony', *American Journal of Sociology* 83: 340–63.

MSI (1991) 'Databrief: short break holidays', Marketing Strategies for Industry London.

Newman, H. (1978) 'Strategic groups and the structure – performance relationship', *Review of Economics and Statistics* 60 (3): 417–27.

Olsen, M.D. (1993) 'International growth strategies of major US hotel companies', *EIU Travel and Tourism Analyst* 3: 51–64.

Overstreet, G.A. (1993) 'Creating value in oversupplied markets', *The Cornell Hotel and Restaurant Administration Quarterly* October: 68–96.

Porter, M.E. (1980) *Competitive Strategy* New York: Free Press.

Porter, M.E. (1985) *Competitive Advantage: Creating and Sustaining Superior Performance*. New York: Free Press.

Potier, F. and Cockerell, N. (1992) 'The European international short break market', *EIU Travel and Tourism Analyst* 5: 45–65.

Reger, R. and Huff, A. (1993) 'Strategic groups: a cognitive perspective', *Strategic Management Journal* 14: 103–24.

Scherer, F. (1980) *Industrial Market Structure and Economic Performance*, 2nd edn, Houghton Mifflin, Boston.

Schidhauser, H. (1992) 'The distinction between short and long holidays', *Tourist Review* 2: 10–17.

Senior, M. and Morphew, R. (1990) 'Competitive strategies in the budget hotel sector', *International Journal of Contemporary Hospitality Management* 2(3): 2–9.

Tang, M.J. and Thomas, H. (1992) 'The concept of strategic groups: theoretical construct or analytical convenience', *Management and Decision Economics* 13: 3239.

Tarrant, C. (1989) 'UK hotel industry – market restructuring and the need to respond to customer demands', *Tourism Management* 10(3): 187–91.

Teare, R., Mazanec, J.A., Crawford-Welch, S. and Calver, S. (1994) *Marketing in Hospitality and Tourism: A Consumer Focus*, Cassell, London.

Teece, D. (1984) *The Competitive Challenge*, Boston, MA: Ballinger.

Thomas, H. and Venkatraman, N. (1988) 'Research on strategic groups: progress and prognosis', *Journal of Management Studies* 25: 537–55.

United Kingdom Tourism Survey (UKTS) (1991) 'The UK tourist statistics 1991', ETB/STB/WTB/NITB, London.

Wright, P., Kroll, K., Chan, C. and Hamel, M. (1991) 'Strategic profiles and performance: an empirical test of select key propositions', *Journal of the Academy of Marketing Science* 19(3): 245–54.

Vierich, W. and Calver, S. (1991) 'Hotels and the Leisure Sector: A Product Differentiation Strategy for the 1990s', *International Journal of Contemporary Hospitality Management* 3(3): 10.

21

AN INVESTIGATION OF FACTORS INFLUENCING THE DECISION TO TRAVEL

A case study of Japanese pleasure travellers

Sheauhsing Hsieh and Joseph T. O'Leary

Tourism has become one of the world's most powerful components of economic development. Many countries have established tourism development as a high priority concern. However, increases in discretionary time and money, as well as the variety of vacation choices, have given the potential traveller more flexibility of choice. As a result, the factors influencing travel decisions are becoming more complex. If a travel or tourism organisation wants to influence a travel decision, it needs to understand who is making the decision and how that decision is made.

The 'decision process' has been described as a simple association between a stimulus and a response. It has also been described as a very complex interaction among many behaviour determinants (Nicosia, 1966). An effective travel decision-making model must incorporate important factors affecting the decision (such as sociodemographic, sociological and psychographic characteristics) and provide an understanding of the relationships between these variables. Travel decisions may also be affected by factors such as travel characteristics, destination attributes and past travel patterns. One of the key steps in tourism planning and marketing is to develop travel behaviour choice models by analysing these travel factors. Traditionally, sociodemographic variables and travel characteristics have been used to understand the choice of vacation types, destination and accommodation, or to predict the demand for travel (Silberman, 1985; Sheldon and Mak, 1987; Witt and Martin, 1987). An understanding of the significance of the variables used to make travel choices in each market can assist travel service providers to understand what potential visitors wish to see, what they wish to do, where they wish to do it, what advertising and promotional vehicles to utilise, and when to place various advertisements.

To investigate these ideas, a secondary analysis of data from the 1989 Pleasure Travel Market Survey for Japanese travellers was conducted to develop a model of travel decision-making in order to develop new knowledge and analysis of market characteristics. The purpose of this study was to understand travel markets by developing a travel decision model using sociodemographics, travel characteristics and psychographic factors, apply this model to the Japanese travel market and identify the important factors affecting the Japanese traveller's decision.

Literature review

Consumer behaviour can be viewed as the study of how, why and how often individuals make decisions to spend their available resources (such as money, time or energy) on consumption-related items (Schiffman and Kanuk, 1990). Many factors affect the consumer decision process. Factors can come from marketing (including such things as perceived product quality, price and distinctiveness), social sources (which includes family and reference groups), individual differences (such as sociodemographics, lifestyle and personality types), or psychological processes (like motivation, or destination perceptions) (Engel and Blackwell, 1990; Schiffman and Kanuk, 1990).

Sociodemographics, travel characteristics and psychographic variables have all been recognised as important factors influencing travel decisions. One of the strategic approaches in tourism planning and decision-making is to develop travel choice models by analysing these travel factors. Models can be defined as 'systems of hypotheses relating one or more dependent variables . . . to several independent variables' (Mazanec, 1989: 63). In studying travel and tourism, dependent variables could be the choice of a tourist destination, hotel or accommodation, the likelihood of taking a future trip, the length of stay or the total number of visits. On the other hand, independent variables could include factors such as sociodemographics, psychographics, travel characteristics, destination attributes or economic variables. In summary, many variables have been identified as important indicators of travel decisions.

Consumer characteristics

Consumer characteristics such as demographics, lifestyle, and personality have often been described as influencing consumer needs and attitudes towards product choice (Assael, 1987; Engel and Blackwell, 1990; Schiffman and Kanuk, 1990). These characteristics are generally easy to identify and to measure. Furthermore, they can often be associated with the use of specific products and with media (Schiffman and Kanuk, 1990). For example, Schul and Crompton (1983) used sociodemographics including age, marital status, sex, education, place of residence and travel-specific lifestyles to predict and explain the external travel information search behaviour of a sample of international vacationers. Sheldon and Mak (1987) presented a model that explained a traveller's choice of independent travel versus travel on package tours to the Hawaiian Islands by using logistic analysis of survey data on travel. Their results indicated that vacation mode decisions were related to certain sociodemographic attributes. For example, purchasers of package tours were likely to be older, intended visiting several destinations, contained fewer people in the party, made short visits, and tended to be first-time visitors to Hawaii. Later, Hsieh *et al.* (1992) found that in the choice of activity participation by the Hong Kong travel market, younger and better-educated travellers tended to participate in the 'full-house activity set' (participating in the broadest variety of activities) and older travellers focused on sightseeing activities. Further, Hsieh *et al.* (1994) found for British travellers that female travellers, compared to male travellers, tended to take more overseas package tours and travellers in the lower income levels sought independent travel more often than travellers in the high income group.

Travel characteristics

The travel and hospitality industry is different from the retails sales industry because of the intangible nature of service and on-site consumption of these services. Travel characteristics such as package travel, destinations, travel party size, and lengths of trips, may affect travellers' choices. Witt and Martin (1987) examined econometric models for forecasting international tourism demand. These models were developed to predict visits from West Germany and the United Kingdom to various tourist destinations (e.g. Austria, France and Greece) and used sociodemographic and price variables. The results indicated that differences existed between Germans and British travellers in their international vacation behaviours. The British were more likely to regard foreign holidays as 'luxuries' whereas the Germans were more likely to regard them as 'necessities'. Since German residents have a lower level of brand loyalty, destinations could compete more effectively on the basis of price and quality. Silberman (1985) estimated the effects of demographic, economic, vacation and destination charac-teristics on the length of stay of individuals on summer vacation trips to Virginia Beach, Virginia. Variables such as cost, the impact of an economic recession, and the number of trips to Virginia Beach were negatively associated with the length of stay. Variables such as distance and income, however, were positively related to the length of stay. Since many resorts are confronted with management issues such as the direction and targeting of advertising campaigns, the possibility of changing sales taxes paid primarily by tourists, and the planning of capital improvement projects, Silberman suggested that information about travel characteristics such as length of stay can help resort managers make effective operating and planning decisions.

In international travel, Hsieh *et al.* (1992) found that the travel party size had an impact on the choice of activity participation among Hong Kong travellers. Hong Kong travellers who liked to participate in the 'full house' (participating in the broadest variety of activities) and 'entertainment' (nightlife, gambling/casinos, and theme park) activity sets tended to travel in larger party sizes. Lang (1991) examined the overseas activities of the Australia travel market and found that active travellers participating in the 'combo' (many different activities) activity set tended to take longer trips than other groups. Dybka (1988) compared overseas travellers from four countries (Japan, West Germany, United Kingdom and France) to Canada. Dybka's findings indicated that an ideal overseas destination for the Japanese would probably combine the main elements that made for an enjoyable touring trip experience – sightseeing in cities, shopping, dining out, guided tours and visiting scenic landmarks, with a place where they can enjoy beautiful sights. In addition, Japanese travellers liked to take long-haul pleasure trips using all-inclusive packages. On the other hand, West German travellers were more likely to embark on resort trips involving beaches, skiing, golf and/or tennis, as well as visiting friends and relatives. Destinations could possibly attract more tourists by carefully developing vacation packages according to specific travel characteristics of various target markets.

Psychographic (travel benefits sought) variables

In the past, survey research has focused on demographic factors because they were easy to identify, useful for interpretation, and readily understood by most people. Nevertheless, people having the same demographic attributes may make different vacation choices in terms of destinations, transportation or accommodation. The only way to find out why they choose different vacation styles may be to understand their psyches (Plog, 1987). Thus an understanding of the psychological factors influencing travel, may assist tourism organisations to serve travellers more effectively and profitably. Psychological factors can influence whether individuals will travel, the specific destination to which they will travel, how they will get there and what they will do on arrival (Mayo and Jarvis, 1981; Plog, 1987). In addition, psychographic variables may be more predictive and can be used to support such tourism decisions as the development of destinations and supporting services, product positioning, advertising, promotion and packaging of products (Plog, 1987). Psychographic researchers assume that people have different sets of motives and behaviours in the market so that unique appeals can be developed for each of the separate psychographic groups (Plog, 1987).

An early study conducted by Woodside and Pitts (1976) found that lifestyle information may be more important in predicting foreign and domestic travel behaviour than demographic information. Similarly, Abbey (1979) concluded that tour travellers preferred tours designed around vacation lifestyle information to tour designs based on demographic information. Later, Schul and Crompton (1983) used two separate multiple regression procedures to examine the relative effects of six psychographic variables and sociodemographics on two measures of external search behaviour, travel planning time and the number of external travel organisations consulted by British travellers. They found that the travel-specific psychographics were more effective than sociodemographics in predicting external search behaviour. Thus they suggested that the use of psychographic information by tour suppliers and marketers should be given a higher priority in the development of effective copy and promotional themes, as well as in the selection of appropriate media for advertising.

In terms of studies of travel benefits sought, Goodrich (1977) found that four major sets of benefits (entertainment, purchase opportunities, climate for comfort and cost) influenced American Express international travellers. Crompton (1979) identified seven socio-psychological motives: escape from a perceived mundane environment, exploration and evaluation of self, relaxation, prestige, regression (less constrained behaviour), enhancement of kinship relationships and facilitation of social interaction. These motives influenced the selection of a particular type of vacation or destination in preference to all the alternatives of which the tourist is aware. McIntosh (1990) suggested that basic travel motivation could be divided into four categories: physical, cultural, interpersonal, and status and prestige motivators. In a more specific study, Cha et al. (1995) identified three distinct groups of Japanese outbound travellers. These groups were labelled sports seekers, novelty seekers and family/relaxation seekers. The advantages of segmenting markets by benefits were that this approach offered a competitive strategy and a guide to market planning and promotional strategies.

Relationship between sociodemographics, travel characteristics and psychographic variables

Van Raaij and Francken (1984) stated that sociodemographic factors may affect psychological factors (e.g. attitudes, expectations, aspirations, values, needs, experiences) of individuals and their households (e.g. lifestyle, traditional/ modern, time orientation, decision-making style, role power structure). In addition, these individual and household factors exert an influence on vacation choices. Woodside and Lysonski (1989) presented a general model of travel destination awareness and choice. This model included two groups of exogenous variables, travel characteristics and marketing variables, which could influence traveller destination awareness. In addition, psychological processes were also considered to be important factors in consumer behaviour in their study. Um and Crompton (1990) also proposed a tourism destination choice model. In this model, external inputs consisted of social interactions and marketing commu-nications to which potential pleasure travellers are exposed. Internal inputs, derived from the socio-psychological set of the potential traveller, included personal characteristics, motives, values and attitudes. It was argued that beliefs about a destination's attributes are formed by the traveller being exposed to the various media, but the nature of these beliefs varies according to the traveller's socio-psychological background.

Conclusions

The study of consumer behaviour is concerned not only with how travellers behave but with why they behave as they do. Only a few researchers (Van Raaij and Francken, 1984; Woodside and Lysonski, 1989; Um and Crompton, 1990) have developed travel destination or activity choice models by examining the travel decision process. Factors such as sociodemographic and travel characteristics have frequently been used to predict the choice of vacation type, but not to predict travel decisions or behaviour. Recently, psychographic factors have received more attention and been suggested as key additions to better explain the decision-making process. Therefore, the opportunity appears to be ripe to develop travel decision models by examining the relationships between sociodemographics, travel characteristics, psychographics (travel benefits sought) and travel decisions. The information regarding the relative importance of these travel factors and decisions may provide new insights for tourism planning and marketing strategies and for the development of travel decisions.

Methodology

Background

In 1986 the US Travel and Tourism Administration and Tourism Canada made a five-year agreement to undertake a jointly funded market research program of mutual interest in overseas countries. By combining resources, the national tourism organisations of the two countries foresaw an ability to produce better market information. Over fourteen potential travel markets were investigated up to 1989. These investigations are called the Pleasure Travel Markets Surveys. Japan is the country chosen in this study.

Japan has emerged as one of the strongest countries in terms of current account surpluses and has led the countries of the Asia-Pacific region in becoming the fastest-growing economies in the world. In the 1980s, Japanese overseas travel experienced tremendous growth. Japan became one of the world's top spenders for international tourism in 1988 following the United States and West Germany (Waters, 1990). The Travel and Leisure World Travel Overview (1990) predicted that Japanese international trips would double from 7.1 to 15.6 million and expenditures increase from US$6.8 billion to US$12.5 billion in the decade from 1986 to 1995. There is no visible trend or phenomenon which is likely to halt the rapid expansion of international travel by Japanese in the future. The Japanese government further announced a new program, 'Two-way Tourism 21' for Japanese tourism in the twenty-first century in 1991. This new program is aimed at facilitating tourists flow to and from Japan to enhance mutual understanding between the Japanese and people from all over the world (Nozawa, 1992). Japanese travellers, therefore, offer an excellent case for the study of travel behaviour and the development of travel decision models.

Data collection and sampling

Data from the Pleasure Travel Markets Survey for Japan for this study was collected in 1989. A total of 1,199 personal in-home interviews averaging 50 minutes in length were conducted. All respondents were people 18 years of age or older who had taken overseas vacations in the past three years or who intended to take such a trip in the next two years. However, only respondents who had taken overseas vacations in the past three years were analysed in this study (916 respondents). In Japan, interviewing was conducted in seven major centres (Tokyo and vicinity, Osaka and vicinity, Nagoya, Fukuoka/Kitakyusha, Sapporo, Hiroshima, Sendai) proportionate to the population (MFCL Japan, 1989: 3).

The questionnaire, designed originally in English, was translated into Japanese. The survey collected information on sociodemographics, travel characteristics, destinations, modes of transportation, activities engaged in on the most recent trip, the most important information sources for planning overseas trips, travel philosophy, benefit sought, media habits, and perceptions of the USA and Canada as travel destinations.

Model development

The travel decision model was developed using the Linear Structural Relation (LISREL VII) program (Joreskog and Sorbom, 1989). LISREL is a versatile and powerful method that combines features of factor analysis and multiple regression to estimate a series of interrelated dependence relationships simultaneously (Lavee, 1988, Hair *et al.*, 1995). That is, LISREL estimates a series of separate, but interdependent, multiple regression equations simultaneously by specifying the structural model used by the statistical program (Hair *et al.*, 1995).

The LISREL program allowed the researchers to investigate two major sets of relationships. First, the model looked at the relative contributions of a large set of variables, including sociodemographics, travel characteristics and travel benefits sought, to a key travel decision variable. Second, the program offered the opportunity to investigate relationships within the sets of variables analysed. Specifically, the analyses concentrated on the connections between socio-demographics and travel characteristics and travel benefits sought. In assessing the overall model fit (that is, how useful the variables were for predicting the travel decision variable) a Chi-square measure that is statistically significant indicates a good fit of the model to the data. A Goodness of Fit Index (GFI) score greater than 0.90 also indicates a good fit (Lavee, 1988). The relationships within the variables were tested by assessing the statistical significance of each parameter. The statistical significance of each parameter was determined by a t statistic at the 0.05 alpha level.

Selection of variables

The sociodemographic variables examined included age, income, education, marital status, sex and language ability (Table 21.1). The indicators of travel characteristics were the number of trips taken before, the party size and the length of trip (Table 21.2). These variables are referred to as endogenous variables.

Table 21.1 Sociodemographic variables used in the analysis

Income	*Sex***
Under 4 million Yen	Male
4–6 million Yen	Not male (female)
6–8 million Yen	
8–10 million Yen	*Language ability*
10–15 million Yen	Yes
15–20 million Yen	No (not good)
Over 20 million Yen	
	*Marital status**
Education	Single
Junior high school	Not single
Senior high school	
Technical/ vocational	*Age group*
Junior college	18–24 yrs
University/ postgraduate	25–34 yrs
	35–44 yrs
	45–54 yrs
	55–64 yrs
	65 yrs or older

* Marital status was coded as 'single' and 'not single' in the model of a travel decision.
** Sex was coded as 'male' and 'not male (female)' in the model of a travel decision.

Table 21.2 Travel characteristics used in the analysis

Party size
1
2
3
4
5
6
7
8
9 or more

Number of trips taken before
1–99 trip(s)

Number of nights away from home
4–365 nights

LISREL requires that these variables be measured at an interval or ratio scale and that the relationships among the variables are linear. However, these restrictions are not absolute, nominal variables can be incorporated into LISREL through the use of 'dummies', and nonlinear and linear relationships can be handled through transformation of variables (Kim and Kohout, 1975). Among these variables, 'education level', 'income', 'age group' and 'party size' were treated as an 'interval-like variable'. Other nominal variables such as 'sex' , 'marital status' and 'language ability' were defined as 'dummy variables'. It is relevant to note that a careful interpretation of 'interval-like' variables above should be made because the change of each level of variable does not produce the same effect on dependent variables. Multicollinearity problems between exogenous variables were detected by examining a matrix of bivariate correlations.

The travel decision variable examined was the likelihood of taking a future trip to fly to somewhere outside of the country, entirely or in part for vacation or pleasure, and stay away from home at least four nights. Respondents were asked to respond to the likelihood on a 1–5 Likert scale (ranging from 5: 'definitely take such a trip', 4: 'very likely to take such a trip', 3: 'somewhat likely to do so', 2: 'might or might not take such a trip', to 1: 'not likely to take such a trip'). Finally,

Table 21.3 Development of travel benefits scales

Travel benefits scales were based on a set of reasons why people might want to go on a vacation. Twenty-five reasons were categorised into six sub-scales of travel benefit sought. The six sub-scales constructed for the Travel Benefit Sought Scales were: being and seeing, sport interest, show and tell, heritage, social escape, and adventure.

Being and seeing Travellers with high scores are interested in travelling to places important in history, experiencing a foreign destination, and generally seeing as much as possible in the time available. A low scorer is more conservative and cautious towards travel.

Sport interest A high scorer on this scale is likely to be a sports participant or a spectator. Travellers with low scores may not be very interested in physical sports activities.

Show and tell Travel is usually considered as a symbol of wealth or social class. Some people desire the recognition, attention and good reputation associated with taking trips. High scorers think that travel to places friends haven't been and talking about their trips after returning home are the major benefits of travelling.

Heritage Visiting friends and relatives usually is a major travel motivation for people, especially those who are from Europe and North America. Heritage scales measure the strength of the heritage orientation in terms of the benefits travellers seek. Travellers who score high like to visit friends, relatives and places their family came from. They enjoy being together as a family and reliving past good times.

Social escape Travel behaviour reflects the individual's way of life. Some travellers seek a break from the stresses of daily life while travelling. Social escapees with high scores want to get away from the demands of home, get a change from a busy job and generally escape from the ordinary. Travel for them means relaxing and doing nothing at all as well as the chance to be free to act the way they feel.

Adventure People desire to explore the unknown. Travel for some people is to have a new experience, a different lifestyle and excitement. Travellers who score high are looking for thrills and excitement, and want to be daring and adventuresome. Low scorers consider safe travel as an important factor in terms of travel benefits.

---

a set of 25 items relating to the benefits and experiences sought from vacation were analysed. For each of these items the respondents were required to rate their importance on a 4-point scale ranging from 1: not at all important to 4: very important. The 25 travel benefits sought were grouped into six categories described in Table 21.3.

Results and conclusions

The measures of the overall fit of the Japanese travel decision model to the data were good. The Chi-square value with 64 degrees of freedom was 80.56 (p = 0.08). The Goodness of Fit Index (GFI) was 0.99. That is, the travel decision was affected by sociodemographics, travel characteristics and psychographic variables. The structural model that was derived from the data is depicted in Figure 21.1.

LISREL (standardised and unstandardised) estimates, standard error and t-test of the structural parameters in the Japanese model are presented in Tables 21.4 and 21.5. The travel decision variable (the likelihood of taking a trip in the next two years) was related to sociodemographic variables for Japanese travellers. An examination of Table 21.4 indicates that age, income, marital status and language ability had a direct influence on the travel decisions in the Japanese travel market. That is, Japanese travellers who were older, single, and with better income and language ability were more likely to choose to travel. Besides sociodemographics, travel characteristics were also causally related to the travel decision. Travel characteristics such as 'number of trips taken before' were positively related to Japanese travel decisions. In regard to the relationships among the endogenous variables (travel decisions and the six categories of travel benefits sought), the 'sport interests' benefit was positively related to the likelihood of taking future trips while the 'show and tell' benefit was negatively associated with the travel decision (Table 21.5). It is worth noting from Table 21.5 that there were significant positive relationships between the 'being and seeing' , 'sport interests' and 'show and tell' benefits packages. These relationships indicate that there exist some indirect effects from other endogenous variables on the travel decision.

In addition to examining the variables related to the likelihood of future travel, the model provided profiles of the Japanese travellers seeking particular benefit or holiday experience packages. For example, Japanese travellers who were older and with better language ability were likely to pursue the 'being and seeing' benefit package (Table 21.4), while the 'sports interest' benefit package was mostly likely to be sought by Japanese who were male and young. The 'show and tell' benefits motivated Japanese travellers who were single, with lower incomes and travelled with a larger party size. For the 'heritage' benefit, Japanese travellers tended to be female, married and with lower incomes. Furthermore, married and female Japanese tended to be 'social escapees' in trips. Finally, 'adventure trips' attracted Japanese, who were young, male, with a better language capacity, and on longer trips.

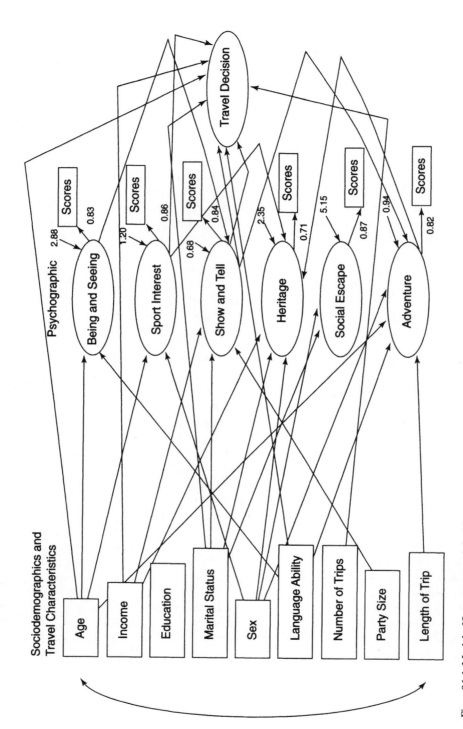

Figure 21.1 Model of Japanese travel decisions
Note: Each line represents a significant relationship between variables

Table 21.4 Maximum likelihood estimates (ML est.) of the effects of the exogenous variables on the endogenous variables.

Dependent variables (endogenous variables)	Travel decision			Being and seeing			Sports interest			Show and tell			Heritage			Social escape			Adventure		
	ML est.	SE	T-test	ML est.	SE	T-test	ML est.	SE	T-test	ML est.	SE	T-test	ML est.	SE	T-test	ML est.	SE	T-test	ML est.	SE	T-test
Age	0.25 (0.30)	0.04	6.72*	0.21 (0.10)	0.09	2.34*	-0.60 (-0.42)	0.06	-9.52*	—	—	—	—	—	—	—	—	—	-0.42 (-0.40)	0.04	10.03*
Income	0.08 (0.11)	0.02	3.18*	—	—	—	—	—	.	-0.11 (-0.12)	0.04	-2.99*	-0.14 (-0.11)	0.05	-2.65*	—	—	—	—	—	—
Education	—	—	—	—	—	—	—	—	—	—	—	—	—	—	—	—	—	—	—	—	—
Single	0.78 (0.28)	0.11	7.10*	—	—	—	—	—	—	0.30 (0.09)	0.14	2.17*	—	—	—	-1.30 (-0.13)	0.38	-3.44*	—	—	—
Male	—	—	—	—	—	—	-0.35 (0.08)	0.16	2.19*	—	—	—	-0.50 (-0.11)	0.22	-2.27*	-1.54 (-0.17)	0.38	-4.02*	0.50 (0.15)	0.13	3.86*
Language ability	0.22 (0.09)	0.08	2.57*	0.95 (0.16)	0.23	4.07*	—	—	—	—	—	—	—	—	—	—	—	—	0.27 (0.08)	0.11	2.40*
Number of trips taken before	0.24 (0.25)	0.03	7.66*	—	—	—	—	—	—	—	—	—	—	—	—	—	—	—	—	—	—
Travel party size	—	—	—	—	—	—	—	—	—	0.05 (0.09)	0.02	2.14*	—	—	—	—	—	—	—	—	—
Length of trip	—	—	—	—	—	—	—	—	—	—	—	—	—	—	—	—	—	—	0.00 (0.07)	0.00	2.36*

Note. Standardised solutions are in parentheses.
Only significant relationships are presented; '——' represents an insignificant relationship between variables.
* p < 0.05.

Table 21.5 Maximum likelihood estimates (ML est.) of the effects of the endogenous variables.

Dependent variables (endogenous variables)	Travel decision			Being and seeing			Sports interest			Show and tell			Heritage			Social escape			Adventure		
	ML est.	SE	T-test	ML est.	SE	T-test	ML est.	SE	T-test	ML est.	SE	T-test	ML est.	SE	T-test	ML est.	SE	T-test	ML est.	SE	T-test
Travel decision	—	—	—	—	—	—	0.09 (0.16)	0.03	3.17*	-0.08 (-0.10)	0.04	-2.10*	—	—	—	—	—	—	—	—	—
Being and seeing	—	—	—	—	—	—	—	—	—	—	—	—	—	—	—	—	—	—	—	—	—
Sports interest	—	—	—	0.31 (0.44)	0.07	4.40*	—	—	—	0.61 (0.43)	0.07	9.17*	—	—	—	—	—	—	—	—	—
Show and tell	—	—	—	0.11 (0.23)	0.03	4.59*	—	—	—	—	—	—	—	—	—	—	—	—	—	—	—
Heritage	—	—	—	—	—	—	0.38 (0.38)	0.07	5.45*	—	—	—	—	—	—	—	—	—	0.73 (0.57)	0.14	5.38*
Social escape	—	—	—	—	—	—	—	—	—	—	—	—	—	—	—	—	—	—	2.68 (0.99)	0.18	14.71*
Adventure	—	—	—	—	—	—	—	—	—	0.28 (0.25)	0.04	6.47*	—	—	—	—	—	—	—	—	—

Note: Standardised solutions are in parentheses.
Only significant relationships are presented; '——' represents an insignificant relationship between variables.
* p < 0.05.

In sum, the travel decision model was found to have an adequate fit for the Japanese travel market. That is, the travel decision studied was found to be significantly influenced by sociodemographics, travel characteristics and psychographic variables. In particular, the model also indicated that psychographic variables had a significant impact on travel decisions. This finding supports the model of tourism destination choice tested by Um and Crompton (1990).

Marketing implications

This study suggests that travellers have unique sets of characteristics which are both related to their travel decisions and which can be used to develop marketing strategies, including effective media selection and communication. The travel characteristics of each group can also provide useful information for developing the format of future tours. Travel organisations and businesses can use travellers' psychographic information to design specific advertisements and tour packages to suit the specific sets of travel benefits sought. For example, if Japanese travellers who are older, with better language ability and wishing to pursue the 'being and seeing' benefit is the target market, then planners and marketers should design promotions which focus on 'experiencing various cultures and scenic areas'. In addition, since the benefit 'being and seeing' had significantly positive relationships with the other benefit categories of 'sport interest' and 'show and tell', the promotion might include some of these themes to attract Japanese travellers.

The results also showed that language ability had a significant effect on the travel decision. In general, not many Japanese are familiar with English, especially older travellers. Brochures and pamphlets should be written in English and Japanese. Major attractions and airports might provide bi-language services. In addition, since the variable 'number of trips taken before' was positively related to the travel decisions of Japanese, these repeat travellers need additional attention. Special programs or promotions can be designed for them. For example, an airline company can develop cooperative relationships with hotels and car rental companies to provide a frequent travel program.

A travel decision model can also provide valuable information for various travel organisations when designing appealing travel packages and experiences. Most travellers do not have the time, experience, knowledge, or financial resources to assemble their own travel trip activities and experiences in foreign destinations. According to the unique sociodemographic background, travel characteristics and travel benefits sought by different groups of Japanese travellers, it is suggested that all host countries should put more effort into designing vacation packages around specific groups of activities rather than producing literature or collateral materials that present a 'laundry list' of every conceivable activity at their destination.

Theoretical implications

An important finding from the research was that there were significant relationships between sociodemographics, travel characteristics and psychographics. That is, sociodemographic factors and travel characteristics affect psychological factors. All these findings support the early studies conducted by Woodside and Pitts (1976), Abbey (1979) and Schul and Crompton (1985). That is, travel-specific psychographics are affected by sociodemographic characteristics and past travel experience. However, compared with previous research, this study provides a unique way to further examine the interrelationship between these variables using a multivariate method. In addition, it is relevant to note that the decision to take a trip is complicated and affected by many factors.

Another unique finding is that sociodemographic variables and travel characteristics not only have direct, but also indirect, impacts on travel decisions. Earlier research has most often focused on the direct relationship between sociodemographics, travel characteristics and the travel decision. The interrelationships among variables described in the present model suggest, however, that travel decisions can be affected by sociodemographics and travel characteristics through psychographic variables. In the example of the Japanese travel market, age has an impact on travel decisions directly and indirectly through its association with the sport interest benefit. Further, psychographic variables may have an impact on travel decisions through other psychographic variables. That is, the relationships among the travel decision models are not single effects but related to each other. These findings should encourage researchers and marketers to view the travel decision as a series of complicated relationships between many variables.

Limitations and future research directions

One of the concerns with the results presented in this chapter is that only travellers from Japan were examined. If the analysis was done for travellers from another country would the same model emerge? Are the sociodemographics and travel characteristics affecting travel decisions in this case also significant for travellers from other countries or for domestic tourists? In addition, since the study is based on a secondary data set, the model of travel decisions could be explained better by adding other important variables. It is thus suggested that additional sociodemographics and other household factors (e.g. lifestyle, family life cycle) as well as psychological factors (e.g. values, perception and image of destinations, expectations) could be added to the model of travel decisions for future studies.

In sum, the structural model provides a new approach to examining a model of travel choice behaviour. The results of this study provide important new information for marketing strategies. By understanding the travel decision process of potential travellers, marketing managers can be in a better position to develop promotional and planning strategies.

References

Abbey, J.R. (1979) 'Does life-style profiling work?' *Journal of Travel Research* 18 (summer): 8–14.

Assael, H. (1987) *Consumer Behaviour and Marketing Action*, Boston, MA: Purs-Kent Publishing Company.

Cha, S., McCleary, K.W. and Uysal, M. (1995) 'Travel motivations of Japanese overseas travellers: a factor-cluster segmentation approach', *Journal of Travel Research* 34(1): 33–9.

Crompton, J. L. (1979) 'Motivations for pleasure vacation', *Annuals of Tourism Research* 6(4): 408–24.

Dybka, J. (1988) 'Overseas travel to Canada: new research on the perceptions and preferences of the pleasure travel market', *Journal of Travel Research* 26(4): 12–15.

Engel, J. F. and Blackwell, R. D. (1990) *Consumer Behaviour*, 6th edn, New York: Dryden Press.

Goodrich, J. N. (1977) 'The relationship between preferences for and perceptions of vacation destinations: application of a choice model', *Journal of Travel Research* 17: 8–13.

Hair, J.F., Abdersibm, R.E., Tatham, R.L. and Black, W.C. (1995) *Multivariate Data Analysis with Readings*, 4th edn, Englewood Cliffs, NJ: Prentice-Hall.

Hsieh, S., O'Leary, J. T. and Morrison, A. M. (1992) 'Segmenting the international travel market by activity', *Tourism Management* 13(2): 209–23.

Hsieh, S., O'Leary, J. T. and Morrison, A. M. (1994) 'A comparison of package and non-package travellers from the United Kingdom', *Journal of International Consumer Marketing* 6(3): 79–100.

Joreskog, K. G. and Sorbom, D. (1989) *LISREL 7: A Guide to the Program and Applications*, 2nd edn, Chicago: SPSS.

Kim, J. and Kohout, F.J. (1975) 'Multiple regression analysis: subprogram regression', *Statistical Package for the Social Science*, 2nd edn, New York: McGraw-Hill Book Co.

Lavee, Y. (1988) 'Linear Structural Relationships (LISREL) in family research', *Journal of Marriage and the Family* 50: 937–48.

Lang, C. (1991) *Types of Activity Participation of Australian Overseas Travellers to National Parks/Forests*. Unpublished master thesis, Purdue University, West Lafayette, IN., USA.

Market Facts of Canada Limited (1989) *Pleasure Travel Markets to North America: Japan*, Toronto.

Mayo, E. J. and Jarvis, L. P. (1981) *The Psychology of Leisure Travel*, Boston, MA: CBI Publishing Company, Inc.

Mazanee, J. A. (1989) 'Consumer behaviour in tourism', in S. F. Witt and L. Moutinho (eds.) *Tourism. Marketing and Management Handbook*, pp. 63–8. New York: Prentice-Hall.

McIntosh, R. W. (1990) *Tourism Principles, Practices, Philosophies*, 6th edn, Columbus, OH: Grid Inc.

Nicosia, F. M. (1966) *Consumer Decision Processes: Marketing and Advertising Implications*, Englewood Cliffs, N.J.: Prentice-Hall, Inc.

Nozawa, H. (1992) 'A marketing analysis of Japanese outbound travel', *Tourism Management*, 13(2): 226–34.

Plog, S. C. (1987) 'Understanding psychographics in tourism research', in J. R. B. Ritchie and C. R. Goeldner (eds) *Travel, Tourism and Hospitality Research*, pp. 203–14. New York: Wiley.

Schiffman, L.G. and Kanuk, L. L. (1990) *Consumer Behaviour*, Englewood Cliffs, NJ: Prentice-Hall.

Schul, P. and Crompton, J. L. (1983) 'Search behaviour of international vacationers: travel-specific lifestyle and sociodemographic variables', *Journal of Travel Research* 22(2): 25–30.

Sheldon, P. J. and Mak, J. (1987) 'The demand for package tours: a mode choice model', *Journal of Travel Research*, 25(3): 13–17.

Silberman, F. (1985) 'A demand function for length of stay: the evidence from Virginia beach', *Journal of Travel Research* 23(9): 16–23.

Travel and Leisure's World Travel Overview (1988/1989). New York: American Express Publishing Corporation.

Um, S. and Crompton, J. L. (1990) 'Attitude determinants in tourism destination choice', *Annuals of Tourism Research* 17: 432–48.

Van Raaij, W. F. and Francken, D. A. (1984) 'Vacation decisions, activities, and satisfactions', *Annals of Tourism Research* 11(1): 101–112.

Waters, S. R. (1990) *Travel Industry World Yearbook – The Big Picture 1990*, New York: Child and Water Inc.

Witt, S. F. and Martin, C. A. (1987) 'Econometric models for forecasting international tourism demand', *Journal of Travel Research* 25(3): 23–30.

Woodside, A. G. and Pitts, R. E. (1976) 'Effects of consumer life styles, demographics and travel activities on foreign and domestic travel behaviour', *Journal of Travel Research* 14(Winter): 13–15.

Woodside, A. G. and Lysonski, S. (1989) 'A general model of traveller destination choice', *Journal of Travel Research* 27(4): 8–14.

22

AN ANALYSIS OF THE PROMOTION OF MARINE TOURISM IN FAR NORTH QUEENSLAND, AUSTRALIA

Diane Burns and Laurie Murphy

A basic premise of marketing is that, 'satisfaction with a product is determined by how well the product meets the customer's expectations for that product' (Kotler *et al.* 1996: 26). Customers whose expectations are not met may become dissatisfied, are unlikely to use the product again and may complain (Peter and Olsen, 1987) whereas, 'a satisfied customer normally provides a company with repeat business and also creates goodwill that is expressed by positive word of mouth recommendations' (Duke and Persia, 1996: 78).

Given the importance of expectations in determining satisfaction, it is not surprising that there exists a substantial and growing literature which focuses on understanding the factors involved in creating tourists' expectations. In various models of destination choice and image formation (see Um and Crompton, 1990; Woodside and Lysonski, 1989; Gunn, 1988) promotional activities undertaken by tourist organisations are a major source of information about destinations and one way in which tourism managers can influence visitor expectations. One component of promotion is advertising and the brochure is a commonly used advertising medium in tourism. For the tour operator, the brochure is one of the most widely used and efficient forms of advertising available (Holloway and Plant, 1988; Dann, 1993). Images in brochures influence the tourists' expectations of the product and where the image of the product differs from the actual reality or where the expectations are not met by the experience, the tourist is likely to become dissatisfied. It therefore becomes important that the images portrayed in the brochures match the experience provided to the visitor. The images presented in brochures may also vary in terms of their appeal to different markets and thus may differ in their effectiveness at matching markets to products (Morrison, 1996).

The aim of the case study presented in this chapter is to examine the link between the products or experiences provided for tourists, the expectations

415

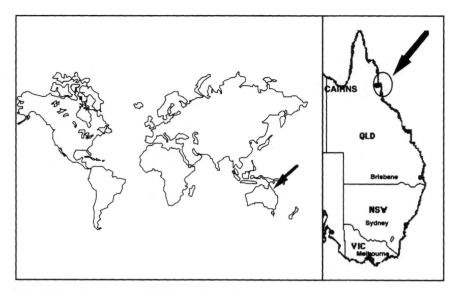

Figure 22.1 Cairns section of the Great Barrier Reef

created in brochures and the markets currently buying day trip tours to the Great Barrier Reef. Figure 22.1 shows the position of the case study region, located in the tropical north east of Australia and having the city of Cairns as the focal point of the tourism industry in the region. The research had two major components. One sought to identify and profile the image of the marine tourism product in the Great Barrier Reef region through a content analysis of brochures, while the other sought to determine visitor experience of the product through a market survey. An examination of the similarities and inconsistencies between the image and the reality allowed the researchers to develop a better understanding of the present situation and provide directions for future marketing.

The case study background

It is estimated that 70 per cent of visitors (Pearce *et al.*, 1996; Coopers & Lybrand, 1996) come to the Far North Queensland region primarily for a reef experience and, according to the Great Barrier Reef Marine Park Authority, 947,000 visitor days were spent at the Great Barrier Reef in 1995/96. It was also estimated that marine tourism generated approximately Aus.$665 million for the region's economy during that year (based on Coopers & Lybrand, 1996: 49). The potential for employment within this industry, combined with revenue generated, places the Great Barrier Reef and its associated marine tourism as a driving force for development of the regional economy.

This research was undertaken under the direction of the Cooperative Research Centre for Ecologically Sustainable Development of the Great Barrier Reef (CRC Reef Research) and the Reef Tourism 2005 project, with the broad aim of developing a profile of Marine Tourism Marketing in the Far North Queensland region. Reef Tourism 2005 is a marine tourism industry, federal government and state government project, initiated to develop a ten-year strategic plan for the sustainable development of marine tourism in the region. The industry component of this organisation represents a variety of operations to the Great Barrier Reef, including general day trip operators, overnight trips, specialist dive trips, extended dive trips and game fishing trips with the capacities of the operations varying from over 300 passengers to fewer than 10 passengers.

The impetus for the establishment of the project was provided by a number of factors that concerned marine tourism operators in the region. Some of these factors can be more clearly outlined with reference to Figure 22.2 which displays Butler's model of the life cycle of a tourist destination (1980). Butler proposed that a tourist destination experienced a life cycle similar to any other marketable product and that by determining where a destination was positioned on this life cycle, planning agencies could adjust development plans to ensure that the destination would avoid the point of 'Stagnation' or 'Decline'. The marine tourism operators associated with Reef Tourism 2005 believed that the Far North Queensland region was at the 'Consolidation' stage and that coordinated

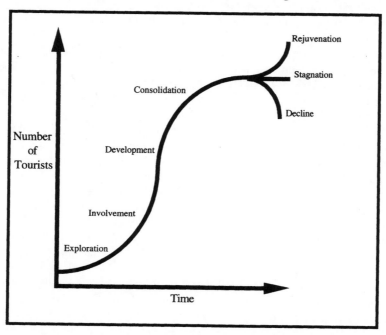

Figure 22.2 Butler's developmental stages of the life cycle of a tourist destination

plans for development and marketing needed to be developed to bring about a 'Rejuvenation' of the region. It was decided that one of the key research projects required to support this planning exercise was that of evaluating promotional activities and examining market characteristics.

The methods used

The approach taken in the case study was to compare information obtained from an analysis of the content of brochures used to promote marine tourism in the area, to the data collected in a market survey. The overall aim was to identify areas where the market profile did not match promotional images, thereby providing information valuable to future marketing efforts.

Brochure analysis

An audit of marine tourism promotional material was undertaken by conducting a content analysis of brochures available from both within and outside the Far North Queensland region. This was carried out with the aim of developing a profile of the marine tourism product, as promoted in its advertising. The brochures were divided into two groups of promotional material, those that promoted the product of a single operator and those that promoted the tourism product of the region as a whole.

Brochures were obtained within the region from the major tourist information centre adjacent to the departure point of Great Barrier Reef trips from Cairns. This resulted in a sample of promotional material from 61 operators. Brochures pertaining to the region as a whole were obtained from the Australian Tourist Commission, the Queensland Tourist and Travel Corporation and the Far North Queensland Promotion Bureau by requesting all brochures that contained information relating to the region. This resulted in a sample of 35 brochures which was reduced to a sample of 22 after elimination of duplications. Analysis of text was further reduced to 19 brochures written in English.

Content analysis has been used previously in tourism, and can be defined as, 'a research technique for making replicable and valid inferences from data to their context' (Krippendorf, 1980: 21). These inferences are generally about forms of communication and can provide information 'about the sender of the message, the message itself, or the audience of the message' (Weber, 1985: 9). The history of content analysis dates back to the 1600s when newspaper articles were thought to be affecting people's religious beliefs, and newspapers and their influence on behaviour became a major topic of investigation at the turn of this century (Krippendorf, 1980). Content analysis has also been recognised as having limitations of reliability due to the context and interpretation of the data where many words and images are classified into fewer content categories (Weber, 1985), but it has been recommended as an effective tool for media research and a starting point for more thorough analysis of images (Barrat, 1986).

The analysis in the present study was divided into two stages, the first stage involved coding of the images/messages presented in the operators' brochures. Stage 2 required the coders to determine initially the proportion of the regional brochure material which contained reef content and then analyse the images presented in the reef component of these brochures.

Coding of the material was broken down to record images featured on;

1   the front page of the brochures,
2   photographs within the body of the brochure, and
3   text within the body of the brochure.

The content of each brochure was analysed using a predetermined set of variables that enabled comparison with relevant questions in the visitor survey. Coders recorded the number of times images occurred in the brochures. Key words from the front page of each brochure were coded. Coders were also asked to note the types of locations, activities, transport, travel parties and marine life depicted in the brochures. Further details obtained from the brochures included information relating to the type of tour being promoted by the brochure and whether the brochure included any ecological/educational messages.

Market survey

The data used in the present study was collected from Great Barrier Reef visitors on fourteen different day trip operations of various sizes, departing from ports in the Far North Queensland region. Visitors were approached on the boat and asked to complete a 6-page questionnaire written in English. Approaches were made to 1,783 visitors with a response rate of 80 per cent. The survey included questions on the nature of the market such as: general holiday travel motivations, specific reasons for the current visit, activity participation and length of stay in the region. The questionnaire also examined aspects of the reef experience such as; the importance of various benefits expected from a reef trip; open-ended responses detailing the best feature of the trip and suggestions for improvement and trip enjoyment rating scales. Data was also collected regarding the visitors' present and future travel patterns. Demographic details collected included country of origin, gender and age of respondent. Other questions included source of information used for the region, and type of transport both to and within the region.

Of the 1,423 respondents to the questionnaire, 48 per cent were male and 52 per cent female, 62 per cent indicated that they were international visitors and 38 per cent were Australian visitors. Just over half of the visitors (58 per cent) were surveyed on larger boats (with a capacity for more than 100 passengers). Most respondents were aged between 18 and 55 years (81 per cent), with 49 per cent being aged between 18 and 35; 57 per cent of respondents were staying in the region for one week or less and 90 per cent stayed in the region for three weeks or less.

Table 22.1 Use of information sources

Source	Respondents (%)
Word of mouth	54.3
Travel agent	34.2
Brochures (outside region)	28.8
Brochures (inside region)	28.8
Books	25.3
Media article	21.4
Tour operator	19.3
Accommodation	15.2
Been before	13.4
Not applicable	6.3
Other sources	5.9
No information	2.6
Clubs/assoc./govt	1.5

Respondents were asked to list the three main sources they used to gain information about the Far North Queensland region. Table 22.1, which provides a summary of these responses, indicates that 29 per cent of visitors used brochures from outside of the region and 29 per cent used brochures from within the region as a main source of gaining information about Far North Queensland. In combination, brochures were the most common source of information about the region. This high level of use supports the analysis of brochures as a basis for determining the promotional image of the region. The high level of use of other people as sources of information also supports the assumption that visitor evaluations of their experience can have substantial impact on the generation of further business.

Results

The results section is divided into three main sections, with the first section describing the images of reef tourism products presented in the brochures examined. The second section examines the responses of reef visitors and discusses the consistency between the images presented and the actual experiences of visitors, while the third section looks at overall satisfaction of reef tourists.

Brochure analysis

The front cover of a brochure is important in attracting the attention of a potential customer. The first step in the content analysis was to examine the keywords and visual images used on the front pages of both regional and marine tour operators. The major physical locations depicted on the front page of the brochures are described in Table 22.2. Marine environments feature in almost 45 per cent of the regional covers whilst reef-specific locations account for 22 per cent of the

Table 22.2 Images on brochure front pages

Visual image	Operators brochures % of locations featured	Regional brochures % of locations featured
Ocean	46.9	5.6
Reef – no pontoon	30.6	5.6
Island	10.2	
Coral cay	6.1	5.6
Beach	2.0	11.1
Reef – with pontoon	2.0	
Reef, rainforest and outback		11.1
Rainforest and ocean		5.6
Rainforest		16.7
Cultural location		11.1
Outback		5.6
Wildlife sanctuary		5.6
Other setting	2.0	16.7

Table 22.3 Key words used on operator brochure front pages

Key word	Brochures	
	Frequency	%
Dive	22	36.1
Reef	12	19.7
Boat/operator's name	6	9.8
Island	5	8.2
Cruise	5	8.2
Fishing	2	3.3
Sail	2	3.3
Snorkel	1	1.6
Submarine	1	1.6
Junk	1	1.6
Marlin	1	1.6
No key word	3	4.8
Total	61	100.0

images presented on the front pages of regional brochures. Consistent with the importance of the marine environment for tourism in the region, marine and reef images were the most commonly used features on regional brochure front pages. The results also indicated that operator brochures were more likely to present a general image of the marine environment rather than specific reef locations on their front cover. General ocean images accounted for 47 per cent of the locations depicted, whilst reef-specific images were 39 per cent of locations depicted. Nearly 20 per cent of the operator brochures did not picture a specific location on the front page and 18.2 per cent of regional brochures did not picture a specific location on the front page. Table 22.3 lists the key words presented on the front page of the operator brochures. Clearly 'dive' and 'reef' feature as the predominant key words for operator brochures.

The content analysis next considered the physical features or locations presented in the entire brochures, both as visual images and text, and the results are given in Table 22.4. These results are consistent with those presented in the previous tables with the reef and the ocean in general being the physical features or locations most likely to be portrayed visually or mentioned in the text of the brochures. Underwater images were also common in the photographs used in the reef operator brochures.

The brochures were also examined for the images of marine life portrayed (see Table 22.5). Half of the front page photographs for operator brochures depicted marine life and less than half of the photographs in the brochures considered as a whole were depictions of marine life. Those images which did include marine life were most likely to be of coral. The text of the brochures provided a more even distribution of references to coral and various other marine life.

After examining the physical features or locations used on the brochures, the content analysis concentrated on the nature of the experience presented focusing on the types of activities and visitors described or depicted. Of the photographs in operators' brochures, 78 per cent included activities of some kind and 67 per cent of photos in regional brochures included activities. Table 22.6 indicates the types of activities depicted in the brochures. For both types of brochure scuba diving was the main reef-related activity described, followed by snorkelling and sailing. The brochures were also analysed to determine the type of the travel parties

Table 22.4 Physical features/locations depicted/described in brochures

Physical feature / location	Operators' photos (% of photos)	Regional reef photos (% of photos)	Operators' text (% of locations depicted)	Regional reef text (% of locations depicted)
Reef (without pontoon)	24.6	18.3	40.7	22.7
Ocean	21.7	11.1	2.5	2.7
Facilities	20.0	2.4	7.5	6.8
Beach	9.0	15.9	5.0	8.2
Nothing/no location	5.8	2.4	0.4	1.4
Specific site/features – land based	5.5	0.8	10.3	30.6
Underwater	4.6	19.1		
Reef with pontoon	4.0	11.9	3.3	0.5
Coral cay	2.3	5.6	11.2	3.6
Island	2.0	11.9	12.0	7.7
Other	0.6	0.8	0.8	7.8
Specific site/features – marine based			6.1	8.2

Table 22.5 Marine life depicted in brochures

Marine life	Operators' front page % of brochures	Operators' text % of brochures	Regional text % of brochures	Operators' photos % of brochures	Regional photos % of brochures
No marine life depicted	50.0	6.9	8.2	61.3	58.7
Coral	18.3	23.3	24.7	18.2	16.7
Large fish	16.7	25.2	14.1	6.1	6.3
Small fish	11.7	12.4	12.9	6.3	14.3
Marine life general/large	3.4	16.5	13.1	5.6	2.4
Marine life general/small	–	6.0	8.1	–	–
Marine life general	–	9.9	18.9	2.4	1.6

Table 22.6 Activities depicted/described in brochures

Activities	Operators text (% of activities)	Regional reef text (% of activities)	Operators photos (% of activities)	Regional reef photos (% of activities)
Swimming	4.5	5.7	2.2	2.4
Snorkelling	14.3	9.0	11.5	16.7
Diving	22.9	16.5	28.1	25.0
Other reef based	4.5	0.9	0.7	1.2
Glass bot/semi sub	5.4	6.1	4.8	3.6
Fishing	2.7	7.1	4.1	1.2
Sailing	7.1	0.9	8.9	3.6
Visiting islands	2.4	0.9	1.9	3.6
Viewing marine life	4.5	2.4	5.6	6.0
Visiting beaches	1.5		4.1	6.0
Other land based	6.0	13.7	4.1	
Other water based	2.7	4.2	1.5	2.4
Boating	1.8	10.4	3.7	20.2
Education/learning	1.8	2.4	1.5	
Other	5.1	15.1	1.1	3.6
Relaxation	5.4	2.4	10.0	4.8
Picnics/food	7.7	2.4	6.3	

depicted (see Table 22.7). The results suggested that operator brochures concentrated on single travellers in social interaction situations, whilst regional brochures concentrated on couples.

The final component of the content analysis was an examination of the use of educational or ecological messages. The marine tour operators suggested this area as having potential for greater emphasis in promotion of their product. Table

Table 22.7 Travel parties depicted in brochure photographs

Travel party	Operators' photos (% of travel parties depicted)	Regional reef photos (% of travel parties depicted)
Couples	28	31
Groups	37	32
Families	5	9
Individuals alone	28	26
Other parties	2	2

Table 22.8 Educational/ecological messages in brochures

Learning/educational image	Operators' brochures % of brochures	Regional brochures % of brochures
Ecological message	18	28
Educational material	33	39

22.8 indicates that generally brochures were not used to provide this kind of information.

Summary

An examination of marine tour operator brochures indicated that the visual images presented were dominated by photographs of the Great Barrier Reef, the ocean and the facilities available onboard the boats or on the pontoons used by the operators. Just over one third of the photographs used in the operator brochures included marine life with images of coral predominant. 'Scuba diving' was found to be the most commonly used key word on the front pages of the operator brochures and this was consistent with it being the main activity pictured or described in the brochures. It was also found that individuals and groups of individuals were the most common travel party presented in the brochures and that few brochures included educational or environmental messages.

Market survey

The first step in the analysis of the visitors survey data was to examine the actual experience of reef visitors. This was done by looking at the variables of actual travel party composition and activity participation, which match two of the features of the content analysis. The results displayed in Table 22.9 showed that the majority of respondents travelled as couples (58 per cent). Travel parties ranged in size from single travellers to parties of 53 with 83 per cent of respondents travelling in groups of 5 or fewer. These results present quite a contrast to the images in the brochures in which individuals and groups are dominant. Further,

Table 22.9 Travel parties of reef visitor survey sample

Travel party	Market survey % of respondents
Couples	58
Groups	33
Families	19
Individuals alone	10
Other parties	4

Table 22.10 Actual and intended activity participation by reef visitors

Reef activities	Have done and/or intend to do (respondents %)
Snorkelling	81
Swimming	80
Viewing marine animals	62
Glass bottom boat/ semi-sub coral viewing	51
Visiting islands	44
Scuba diving	31
Sailing	21
Cruises of one or more nights	16
Fishing	12

only 5 per cent of the groups portrayed in the brochures were in family groups and this contrasts with 19 per cent of the actual travel parties of visitors. In another visitor survey conducted in the region, Moscardo (1996) found that one of the variables that distinguished non-reef from reef visitors was that visitors who did not go to the reef were significantly more likely to be travelling with children. This suggests at least the possibility that families may be deterred from taking reef trips because the brochures generally exclude children from their images of reef experiences.

Table 22.10 lists the ten activities with the highest actual participation rates. The survey found that the marine activities of snorkelling, swimming, viewing marine life and glass bottom boat or semi submersible rides were the most frequent activities with scuba diving listed by fewer than a third of the visitors. As with the travel party variables this indicated a mismatch between the activities portrayed and the activities which visitors were most likely to participate in.

The other variable studied to determine actual reef experiences were the responses given to a question asking reef visitors to list the best features of their reef experience. The survey results for this question are presented in Table 22.11 and the most common best features of reef visits related to snorkelling, scuba diving, seeing the coral and being amongst marine life. The dominance of seeing

CHANGE MANAGEMENT STRATEGIES

Table 22.11 Ten best features of a Great Barrier Reef visit

Best feature	% responses
Snorkelling	19.6
Diving, diving related	12.1
Coral	8.3
General swimming amongst reef/nature/fish, etc.	8.0
Semi-sub, glass bottom boat, observatory	7.0
Marine life other than fish and coral	6.9
Beautiful environment, water, weather, etc.	6.0
Fish	5.6
Relaxing, peaceful, restful holiday	3.1
Nature, wilderness, natural	3.1
Everything	4.4

Table 22.12 Mean ratings of importance of expected benefits from a reef trip

Expected benefits	Mean rating of importance
Seeing the beauty of the Great Barrier Reef	1.160
Seeing coral in its natural surroundings	1.188
Seeing marine life in detail	1.440
Being close to nature	1.490
Something new and different	1.515
Swimming amongst the marine life	1.636
A learning/educational experience	1.641
An opportunity to rest and relax	1.885
Provides excitement	1.911
Being physically active	1.984
Provides a chance to escape	2.125
An opportunity to be with friends, partners or family	2.181

Note: A lower score indicates greater importance as the scale ranged from 1 'always important' to 4 'Never important'.

the coral as a highlight of reef trips supports the high use of coral images in operator brochures. The importance of diving also lends support to its presentation in brochures.

Two other variables also offered insights into visitor expectations: the benefits they expected from a reef trip and their suggested improvements. Twelve specific benefits sought from a trip to the reef were rated in importance by respondents and the mean scores for each of this is given in Table 22.12. The six most important expected benefits were to see the Great Barrier Reef, to see coral, to see marine life, to get close to nature, to experience something new and different and to have an educational or learning experience. Several of these are consistent with the images given in the reef operator brochures.

426

The final variable examined was suggested improvements for a reef trip and the ten most common responses are given in Table 22.13. The three most common responses were requests for more time at the reef, more information, education or interpretation, and more action on the part of operators to limit visitor impacts on the environment. The importance attached to the environment and learning in both the expected reef benefits and in suggested improvements identify some options for changing both the reef product offered and the brochure designs. These visitor responses suggest that more interpretation and discussion of ways to protect the reef would be welcomed during the reef trips and that the addition of educational and environmental messages in brochures may contribute to the appeal of the reef experiences being advertised.

Table 22.13 Ten most often suggested improvements for a reef visit

Suggestions for improvement	% of responses
No improvement necessary	34.0
More time at reef	13.5
More info/education/interpretation/guides	7.7
Limit impacts, reduce litter, touching, keep control, etc.	5.1
Restrict numbers of visitors, reduce crowding	4.5
Improve lunch/variety	3.7
Improve facilities on pontoon, other hardware (e.g. semi-sub)	2.9
Better quality reef/water	2.5
Reduce prices/costs, discounts, etc.	2.1
Improve quality, speed of boat, seating on boat	1.8

Visitor satisfaction

This chapter has previously noted that a consumer's satisfaction with a product is a reflection of the match between the expectations of the consumer and the reality of the product, or, in the case of marine tourism, the expectation of the visitor and the reality of the experience. The survey asked a number of questions to determine the level of visitor satisfaction with their reef trip. One question asked people to rate the level of enjoyment of their current reef trip on a scale from 0 = 'not at all' to 10 = 'very much', while another asked whether visitors would recommend the trip to friends and family.

Responses to the question about enjoyment indicated that only five people (corresponding to less than 1 per cent) did not enjoy their trip at all, whilst 39 per cent rated their enjoyment as 10 out of 10 and 87 per cent of respondents ranked their level of enjoyment as an '8' or higher. The mean level of enjoyment was 8.8 indicating a high level of satisfaction (see Figure 22.3). Further, a majority of visitors (92 per cent) would definitely or probably recommend the experience to other people with only 1.5 per cent stating that they would not recommend the reef trip.

Enjoyment of GBR Trip

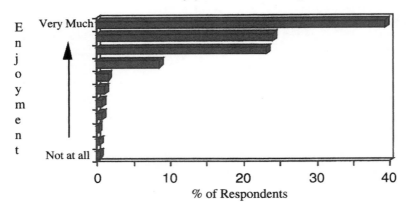

Figure 22.3 Enjoyment levels for Great Barrier Reef trip

Summary and conclusions

The high levels of satisfaction with the reef trip and willingness to recommend indicate that visitors' expectations were largely met. However, some differences were found between the market survey results and the analysis of promotional material which have marketing implications. For example, seeing marine life and learning about the reef were important reasons for taking a trip to the reef. Analysis of promotional material indicated relatively little use of educational and ecological messages and information in brochures, and limited use of marine life in visual images.

With regard to the actual experience of visitors, the presentation of visitors in the promotional material tended to overemphasise individuals on their own, and underemphasise families. The major discrepancy resulting from a comparison between visitor activity participation and the depiction of activities in promotional material was that scuba diving was over-represented in brochures (both text and photographs). Diving ranked sixth overall in terms of marine activity participation, behind snorkelling, swimming, viewing marine life, glass bottom boat trips and visiting islands. Swimming, viewing marine life and visiting islands were particularly under-represented in promotional material.

The following specific recommendations can be made as a result of the analyses reported.

- A range of travel party groups could be presented to encourage a broader range of visitors to the region to participate in reef tours.
- More emphasis could be placed on marine activities other than diving in promotional material. Diving is a very specialised activity requiring substantial time and money commitments. It appears that many visitors do

not participate in diving, so further emphasis should be placed on the range of alternative reef activities such as snorkelling and glass bottom boat trips, which can also result in a satisfying experience.

Brochures were an important source of information for visitors to the Far North Queensland region. By addressing some of the discrepancies just discussed, operators may be able to attract more customers to their operation and ensure that their customers are those who are seeking experiences which closely match the product on offer. It is important to monitor the overall or 'composite' image of marine tourism in the region which is promoted to visitors to ensure that the dynamic needs and characteristics of the market are addressed.

References

Barrat, D. (1986) *Media Sociology*, London: Tavistock.

Butler, R.W. (1980) 'The Concept of a Tourism Area Cycle of Evolution: Implications for the Management of Resources', *The Canadian Geographer* 24: 5–12.

Coopers & Lybrand Consultants (1996) *Reef Tourism 2005. Structure and Economics of the Marine Tourism Industry in the Cairns Section of the Great Barrier Reef. Final Report.* Sydney: Coopers & Lybrand Consultants.

Dann, G.M.S. (1993) 'Advertising in Tourism and Travel: Tourism Brochures', in M.A. Khan, M.D. Olsen and T. Var (eds) *VNR's Encyclopedia of Hospitality and Tourism*, pp. 893–901. New York: Van Nostrand Reinhold.

Duke, C.R. and Persia, M.A. (1996) 'Consumer-defined Dimensions for the Escorted Tour Industry Segment: Expectations, Satisfactions, and Importance', *Journal of Travel and Tourism Marketing* 5 (1/2): 77-99.

Gunn, C.A. (1988) *Vacationscapes: Designing Tourism Regions.* New York: Van Nostrand Reinhold.

Holloway, J.C. and Plant, R.V. (1988) *Marketing for Tourism.* London: Pitman.

Kotler, P., Bowen, J. and Makens, J. (1996) *Marketing for Hospitality and Tourism*, Upper Saddle River, NJ: Prentice-Hall Inc.

Krippendorff, K. (1980). *Content Analysis: An Introduction to its Methodology*, Vol. 5, The Sage COMMTEXT Series. Beverly Hills, CA: Sage Publications.

Morrison, Alastair M. (1996) *Hospitality and Travel Marketing*, 2nd edn. New York: Delmar Publishers.

Moscardo, G. (1996) 'Using Tourism Research to Develop New Tourism Products: Creating Sustainable Tourism Experiences for the Great Barrier Reef', in *It's Showtime for Tourism: New Products, Markets and Technologies*, Proceedings of Travel and Tourism Research Association Conference, Las Vegas, June 1996, pp. 57–65. Lexington, KY: Travel and Tourism Research Association.

Pearce, P. L., Morrison, A., Scott, N., O'Leary, J., Nadkarni, N. and Moscardo, G. (1996) 'The Holiday Market in Queensland: Building an Understanding of Visitors Staying in Commercial Accommodation', *Tourism and Hospitality Research: Australian and International Perspectives*, Proceedings from the Australian Tourism and Hospitality Research Conference, 1996. Australia: Bureau of Tourism Research.

Peter, J.P. and Olsen, J.C. (1987) *Consumer Behaviour: Marketing Strategy Perspectives.* Homeword, IL: Irwin.

Um, S. and Crompton, J.L. (1990) 'Attitude Determinants in Tourism Destination Choice', *Annals of Tourism Research* 17: 432–48.

Weber, R.W. (1985) *Basic Content Analysis*. Beverly Hills: Sage Publications.

Woodside, A.G. and Lysonski, S. (1989) 'A Research Model of Traveller Destination Choice', *Journal of Travel Research*, 17(4): 8–14.

INDEX